Field Manual
No. 3-0

Headquarters
Department of the Army
Washington, DC, 6 October 2017

Operations

Contents

		Page
	PREFACE	vii
	INTRODUCTION	ix
Chapter 1	**OVERVIEW OF ARMY OPERATIONS**	**1-1**
	Large-Scale Combat Operations	1-1
	Challenges for Army Forces	1-2
	Anticipated Operational Environments	1-4
	Multi-Domain Extended Battlefield	1-6
	Threats	1-9
	Joint Operations	1-12
	The Army in Joint Operations	1-14
	The Army's Operational Concept	1-16
	Operational Art	1-19
	Strategic and Operational Reach	1-22
	Operational Framework	1-25
	Sequencing Operations	1-36
	Control Measures	1-37
	Paths to Victory	1-38
Chapter 2	**ARMY ECHELONS, CAPABILITIES, AND TRAINING**	**2-1**
	Section I – Overview of Army Echelons, Capabilities, and Training	**2-1**
	Section II – Army Echelons	**2-1**
	Theater Army	2-1
	Theater Army Organization	2-3
	The Expanded Theater	2-6
	Other Organizations for Theater Support	2-9
	Corps	2-11
	Divisions	2-13
	Brigade Combat Teams	2-14
	Multifunctional and Functional Brigades	2-16
	Section III – Army Capabilities (Combat Power)	**2-21**
	The Elements of Combat Power	2-22
	Mission Command	2-23
	Movement and Maneuver	2-41
	Intelligence	2-42
	Fires	2-45
	Sustainment	2-47
	Protection	2-49

Distribution Restriction: Approved for public release; distribution is unlimited.

Contents

	Section IV – Training for Large-Scale Combat Operations	2-52
	Mission-Essential Tasks	2-54
	Training Techniques	2-54
	Training Considerations by Warfighting Function	2-55
Chapter 3	**OPERATIONS TO SHAPE**	**3-1**
	Overview of Operations to Shape	3-1
	Operations Assessments	3-3
	Threats	3-4
	Shaping Activities	3-4
	Interagency Coordination	3-9
	Army Organizations	3-9
	Consolidate Gains	3-14
Chapter 4	**OPERATIONS TO PREVENT**	**4-1**
	Overview of Operations to Prevent	4-1
	Threats	4-2
	Prevent Activities	4-3
	Sustainment Preparation	4-8
	Deployment	4-11
	Army Forces During Operations to Prevent	4-14
	The Corps	4-19
	The Division	4-24
	Consolidation of Gains	4-25
Chapter 5	**LARGE-SCALE COMBAT OPERATIONS**	**5-1**
	Section I – Overview: Large-Scale Combat Operations	**5-1**
	Joint Large-Scale Combat Operations	5-1
	Army Forces in Large-Scale Combat Operations	5-3
	Threat	5-5
	Stability in Large-Scale Combat Operations	5-6
	Section II – Tactical Enabling Tasks	**5-9**
	Reconnaissance	5-9
	Security Operations	5-12
	Troop Movement	5-13
	Relief in Place	5-15
	Passage of Lines	5-16
	Encirclement Operations	5-16
	Mobility Operations	5-16
	Section III – Forcible Entry	**5-20**
	Section IV – Transition to Consolidate Gains	**5-22**
Chapter 6	**LARGE-SCALE DEFENSIVE OPERATIONS**	**6-1**
	Overview of Large-Scale Defensive Operations	6-1
	Enemy Attack	6-2
	Planning Corps and Division Defensive Tasks	6-8
	Preparing Corps and Divisions for the Defense	6-18
	Defensive Tasks	6-26
	Defending Encircled	6-46
	Consolidation of Gains in the Defense	6-49
Chapter 7	**LARGE-SCALE OFFENSIVE OPERATIONS**	**7-1**
	Overview of Large-Scale Offensive Operations	7-1
	Enemy Defense	7-2
	Planning for the Corps and Division in the Offense	7-5
	Preparing Corps and Divisions for the Offense	7-14

	Forms of Maneuver	7-21
	Offensive Tasks	7-32
	Subordinate Forms of the Attack	7-52
	Other Tactical Considerations	7-53
	Consolidation Of Gains in the Offense	7-58
Chapter 8	**OPERATIONS TO CONSOLIDATE GAINS**	**8-1**
	Overview of Operations to Consolidate Gains	8-1
	Threats to the Consolidation of Gains	8-4
	Consolidation Area Framework	8-5
	Consolidation of Gains Activities	8-7
	Consolidate Gains Responsibilities by Echelon	8-13
	Transition	8-17
Appendix A	**COMMAND AND SUPPORT RELATIONSHIPS**	**A-1**
Appendix B	**RISK CONSIDERATIONS**	**B-1**
	SOURCE NOTES	**Source Notes-1**
	GLOSSARY	**Glossary-1**
	REFERENCES	**References-1**
	INDEX	**Index-1**

Figures

Introductory figure. FM 3-0 logic chart	x
Figure 1-1. The conflict continuum and the range of military operations	1-1
Figure 1-2. Cyberspace in the multi-domain extended battlefield	1-8
Figure 1-3. Notional large-scale combat joint phasing model	1-12
Figure 1-4. Army strategic roles and their relationships to joint phases	1-14
Figure 1-5. Operational maneuver from strategic distance	1-24
Figure 1-6. Joint operational areas within a theater	1-28
Figure 1-7. Corps area of operations within a theater of operations	1-30
Figure 1-8. Contiguous corps area of operations	1-32
Figure 1-9. Noncontiguous corps area of operations	1-33
Figure 2-1. Theater enabler organizations	2-4
Figure 2-2. Example of a theater army organization for large-scale combat operations	2-7
Figure 2-3. Possible tactical corps task organization	2-12
Figure 2-4. Division task organized for large-scale combat operations	2-14
Figure 2-5. Armored brigade combat team	2-15
Figure 2-6. The elements of combat power	2-22
Figure 2-7. The operations process	2-25
Figure 2-8. The commander's role in the operations process	2-26
Figure 2-9. Dynamic continuum of information operations	2-27
Figure 2-10. Example division airspace coordinating measures	2-33
Figure 3-1. Shaping activities within an environment of cooperation and competition	3-2
Figure 4-1. Example of diplomatic, informational, military, and economic FDOs and FROs	4-4

Figure 4-2. An overview of force tailoring 4-7
Figure 5-1. Relief in place 5-15
Figure 6-1. Operations in Malaya 6-3
Figure 6-2. Enemy disruption forces in the attack 6-5
Figure 6-3. Enemy fixing and assault forces breach U.S. defenses 6-6
Figure 6-4. Enemy exploitation forces attack to destroy assigned objectives 6-7
Figure 6-5. Threat army-level integrated fires command 6-8
Figure 6-6. Engagement area 6-9
Figure 6-7. Main battle area 6-10
Figure 6-8. Obstacle control measures 6-11
Figure 6-9. Forward line of own troops 6-12
Figure 6-10. Battle position 6-12
Figure 6-11. The forward edge of the battle area 6-13
Figure 6-12. Target area of interest 6-14
Figure 6-13. Fire support target symbol 6-15
Figure 6-14. Named area of interest 6-22
Figure 6-15. A blue kill box 6-24
Figure 6-16. Example corps contiguous area defense 6-29
Figure 6-17. Example corps noncontiguous area defense 6-30
Figure 6-18. Passage point 6-34
Figure 6-19. Passage lane 6-34
Figure 6-20. Final protective fire 6-35
Figure 6-21. BHL used with other control measures during a rearward passage of lines 6-35
Figure 6-22. Spoiling attack 6-37
Figure 6-23. Projected major counterattacks 6-38
Figure 6-24. Checkpoint 6-41
Figure 6-25. Mobile defense before the striking force's commitment 6-42
Figure 6-26. Mobile defense after commitment of striking force 6-43
Figure 6-27. Trigger line 6-44
Figure 6-28. Perimeter defense when encircled 6-48
Figure 7-1. Enemy battle handover from disruption forces to contact forces 7-3
Figure 7-2. Enemy battle handover from initial contact force to initial shielding force 7-4
Figure 7-3. Enemy continuous maneuver defense in depth within enemy main defense zone 7-5
Figure 7-4. Example penetration followed by exploitation 7-10
Figure 7-5. Example of a single envelopment 7-22
Figure 7-6. Envelopment control measures 7-23
Figure 7-7. Flank attack 7-24
Figure 7-8. Frontal attack 7-25
Figure 7-9. Frontal attack control measures 7-26
Figure 7-10. Penetration: relative combat power 7-28
Figure 7-11. An example of a corps penetration 7-29

Figure 7-12. Penetration control measures 7-30
Figure 7-13. Example of a corps turning movement 7-31
Figure 7-14. Contact point 7-31
Figure 7-15. Example of a corps movement to contact 7-33
Figure 7-16. Route 7-35
Figure 7-17. Target reference point 7-37
Figure 7-18. Example of a corps attack 7-41
Figure 7-19. Line of contact 7-43
Figure 7-20. Attack by fire position 7-44
Figure 7-21. Coordinated fire line 7-46
Figure 7-22. Example of a corps in a pursuit 7-50
Figure 7-23. Pursuit control measures 7-51
Figure 7-24. Free-fire area 7-56
Figure 7-25. No-fire area 7-56
Figure 8-1. Consolidation area during large-scale combat operations 8-6
Figure 8-2. Search and attack 8-9
Figure 8-3. Cordon and search 8-11
Figure 8-4. Consolidating gains after large-scale combat operations 8-15

Tables

Introductory table. Modified Army terms xii
Table 2-1. Additional functional brigades 2-18
Table 2-2. Command post by echelon and type of unit 2-35
Table 4-1. Force projection terms 4-8
Table 4-2. Theater army responsibilities 4-16
Table A-1. Joint support categories A-2
Table A-2. Army command relationships A-4
Table A-3. Army support relationships A-6
Table B-1. Risk considerations B-2

Foreword

Today's operational environment presents threats to the Army and joint force that are significantly more dangerous in terms of capability and magnitude than those we faced in Iraq and Afghanistan. Major regional powers like Russia, China, Iran, and North Korea are actively seeking to gain strategic positional advantage. These nations, and other adversaries, are fielding capabilities to deny long-held U.S. freedom of action in the air, land, maritime, space, and cyberspace domains and reduce U.S. influence in critical areas of the world. In some contexts they already have overmatch or parity, a challenge the joint force has not faced in twenty-five years.

The proliferation of advanced technologies; adversary emphasis on force training, modernization, and professionalization; the rise of revisionist, revanchist, and extremist ideologies; and the ever increasing speed of human interaction makes large-scale ground combat more lethal, and more likely, than it has been in a generation. As the Army and the joint force focused on counter-insurgency and counter-terrorism at the expense of other capabilities, our adversaries watched, learned, adapted, modernized and devised strategies that put us at a position of relative disadvantage in places where we may be required to fight.

The Army and joint force must adapt and prepare for large-scale combat operations in highly contested, lethal environments where enemies employ potent long range fires and other capabilities that rival or surpass our own. The risk of inaction is great; the less prepared we are to meet these challenges, the greater the likelihood for conflict with those who seek windows of opportunity to exploit. The reduction of friendly, forward-stationed forces, significant reductions in capability and capacity across the entire joint force, and the pace of modernization make it imperative that we do everything possible to prepare for worst-case scenarios. We must be ready to win with the forces we have, and having the right doctrine is a critical part of that readiness.

FM 3-0, *Operations*, provides a doctrinal approach for our theater armies, corps, divisions and brigades to address the challenges of shaping operational environments, preventing conflict, prevailing during large-scale ground combat, and consolidating gains to follow through on tactical success. FM 3-0 is about how we deter adversaries and fight a peer threat today, with today's forces and today's capabilities. It addresses operations to counter threats in three broad contexts that account for what the Nation asks its Army to do. Two chapters describe operations to defeat aggression by subversion of U.S. partners and interests, which is fundamental to winning short of war. Three chapters describe operations to defeat enemies during large-scale ground combat operations, which is fundamental to winning wars. The final chapter describes operations to complete the tasks necessary to ensure enduring outcomes, which is fundamental to achieving the ultimate strategic purpose of employing Army forces.

However, doctrine is only one factor in how we fight. Of greater importance is our training and leader development. Building agile and adaptive leaders and units that can prevail in the relentlessly lethal environment of large-scale combat operations requires tough, realistic, and repetitive training. FM 3-0 provides the tactical and operational doctrine to drive our preparation, and when necessary, execution.

MICHAEL D. LUNDY
LIEUTENANT GENERAL, UNITED STATES ARMY
COMMANDING

Preface

FM 3-0 augments the Army's capstone doctrine on unified land operations described in ADP 3-0 and ADRP 3-0. It describes how Army forces, as part of a joint team, shape operational environments (OEs), prevent conflict, conduct large-scale ground combat, and consolidate gains against a peer threat. Together with ADP 3-0 and ADRP 3-0, this manual provides a foundation for how Army forces conduct prompt and sustained large-scale combat operations.

FM 3-0 is applicable to all members of the Army Profession: leaders, Soldiers, and Army Civilians. The principle audience for FM 3-0 is commanders, staffs, and leaders of theater armies, corps, divisions, and brigades. This manual also provides the foundation for training and Army education system curricula and future capabilities development across doctrine, organization, training, materiel, leadership and education, personnel, and facilities (known as DOTMLPF).

To comprehend the doctrine contained in FM 3-0, readers must first understand the fundamentals of unified land operations described in ADP 3-0 and ADRP 3-0. They must understand the language of tactics and the fundamentals of the offense and defense described in ADRP 3-90. Users of FM 3-0 should also understand the fundamentals of stability described in ADRP 3-07. Commanders and staffs of Army headquarters that form the core of a joint task force, joint land component, or multinational headquarters should refer to applicable joint or multinational doctrine. This includes JP 3-33, JP 3-31, and JP 3-16.

Commanders, staffs, and subordinates will ensure their decisions and actions comply with applicable U.S., international, and, in some cases, host-nation laws and regulations. Commanders at all levels will ensure their Soldiers operate in accordance with the law of war and rules of engagement. They also adhere to the Army Ethic as described in ADRP 1. (See FM 27-10.)

FM 3-0 uses joint terms where applicable. Selected joint and Army terms and definitions appear in both the glossary and the text. Terms and definitions where FM 3-0 is the proponent field manual (the authority) are printed in boldface in the text and indicated with an asterisk (*) in the glossary. For other terms defined in the text, the term is italicized and the number of the proponent publication follows the definition.

FM 3-0 applies to the Active Army, the Army National Guard/Army National Guard of the United States, and the United States Army Reserve unless otherwise stated.

Headquarters, United States Army Combined Arms Center is the proponent for this publication. The preparing agency is the Combined Arms Doctrine Directorate, United States Army Combined Arms Center. Send written comments and recommendations on DA Form 2028 (Recommended Changes to Publications and Blank Forms) to Commander, U.S. Army Combined Arms Center and Fort Leavenworth, ATTN: ATZL-MCD (FM 3-0), 300 McPherson Avenue, Fort Leavenworth, KS 66027 2337; by email to usarmy.leavenworth.mccoe.mbx.cadd-org-mailbox@mail.mil; or submit an electronic DA Form 2028.

Preface

ACKNOWLEDGEMENTS

The copyright owners listed here have granted permission to reproduce material from their works. Other courtesy credits listed.

Excerpts from *On War* by Carl von Clausewitz. Edited and translated by Peter Paret and Michael E. Howard. Copyright © 1976, renewed 2004 by Princeton University Press.

Excerpt by permission from Julian S. Corbett, *Some Principles of Maritime Strategy* (Annapolis, MD: Naval Institute Press, 1988).

Quote reprinted courtesy of *The American Presidency Project*. Online by Gerhard Peters and John T. Woolley. Available at http://www.presidency.ucsb.edu/ws/?pid=11340.

Quote reprinted courtesy The Public Papers of President Ronald W. Reagan, Ronald Reagan Presidential Library. Available at https://www.reaganlibrary.archives.gov/archives/speeches/1984/60684a.htm.

Quotes reprinted by permission from *Military Air Power: The CADRE Digest of Air Power Opinions and Thoughts*, compiled by Charles M. Westenhoff (Maxwell Air Force Base, AL: Air University Press, 1990).

Quotes reprinted courtesy *Dictionary of Military and Naval Quotations*, compiled by Robert Debs Heinl, Jr. (Annapolis, MD: United States Naval Institute, 1988).

War as I Knew It by General George S. Patton. Copyright © 1947 by Beatrice Patton Walters, Ruth Patton Totten, and George Smith Totten. Copyright © renewed 1975 by MG George Patton, Ruth Patton Totten, John K. Waters, Jr., and George P. Waters. Reprinted by permission of Houghton Mifflin Company. All rights reserved.

Quote reprinted courtesy the Perseus Digital Library at Tufts. Available at http://perseus.uchicago.edu/.

Wavell quote reprinted courtesy of Martin Van Creveld's *Supplying War: Logistics from Wallenstein to Patton* (New York: Cambridge University Press, 1991).

Quotes reprinted courtesy *The Art of War* by Sun Tzu, translated by Samuel B. Griffith (New York: Oxford University Press, 1963).

Patton quotes courtesy http://www.generalpatton.com/quotes/.

Reprinted by permission Winston S. Churchill, *The Second World War, Volume 5, Closing the Ring* (Boston: Houghton Mifflin Company, 1979).

Quote reprinted courtesy *Quotes for Air Force Logistician*, Volume I (Maxwell Air Force Base, Alabama: Air Force Logistics Management Agency, 2006).

Quote reprinted courtesy Frederick the Great, "The Instruction of Frederick the Great for His Generals," vii, 1747, in *Roots of Strategy: The 5 Greatest Military Classics of All Time*, edited by Thomas R. Phillips (Harrisburg, Pennsylvania: Stackpole Books, 1985).

Quote reprinted by permission Rear Admiral J. C. Wylie, *Military Strategy: A General Theory of Power Control* (Annapolis, MD: Naval Institute Press, 2014).

Churchill quote reprinted courtesy The International Churchill Society. Available at https://www.winstonchurchill.org/resources/speeches/1941-1945-war-leader/the-price-of-greatness-is-responsibility.

Quote reprinted courtesy B. H. Liddell Hart, *Strategy* (New York: Signet Printing, 1974).

Quote reprinted courtesy Richard Nixon, *Six Crises* (New York: Pyramid Book, 1968).

Quote reprinted by permission from General William T. Sherman, *Memoirs of General William T. Sherman by Himself* (Bloomington, IN: Indiana University Press, 1957).

Churchill quote reprinted by permission from National Churchill Museum. Available at www.nationalchurchillmuseum.org/churchill-in-world-war-i-and-aftermath.html.

Polybius quote reprinted by permission, Joseph Callo, *John Paul Jones: America's First Sea Warrior* (Annapolis, MD: Naval Institute Press, 2006).

Introduction

When the Cold War ended, U.S. defense policy postulated that a new era had dawned in which large-scale combat operations against a peer threat were unlikely. This hypothesis was supported by operations throughout the 1990s. While the U.S. military applied the relative conventional superiority it developed in competition with the Warsaw Pact to dominate a large, conventionally armed opponent in OPERATION DESERT STORM, it was an exception. U.S. forces conducted contingency operations at the lower end of the conflict continuum in the Balkans and elsewhere. In 2001 and 2003 the U.S. conducted two offensive joint campaigns that achieved rapid initial military success but no enduring political outcome, resulting in protracted counterinsurgency campaigns in Afghanistan and Iraq. The focus of Army training and equipping shifted from defeating a peer threat to defeating two insurgencies and the global terrorist threat.

Adversaries have studied the manner in which U.S. forces deployed and conducted operations over the past three decades. Several have adapted, modernized, and developed capabilities to counter U.S. advantages in the air, land, maritime, space, and cyberspace domains. Military advances by Russia, China, North Korea, and Iran most clearly portray this changing threat.

While the U.S. Army must be manned, equipped, and trained to operate across the range of military operations, large-scale ground combat against a peer threat represents the most significant readiness requirement. FM 3-0 provides doctrine for how Army forces, as part of a joint team, and in conjunction with unified action partners, do this. FM 3-0 is concerned with operations using current Army capabilities, formations, and technology in today's operational environment (OE). It expands on the material in ADRP 3-0 by providing tactics describing how theater armies, corps, divisions, and brigades work together and with unified action partners to successfully prosecute operations short of conflict, prevail in large-scale combat operations, and consolidate gains to win enduring strategic outcomes.

The logic map for this manual is shown in the introductory figure on page x. The logic map begins with an anticipated OE that includes considerations during large-scale combat operations against a peer threat. Next it depicts the Army's contribution to joint operations through its strategic roles. Within each phase of a joint operation, the Army's operational concept of unified land operations guides how Army forces conduct operations. In large-scale ground combat, Army forces combine offensive, defensive, and stability tasks to seize, retain, and exploit the initiative in order to shape OEs, prevent conflict, conduct large-scale ground combat, and consolidate gains. The philosophy of mission command guides commanders, staffs, and subordinates in their approach to operations. The mission command warfighting function enables commanders and staffs of theater armies, corps, divisions, and brigade combat teams to synchronize and integrate combat power across multiple domains and the information environment. Throughout operations, Army forces maneuver to achieve and exploit positions of relative advantage across all domains to achieve objectives and accomplish missions.

Introduction

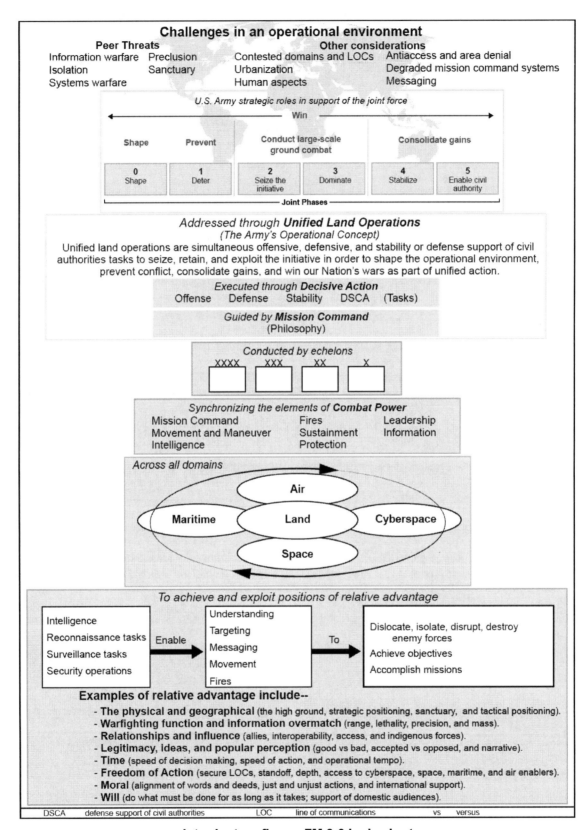

Introductory figure. FM 3-0 logic chart

Overall, the doctrine in FM 3-0 is consistent with ADRP 3-0. The most significant change is the use of the Army strategic roles as an organizing construct to describe how Army forces contribute to joint operations. In addition, FM 3-0—

- Modifies the Army's operational framework by adding a consolidation area to the deep, close, and support area operational framework.
- Adds the physical, temporal, cognitive, and virtual aspects of the operational framework in the context of a multi-domain environment.
- Emphasizes the importance of preparation and training for large-scale combat operations across warfighting functions.
- Recognizes the importance of operations to achieve strategic goals (win) short of armed conflict.
- Emphasizes the roles of echelons, initiative, and maneuver against enemies employing antiaccess and area denial operational approaches to gain and exploit positions of relative advantage.
- Describes corps and divisions as formations (headquarters and subordinate units) rather than just headquarters.
- Emphasizes the importance of consolidating gains to exploit positions of relative advantage and achieve enduring favorable outcomes.
- Establishes a support area command post for corps and division headquarters.

The writing style of FM 3-0 is different from other publications within the Doctrine 2015 construct. FM 3-0 contains fundamentals, tactics, and techniques focused on fighting and winning large-scale combat operations. This manual uses historical vignettes, quotes, and graphics to reinforce the doctrine within.

FM 3-0 contains eight chapters and two appendixes:

Chapter 1 describes large-scale combat operations and associated challenges Army forces face today. It addresses anticipated OEs, threats, joint operations, and the Army's strategic roles in support of joint operations. The chapter then discusses unified land operations and associated topics, including decisive action, operational art, and the operational framework.

Chapter 2 is divided into three sections. Section one provides a general discussion of Army forces in a theater, including theater army, corps, divisions, and brigades. Section two describes Army capabilities by warfighting function. Section three addresses training for large-scale ground combat.

Chapter 3 provides an overview of operations to shape the OE. It discusses operation assessments and describes threat activities prior to armed conflict. A discussion of shaping activities performed by Army forces follows. The chapter then describes Army organizations and their roles as they shape the OE.

Chapter 4 provides an overview of operations to prevent conflict. It addresses assessing OEs in which Army forces conduct activities to prevent war during crisis action, and it provides a description of threats. The chapter continues with a discussion of the major activities within operations to prevent, including planning considerations. The chapter concludes with a description of the roles of the theater army, corps, divisions, and brigades.

Chapter 5 is divided into three sections. Section one provides an overview of large-scale combat operations. Section two addresses tactical enabling tasks that apply to both the defense and the offense. Section three provides a discussion of forcible entry operations from which Army forces may defend or continue the offense.

Chapter 6 begins with a general discussion of the defense, followed by a discussion of how an enemy may attack. It continues with sections on planning and preparing corps and division defenses. It then addresses the three primary defensive tasks.

Chapter 7 begins with a general discussion of the offense, followed by a discussion of how an enemy may defend. It continues with a section on how corps and divisions plan for the offense. This chapter then provides a discussion of forms of maneuver and the four offensive tasks. The chapter concludes with a discussion on the subordinate forms of attack.

Chapter 8 expands upon operations to consolidate gains discussed in previous chapters. It describes how Army forces transition from large-scale ground combat operations to operations that translate tactical and

Introduction

operational success into lasting gains. Next it describes threats to the consolidation of gains. An expanded description of the operational framework and the consolidation area follows. The chapter concludes with a description of consolidation activities and the roles of the theater army, corps, division, and brigade combat teams in consolidating gains.

Appendix A provides doctrine on command and support relationships that form the basis for unity of command and unity of effort.

Appendix B provides commanders with a listing of risk considerations for the planning and execution of large-scale ground combat.

The introductory table outlines different modifications in doctrinal terminology reflected in FM 3-0.

Introductory table. Modified Army terms

Term	Remarks
direct support	FM 3-0 is now the proponent publication.
reinforcing	FM 3-0 is now the proponent publication.

Chapter 1
Overview of Army Operations

War is thus an act of force to compel our enemy to do our will.

Carl von Clausewitz

This chapter describes large-scale combat operations and associated challenges Army forces face today. It addresses anticipated operational environments (OEs), the threat, joint operations, and the Army's strategic roles in support of joint operations. The chapter then discusses unified land operations and associated topics.

LARGE-SCALE COMBAT OPERATIONS

1-1. Threats to U.S. interests throughout the world are countered by the ability of U.S. forces to respond to a wide variety of challenges along a conflict continuum that spans from peace to war as shown in figure 1-1. U.S. forces conduct a range of military operations to respond to these challenges. The conflict continuum does not proceed smoothly from stable peace to general war and back. For example, unstable peace may erupt into an insurgency that quickly sparks additional violence throughout a region, leading to a general war. (See JP 3-0 for the specific types of joint operations conducted across the conflict continuum.)

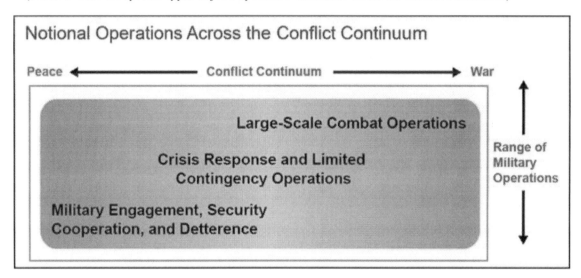

Figure 1-1. The conflict continuum and the range of military operations

1-2. The range of military operations is a fundamental construct that helps relate military activities and operations in scope and purpose within a backdrop of the conflict continuum. All operations along this range share a common fundamental purpose—to achieve or contribute to national objectives. Military engagement, security cooperation, and deterrence activities build networks and relationships with partners, shape regions, keep day-to-day tensions between nations or groups below the threshold of armed conflict, and maintain U.S. global influence. Typically, crisis response and limited contingency operations are focused in scope and scale and conducted to achieve a specific strategic or operational-level objective in an operational area. Large-scale combat operations occur in the form of major operations and campaigns aimed at defeating an enemy's armed forces and military capabilities in support of national objectives.

Chapter 1

1-3. Large-scale combat operations are at the far right of the conflict continuum and associated with war. Historically, battlefields in large-scale combat operations have been more chaotic, intense, and highly destructive than those the Army has experienced in the past several decades. During the 1943 battles of Sidi Bou Zid and Kasserine Pass in World War II, 5,000 American Soldiers were killed over the course of just 10 days; during the first three days of fighting the Army lost Soldiers at the rate of 1,333 per day. Even later in the war, when units were better seasoned, trained, and equipped, casualty rates remained high due to the inherent lethality of large-scale combat operations. In the Hürtgen Forest the Army sustained 33,000 total casualties over 144 days, a loss of 229 Soldiers per day. Similarly, the Battle of the Bulge cost the Army 470 Soldiers per day, for a total loss of 19,270 killed and 62,489 wounded over 41 days of sustained combat.

Close Combat: Hürtgen Forest

In November 1944, the 4th Infantry Division attacked into the Hürtgen Forest in a costly attempt to break through Nazi Germany's "West Wall" and into the Rhine Valley beyond. Opposing them were veteran German divisions, dug into prepared positions consisting of concrete pillboxes and log bunkers, all protected by a carpet of mines. These defenses were skillfully camouflaged in a planted forest that offered perfectly straight fields of fire for machine guns and small arms. On the other hand, the 4th Infantry Division had outstripped its supply lines, resulting in a lack of cold weather gear, especially rubber shoepacs, as the frigid and damp European winter descended.

The division failed to penetrate the German defenses, suffering 4,000 battle casualties and another 2,000 non-battle casualties in less than a month, an average of over 200 per day. In the midst of this ferocious combat, American forces could not rely on artillery support, which had difficulty penetrating the dense forest canopy, or air support, which was likewise limited or grounded by poor weather. The attack degenerated into a close quarters infantry fight, with the Germans using the rugged terrain of high ridges and steep gorges to excellent advantage. Supporting American armor could only use a few cleared trails, which tanks quickly churned into immobilizing, bottomless mud. Exposure to the elements, and especially "trench foot," took a constant toll, leading to high rates of non-battle casualties, even when units were not engaged.

Despite a constant flow of inexperienced replacements, the 4th Infantry Division struggled to reach its objectives. Reduced to less than 50 percent strength, the division had to be withdrawn to a quiet section of the line, where it would hold the southern shoulder of what would soon become the Battle of the Bulge.

1-4. Large-scale combat operations are intense, lethal, and brutal. Their conditions include complexity, chaos, fear, violence, fatigue, and uncertainty. Future battlefields will include noncombatants, and they will be crowded in and around large cities. Enemies will employ conventional tactics, terror, criminal activity, and information warfare to further complicate operations. To an ever-increasing degree, activities in the information environment are inseparable from ground operations. Large-scale combat operations present the greatest challenge for Army forces.

CHALLENGES FOR ARMY FORCES

1-5. Army forces must be organized, trained, and equipped to meet worldwide challenges against a full range of threats. The experiences of the U.S. Army in Afghanistan and Iraq in the early 21st century are not representative of the most dangerous conflicts the Army could face in the future. While the Army conducted combat operations in both locations, for the most part it focused its efforts on counterinsurgency operations and stability tasks. Only a fraction of the forces committed in either theater were engaged in offensive and defensive tasks on any given day. While undoubtedly dangerous and lethal at times, these operations reflected the reality that the enemy operated from positions of disadvantage across all domains. The enemy lacked capabilities in the form of sustained long-range precision fires, integrated air defense systems, robust

Overview of Army Operations

conventional ground maneuver, and electronic warfare. Seldom were friendly units larger than platoons ever at risk of destruction in ground combat. In the future, large-scale combat operations against a peer threat will be much more demanding in terms of operational tempo and lethality.

> **Lethality: Eastern Ukraine**
>
> In July 2014, the Ukrainian Army moved several mechanized brigades into a position near the Russian border to prevent the illegal movement of military equipment across the frontier to rebels in eastern Ukraine. Early on the morning of 11 July, soldiers at the position noticed a drone orbiting above them for some time. Not long after the drone disappeared, rockets fired from 9A52-4 Tornado multiple launch rocket systems began landing on one of the brigades.
>
> The barrage lasted four minutes. Rockets carrying a mixture of high explosive, cluster, and thermobaric munitions smothered the unit's position. Cannon rounds followed the rockets with devastating effect. The Ukrainian soldiers took appalling losses. One battalion was virtually destroyed, and others were rendered combat ineffective due to heavy losses in vehicles and personnel. Casualties quickly overwhelmed army and local medical facilities. In the days that followed, rocket and cannon strikes continued, disrupting the Ukrainian Army's ability to defend that region of eastern Ukraine.

1-6. Army forces participate in operations as part of a joint and multinational force. These operations serve a higher political purpose, and they should be planned and executed at each echelon to support that purpose. Without a clear understanding of the higher purpose, it is difficult to understand what must happen to consolidate gains. How well ground forces consolidate gains determines in large part how enduring the results of operations will be. Failure to effectively consolidate gains in relation to the enemy and the population reduces options for senior leaders and policy makers and contributes to indecisive long-term results.

1-7. Army forces gain, sustain, and exploit control over land to deny its use to an enemy. They do this with combined arms formations possessing the mobility, firepower, and protection to defeat an enemy and establish control of areas, resources, and populations. Army forces are almost entirely dependent upon the capabilities of the other Services to be expeditionary, and Army forces are heavily dependent upon the joint force for fires and information collection in the deep area. Just as the Army requires joint capabilities to conduct operations, the joint force also requires Army capabilities to succeed across the land, maritime, air, space, and cyberspace domains. The Army provides key enabling capabilities to the joint force, including communications, intelligence, protection, and sustainment support.

1-8. Planning and training for most likely scenarios is imperative when all domains can be contested by a peer threat. Many adversaries have systems that facilitate an antiaccess (A2) and area denial (AD) strategy. Massed, multi-layered integrated air defense systems (IADSs), combined with long-range rockets and cruise and ballistic missiles within an integrated fires command, challenge the joint force's ability to project into, and operate within, a theater of operation. Army forces contribute to solving these problems through a multi-domain approach that includes isolating parts of the enemy's integrated fires command to enable or assist joint fires in rendering the system ineffective. (See JP 3-18 for more information on joint forcible entry.)

1-9. Being expeditionary ultimately means deploying on short notice to austere locations and being capable of immediately conducting combat operations. During the initial phases of an operation, Army units may find themselves facing superior threats in terms of both numbers and capabilities. The first deploying units require the capability to defend themselves while they provide reaction time and maneuver space for follow-on forces. Army units with limited joint support may be defending wide frontages as a covering force inside the range of enemy long-range fires. Initial defending forces must exploit the advantages of terrain, both for protection and to impede enemy movement by disrupting the enemy's attack, contesting the enemy's initiative, and establishing the conditions for friendly forces to transition to the offense.

Chapter 1

1-10. The likelihood of the enemy's use of weapons of mass destruction (WMD) increases during large-scale combat operations—particularly against mission command nodes, massed formations, and critical infrastructure. Commanders ensure as much dispersion as tactically prudent. In the offense, Army forces maneuver quickly along multiple axes, concentrate to mass effects, and then disperse to avoid becoming lucrative targets for WMD and enemy conventional fires. Commanders must anticipate that the high tempo of large-scale combat operations will create gaps and seams that create both opportunities and risks as enemy formations disintegrate or displace. Enemy formations may get intermingled with friendly forces or be bypassed, which requires follow-on and supporting units to protect themselves and dedicated forces to secure the consolidation area by defeating or destroying enemy remnants.

1-11. Offensive, defensive, and stability tasks will take place simultaneously throughout the land component's area of operations (AO), although subordinate elements, particularly those at brigade and lower echelons, will generally focus on one type of task at a time. Units engaged in close operations will generally focus on offensive or defensive tasks first, although they will consider long-term stability considerations while executing those tasks. Avoiding civilian casualties and unnecessary destruction of infrastructure are examples of stability considerations that all units account for during planning.

1-12. Fluidity characterizes operations to the rear of forces in the close area. Irregular forces, criminal elements, hostile populations, and the disruptive effects of enemy information warfare will challenge Army forces. These disruptive effects may occur at unit home-stations, ports of embarkation, while in transit to the theater, and upon arrival at ports of debarkation. Army forces may not have the capability, nor the authority, to preempt these attacks.

1-13. Army operations take place in the most complex of environments, on land among humans who have fundamental disagreements. Propaganda, deception, disinformation, and the ability of individuals and groups to influence disparate populations through social technologies reflect the increasing speed of human interaction. Leaders must consider all factors that make up their OE, including social factors initiating and sustaining a conflict. Failure to do so may lead to faulty plans that do not address the desired end state.

1-14. Commanders must recognize the proliferation of cyberspace capabilities (both friendly and threat) and the impact of these capabilities on operations. Threat cyberspace capabilities can disrupt friendly information systems and degrade command and control (C2) across the joint force. Threat operations in cyberspace are often less encumbered by treaty, law, and policy restrictions than those imposed on U.S. forces, which may allow adversaries or enemies an initial advantage. Friendly cyberspace operations enable commanders to defend networks and data and gain advantages in the cyberspace domain through various means. (See FM 3-12 for doctrine on cyberspace operations.)

1-15. The more fluid a situation becomes, the more important and difficult it is to identify decisive points and focus combat power. The philosophy of mission command helps mitigate this uncertainty. Mission command requires mutual trust and shared understanding, and it emphasizes the importance of ensuring that the commander's intent is understood at each level. This mitigates risk when communications are interrupted and allows subordinates to take initiative while out of contact with their higher headquarters. (See ADRP 6-0 for doctrine on mission command.)

ANTICIPATED OPERATIONAL ENVIRONMENTS

> *No matter how clearly one thinks, it is impossible to anticipate precisely the character of future conflict. The key is to not be so far off the mark that it becomes impossible to adjust once that character is revealed.*
>
> Sir Michael Howard

1-16. Factors that affect operations extend far beyond the boundaries of a commander's assigned AO. As such, commanders, supported by their staffs, seek to develop and maintain an understanding of their OE. An *operational environment* is a composite of the conditions, circumstances, and influences that affect the employment of capabilities and bear on the decisions of the commander (JP 3-0). An OE encompasses physical areas of the air, land, maritime, space, and cyberspace domains; as well as the information environment (which includes cyberspace); the electromagnetic spectrum (EMS), and other factors. Included within these are adversary, enemy, friendly, and neutral actors that are relevant to a specific operation.

Commanders and staffs analyze an OE using the eight operational variables: political, military, economic, social, information, infrastructure, physical environment, and time (known as PMESII-PT). (See FM 6-0 for more information on the operational variables.)

1-17. How the many entities behave and interact with each other within an OE is difficult to discern and always results in differing circumstances. No two OEs are the same. In addition, an OE is not static; it continually evolves. This evolution results from opposing forces and actors interacting and their abilities to learn and adapt. The complex and dynamic nature of an OE makes determining the relationship between cause and effect difficult and contributes to the uncertain nature of military operations.

1-18. Understanding an OE is essential to effective decision making. However, commanders realize that uncertainty and time preclude achieving complete understanding before deciding and acting. In addition to the operational variables, there are several tools that assist commanders and staffs with understanding, visualizing, and describing an OE, including—

- Running estimates (described in FM 6-0).
- Army design methodology (described in ATP 5-0.1).
- The military decision-making process (described in FM 6-0).
- Intelligence preparation of the battlefield (described in ATP 2-01.3).
- Sustainment preparation of the OE (described in ADRP 4-0).

LEVELS OF WARFARE

1-19. Operational environments include considerations at the strategic, operational, and tactical levels of warfare. At the strategic level, national leaders develop an idea or set of ideas for employing the instruments of national power (diplomatic, informational, military, and economic) in a synchronized and integrated fashion to achieve theater, national, and multinational objectives. The operational level links the tactical employment of forces to national and military strategic objectives, with the focus being on the design, planning, and execution of operations using operational art. Joint force commanders (JFCs) and their component commanders use operational art to determine how, when, where, and for what purpose military forces will be employed. The tactical level of warfare involves the employment and ordered arrangement of forces in relation to each other using the art and science of tactics. Tactics includes the planning and execution of battles, engagements, and activities to achieve military objectives assigned to tactical units.

1-20. The levels of warfare model the relationship between national objectives and tactical actions. There are no fixed limits or boundaries between these levels, but they help commanders visualize a logical arrangement of operations, allocate resources, and assign tasks to appropriate commands. Echelon of command, size of units, types of equipment, and types and location of forces or components may often be associated with a particular level, but the strategic, operational, or tactical purpose of their employment depends on the nature of their task, mission, or objective.

TRENDS

1-21. Several trends will continue to affect future OEs. The competition for resources, water access, declining birthrates in traditionally allied nations, and disenfranchised groups in many nations contribute to the likelihood of future conflict. Populations will continue to migrate across borders and to urban areas in search of the employment and services urban areas offer. Adversarial use of ubiquitous media platforms to disperse misinformation, propaganda, and malign narratives enables adversaries to shape an OE to their advantage and foment dissention, unrest, violence, or at the very least, uncertainty.

1-22. Proliferating technologies will continue to present challenges for the joint force. Unmanned systems are becoming more capable and common. Relatively inexpensive and pervasive anti-tank guided missiles and advanced rocket propelled grenades can defeat modern armored vehicles. Sensors and sensing technology are becoming commonplace. Adversaries have long-range precision strike capabilities that outrange and outnumber U.S. systems. Advanced integrated air-defense systems can neutralize friendly air power, or they can make air operations too costly to conduct. Anti-ship missiles working in concert with an IADS can disrupt access to the coastlines and ports necessary for Army forces to enter an AO. Adversary cyberspace and space control capabilities can disrupt friendly information systems and degrade C2 across the joint force. Use of WMD and the constant pursuit of the materials, expertise, and technology to employ WMD will increase in the future. Both state and non-state actors continue to develop WMD programs to gain advantage against the United States and its allies. These trends mean that adversaries can contest U.S. dominance in the air, land, maritime, space, and cyberspace domains.

MULTI-DOMAIN EXTENDED BATTLEFIELD

Since men live upon the land and not upon the sea, great issues between nations at war have always been decided—except in the rarest of cases—either by what your army can do against your enemy's territory and national life, or else by fear of what the fleet makes it possible for your army to do.

Sir Julian Corbett

1-23. The interrelationship of the air, land, maritime, space, and the information environment (including cyberspace) requires a cross-domain understanding of an OE. Commanders and staffs must understand friendly and enemy capabilities that reside in each domain. From this understanding, commanders can better identify windows of opportunity during operations to converge capabilities for best effect. Since many friendly capabilities are not organic to Army forces, commanders and staffs plan, coordinate for, and integrate joint and other unified action partner capabilities in a multi-domain approach to operations.

1-24. A multi-domain approach to operations is not new. Army forces have effectively integrated capabilities and synchronized actions in the air, land, and maritime domains for decades. Rapid and continued advances in technology and the military application of new technologies to the space domain, the EMS, and the information environment (particularly cyberspace) require special consideration in planning and converging effects from across all domains.

SPACE DOMAIN

1-25. The *space domain* is the space environment, space assets, and terrestrial resources required to access and operate in, to, or through the space environment (FM 3-14). Space is a physical domain like land, sea, and air within which military activities are conducted. Proliferation of advanced space technology provides more widespread access to space-enabled technologies than in the past. Adversaries have developed their own systems, while commercially available systems allow almost universal access to some level of space enabled capability with military applications. Army forces must be prepared to operate in a denied, degraded and disrupted space operational environment (D3SOE).

1-26. Space capabilities provide information collection; early warning; environmental monitoring; satellite based communication; and positioning, navigation, and timing. Activities in the space domain enable freedom of action for operations in all other domains, and operations in the other domains can create effects in and through the space domain.

1-27. Space operations are inherently joint. Army forces rely on space-based capabilities to enhance each warfighting function and effectively conduct operations. Space capabilities enable Army forces to communicate, navigate, accurately target an enemy, and protect and sustain the force. To achieve optimal military utility from space assets, a basic understanding of space capabilities and the ability to coordinate activities between involved agencies and organizations is required. Space support elements (SSE) at theater army, corps, and division headquarters assist with planning, integrating, and coordinating space capabilities into all aspects of operations. Commanders cannot assume that U.S. forces will have unconstrained use of space-based capabilities. (See FM 3-14 for doctrine on Army space operations.)

INFORMATION ENVIRONMENT

Print is the sharpest and strongest weapon of our party.

Joseph Stalin

1-28. The *information environment* is the aggregate of individuals, organizations, and systems that collect, process, disseminate, or act on information (JP 3-13). The information environment is not separate or distinct from the OE but is inextricably part of it. Any activity that occurs in the information environment

Overview of Army Operations

simultaneously occurs in and affects one or more of the physical domains. Most threat forces recognize the importance of the information environment and emphasize information warfare as part of their strategic and operational methods (see paragraphs 1-41 to 1-43 for a discussion of the threat's use of information warfare).

1-29. The information environment is comprised of three dimensions: physical, informational, and cognitive. The physical dimension includes the connective infrastructure that supports the transmission, reception, and storage of information. The informational dimension contains the content (or data) itself. It refers to the content and flow of information, such as text or images, or data that staffs can collect, process, store, disseminate, and display. The informational dimension provides the necessary link between the physical and cognitive dimensions. The cognitive dimension refers to the minds of those who are affected by and act upon information. This dimension focuses on the societal, cultural, religious, and historical contexts that influence the perceptions of those producing the information and of the targets and audiences receiving the information. In this dimension, decision makers and target audiences are most prone to influence and perception management.

1-30. Across the globe, information is increasingly available in near-real time. The ability to access this information, from anywhere, at any time, broadens and accelerates human interaction across multiple levels, including person to person, person to organization, person to government, and government to government. Social media, in particular, enables the swift mobilization of people and resources around ideas and causes, even before they are fully understood. Disinformation and propaganda create malign narratives that can propagate quickly and instill an array of emotions and behaviors from anarchy to focused violence. From a military standpoint, information enables decision making, leadership, and combat power; it is also key to seizing, gaining, and retaining the initiative, and to consolidating gains in an OE. Army commanders conduct information operations to affect the information environment. (See FM 3-13 for doctrine on the information environment and the various information-related capabilities available to commanders.)

CYBERSPACE AND THE ELECTROMAGNETIC SPECTRUM

1-31. *Cyberspace* is a global domain within the information environment consisting of interdependent networks of information technology infrastructures and resident data, including the Internet, telecommunications networks, computer systems, and embedded processors and controllers (JP 3-12[R]). Friendly, enemy, adversary, and host-nation networks, communications systems, computers, cellular phone systems, social media, and technical infrastructures are all part of cyberspace.

1-32. Cyberspace is an extensive and complex global network of wired and wireless links connecting nodes that permeate every domain. Networks cross geographic and political boundaries connecting individuals, organizations, and systems around the world. Cyberspace is socially enabling, allowing interactivity among individuals, groups, organizations, and nation-states. A way to describe cyberspace is in terms of three layers:
- Physical network layer—geographic locations in land, air, maritime, or space where elements of the network reside.
- Logical network layer—components of the network related to one another in a way abstracted from the physical network.
- Cyber-persona layer—digital representations of individuals or entities identity in cyberspace.

1-33. Cyberspace is highly vulnerable for several reasons, including ease of access, network and software complexity, lack of security considerations in network design and software development, and inappropriate user activity. Access to cyberspace by an individual or group with a networked device is easy, and an individual with a single device may be able to disable an entire network. Vulnerabilities in the systems that operate in cyberspace contribute to a continuous obligation to manage risk and protect portions of cyberspace.

1-34. The *electromagnetic spectrum* is the range of frequencies of electromagnetic radiation from zero to infinity. It is divided into 26 alphabetically designated bands (JP 3-13.1). The EMS crosses all domains, and it provides a vital link between the space and cyberspace domains. Space operations depend on the EMS for the transport of information and the control of space assets. Space operations provide a specific capability of transport through the space domain for long haul and limited access communications. Space assets provide a key global connectivity capability for cyberspace operations. Conversely, cyberspace operations provide a capability to execute space operations. This interrelationship is an important consideration across cyberspace operations, and it is particularly important when conducting targeting in cyberspace.

Chapter 1

1-35. Cyberspace and the EMS will grow increasingly congested, contested, and critical to successful operations. Army forces must be able to effectively operate in cyberspace and the EMS, while controlling the ability of others to operate there. Rapid developments in cyberspace and the EMS present continuous challenges. While Army forces cannot defend against every kind of intrusion, commanders and staffs must take steps to identify, prioritize, and defend their most important networks and data. They must also adapt quickly and effectively to enemy and adversary presence inside cyberspace systems. Figure 1-2 displays friendly, threat, and neutral (or non-attributed) networks both within an operational area and worldwide. It depicts the global nature of cyberspace and the extended battlefield.

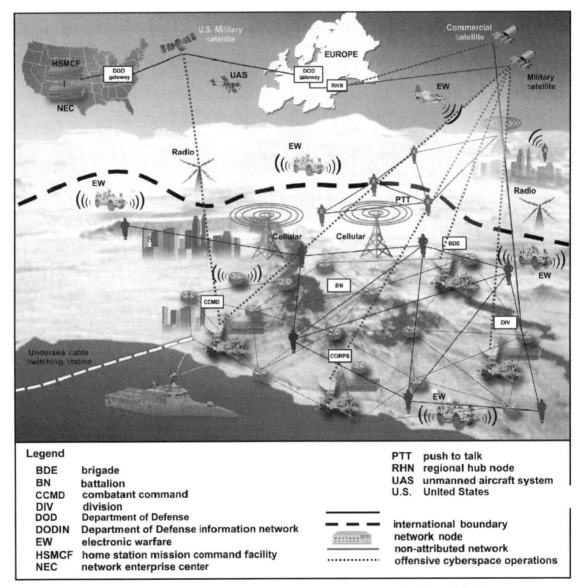

Figure 1-2. Cyberspace in the multi-domain extended battlefield

1-36. Advantages in cyberspace and the EMS result from effectively synchronizing Department of Defense information network (DODIN) operations, offensive cyberspace operations (OCO), defensive cyberspace operations (DCO), electronic attack, electronic protection, electronic warfare support, and spectrum management operations. (See FM 3-12 for more details on requesting cyberspace effects and electronic attack.)

THREATS

I know from personal experience that evil forces are loose in the world, and that their sinister aim is to destroy our free way of life by subversion if possible, by overt hostility if necessary.

General Mark Clark

1-37. A *threat* is any combination of actors, entities, or forces that have the capability and intent to harm United States forces, United States national interests, or the homeland (ADRP 3-0). Threats may include individuals, groups of individuals, paramilitary or military forces, nation-states, or national alliances. In general, a threat can be categorized as an enemy or an adversary:

- An *enemy* is a party identified as hostile against which the use of force is authorized (ADRP 3-0). An enemy is also called a combatant and is treated as such under the law of war.
- An *adversary* is a party acknowledged as potentially hostile to a friendly party and against which the use of force may be envisaged (JP 3-0).

1-38. While ADRP 3-0 addresses various threats across the range of military operations, FM 3-0 is focused on peer threats in large-scale combat operations. A peer threat is an adversary or enemy with capabilities and capacity to oppose U.S. forces across multiple domains world-wide or in a specific region where they enjoy a position of relative advantage. Peer threats possess roughly equal combat power in geographical proximity to a conflict area with U.S. forces. A peer threat may also have a cultural affinity to specific regions, providing them relative advantages in terms of time, space, and sanctuary. Peers threats generate tactical, operational, and strategic challenges that are an order of magnitude more challenging militarily than those the U.S. Army has faced since the end of the Cold War.

1-39. Peer threats employ strategies that capitalize on their capabilities to achieve their objectives. When these objectives are at odds with the interests of the United States and its allies, conflict becomes more likely. Peer threats prefer to achieve their goals without directly engaging U.S. forces in combat. They often employ information warfare in combination with conventional and irregular military capabilities to achieve their goals. During a conflict, peer threats will try to weaken the resolve of the United States and its partners to sustain conflict. They will exploit friendly sensitivity to world opinion and attempt to exploit American domestic opinion and sensitivity to friendly casualties. Peer threats believe they have a comparative advantage because of their willingness to endure greater hardship, casualties, and negative public opinion.

1-40. Peer threats employ their resources across multiple domains to attack U.S. vulnerabilities. They use their capabilities to create lethal and nonlethal effects throughout an OE. During combat operations, threats seek to inflict significant damage across multiple domains in a short period of time. They seek to delay friendly forces long enough to achieve their goals and end hostilities before friendly forces reach culmination. Peer threats will employ various methods to employ their national elements of power to render U.S. military power irrelevant. Five broad peer threat methods, often used in combination, include—

- Information warfare.
- Preclusion.
- Isolation.
- Sanctuary.
- Systems warfare.

INFORMATION WARFARE

1-41. Information warfare is a term referring to a threat's orchestrated use of information activities (such as cyberspace operations, electronic warfare, and psychological operations) to gain advantage in the information environment. Peer threats conduct information warfare aggressively and continuously across the range of military operations. During a conflict, peer threats will use information warfare in conjunction with other methods to achieve strategic and operational objectives. Peer threats typically have fewer political and legal restrictions than U.S. forces on how they use information activities, providing them an initial advantage. Peer threats seek to manipulate the information environment and will use all their means against a wide range of audiences, both civilian and military, and domestically and internationally, in support of their goals. Peer threats will exploit shared cultural norms, historical grievances, friendly political decision making, and self-serving interpretation of international law to limit U.S. military actions and degrade U.S. political will. Peer threats use diverse means to conduct information warfare, and these means may include—

- Cyberspace operations.
- Perception management.
- Deception.
- Electronic warfare.
- Physical destruction.
- Operations security.

1-42. A peer threat will systematically and continuously combine all of these means to attack friendly forces and create specific effects within the information environment. Peer threats will use misinformation, disinformation, propaganda, and information for effect to create doubt, confuse, deceive, and influence U.S. and partner decision makers, forces, and audiences. (See FM 3-53 for a discussion on threat information categories).

1-43. The capabilities that facilitate the manipulation of the information environment are in a constant state of change. Threat information warfare will be conducted through a variety of platforms, including social media, face-to-face communication, denial of service, sabotage, and false stories in media outlets. U.S. forces should expect that peer threats will use all capabilities at their disposal to systematically and continuously target U.S. decision-making processes. Adversaries and enemies seek to shape the information environment in ways that cause U.S. commanders to make decisions that ultimately support their own goals. Peer threats will use cyberspace operations to manipulate the information friendly forces receive from information systems. They will leverage disinformation to manipulate media story narratives to distort the truth. Threat deception activities, combined with propaganda, will create the impression that threat misinformation is true when it is not.

PRECLUSION

1-44. To preclude is to keep something from happening by taking action in advance. Peer threats use a wide variety of capabilities to preclude a friendly force's ability to shape the OE and mass and sustain combat power. A2 and AD are two such activities. A2 includes those usually long-range actions and capabilities designed to prevent an enemy force from entering an operational area. For example, A2 activities can deny forces the ability to conduct operations in a desired area. Usually AD includes shorter range actions and capabilities that are not designed to keep friendly forces out, but rather to limit their freedom of action within an operational area to the point that their mission is severely limited or unachievable.

ISOLATION

1-45. Isolation is the containment of a force so that it cannot accomplish its mission. Peer threats will attempt to isolate U.S. forces in several ways. Some examples include—

- Preventing or limiting communications with other units.
- Deceiving friendly forces about the current situation and their role in the OE.
- Deceiving the public about the current situation to reduce its support of friendly operations that counter threat goals.
- Exploiting inadequate friendly understanding of the OE or cultural affinity in an area or region.
- Decisively engaging or fixing units in combat.
- Denying mobility by exploiting complex terrain, obstacles, or WMD.

1-46. Peer threats believe that the defeat of the U.S. military lies not at the end of a substantial battle, but through the culmination of U.S. efforts before U.S. goals are reached. In large-scale combat operations, peer threats will seek to isolate U.S. tactical forces and prevent their mutual support, facilitating threat campaign objectives.

SANCTUARY

1-47. Sanctuary is a threat method of putting threat forces beyond the reach of friendly forces. It is a form of protection derived by some combination of political, legal, and physical boundaries that restrict freedom of action by a friendly force commander. Peer threats will use any means necessary, including sanctuary, to protect key elements of their combat power from destruction, particularly by air and missile capabilities. Peer threats will also protect their key interests, whether these interests reside in their homeland or in another country. To create a sanctuary, both physical and non-physical means can be used in combination to protect key interests, including—

- Complex terrain.
- Hiding among noncombatants and culturally sensitive structures.
- Counter-precision techniques, including camouflage, concealment, and deception.
- Electronic warfare (EW) capabilities like Global Positioning System (GPS) jamming.
- Countermeasure systems, including decoys, hardened and buried facilities, IADSs, and long range fires.
- Dispersion.
- Engineer efforts and fortifications.
- Information warfare.
- International borders.
- International law and treaty agreements.

1-48. Generally speaking, most means of sanctuary cannot protect an entire force for an extended time period. Rather, a threat will seek to protect selected elements of its forces for enough time to gain the freedom of action necessary to pursue its strategic goals or diplomatic ends. Threat forces will seek to protect their conventional forces, advanced aircraft, and extended-range fires systems. Many peer threats have invested in long-range rocket and missile systems, such as the Russian Smerch (BM-30) and Chinese PHL-03, capable of counter-fires at extreme ranges to allow sanctuary beyond international borders. Improved air defense systems, including counter-tactical ballistic missile systems, often provide protection for these advanced fires capabilities.

SYSTEMS WARFARE

1-49. Peer threats view the battlefield, their own instruments of power, and an opponent's instruments of power as a collection of complex, dynamic, and integrated systems composed of subsystems and components. Peer threats use systems warfare to identify specific critical capabilities for disruption or destruction in order to cause failure of a larger friendly system. Simple examples of systems warfare would be to use electronic warfare to disable the links between unmanned aircraft system (UAS) controllers and the aircraft in a specific area or the emplacement of layered IADSs from a position of sanctuary to prevent the integration of opposing airpower with ground operations. Peer threats employ systems warfare to identify and isolate critical subsystems or components that give opposing forces the capabilities necessary to accomplish their mission.

1-50. A peer threat's use of systems warfare during the shape and deter phases of a joint operation focuses primarily on the systems within each instrument of national power, seeking vulnerabilities that can be exploited to preclude the achievement of U.S. objectives. During subsequent phases, peer threats use systems warfare in combat to assist threat commanders in decision making and the planning and execution of their mission. Peer threats believe that a qualitatively or quantitatively weaker force can defeat a superior foe, if the weaker force can dictate the terms of combat. Peer threats believe that the systems warfare approach allows them to move away from the traditional attrition-based approach to combat. Systems warfare makes it unnecessary to match an opponent system-for-system or capability-for-capability. Threat commanders and staffs will locate the critical components of the enemy combat system, patterns of interaction, and opportunities to exploit this connectivity. (See TC 7-100.2 for a more in depth discussion of systems warfare.)

JOINT OPERATIONS

[S]eparate ground, sea, and air warfare is gone forever. If ever again we should be involved in war, we will fight it in all elements, with all services, as one single concentrated effort.

President Dwight D. Eisenhower

1-51. *Joint operations* are military actions conducted by joint forces and those Service forces employed in specific command relationships with each other, which of themselves, do not establish joint forces (JP 3-0). Traditionally, campaigns are the most extensive joint operations, in terms of the amount of forces and other capabilities committed and the duration of operations. In the context of large-scale combat operations, a campaign is a series of related major operations achieving strategic and operational objectives within a given time and space. A major operation is a series of tactical actions, such as battles, engagements, and strikes, and it is the primary building block of a campaign. Army forces conduct supporting operations as part of a joint campaign.

1-52. Most joint operations share certain activities or actions in common. There are six general groups of military activities that typically occur in preparation for and during a large-scale joint combat operation. These six groups are shape, deter, seize initiative, dominate, stabilize, and enable civil authorities. These six general groups of activity provide a basis for thinking about a joint operation in notional phases. These phases often overlap, and they are not necessarily sequential, as shown in figure 1-3.

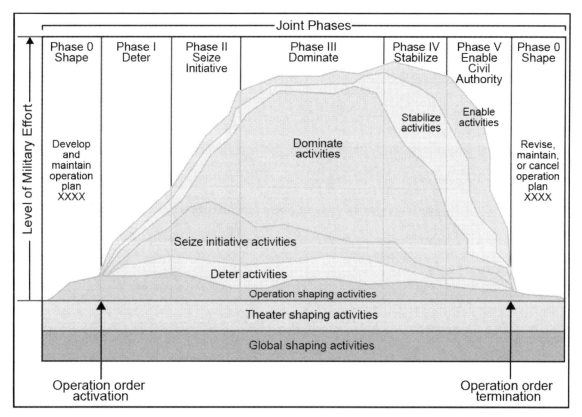

Figure 1-3. Notional large-scale combat joint phasing model

1-53. A phase is a definitive stage or period during a joint operation in which a large portion of the forces and capabilities are involved in similar or mutually supporting activities for a common purpose that often is achieved by intermediate objectives. Phasing helps commanders and staffs to visualize, plan, and execute an entire operation and define requirements in terms of forces, resources, time, space, and purpose. The actual phases of an operation will vary according to the nature of the operation and the JFC's decisions.

Overview of Army Operations

Commanders may compress, expand, or omit a phase entirely. Phases may be conducted sequentially, but many activities from a phase may begin in a previous phase and continue into subsequent phases.

1-54. The nature of operations and activities during large-scale combat operations will change from the beginning (at operation order activation) to the operation's end (at operation order termination). Figure 1-3 shows that change from the deter phase through the enable civil authority phase. The operations and activities in these groups vary in magnitude, time, and force as an operation progresses. Shaping activities precede operation order activation and continue during and after an operation order is terminated. Theater and global shaping activities occur continuously to support specific joint operations and to meet theater and global requirements. Such requirements are defined in theater-specific campaign plans for each area of responsibility (AOR) and the Unified Command Plan that encompasses the six AORs.

SHAPE

I should like to know what advantage it would be to you to lose the support of a man like me.

Saladin

1-55. Activities in the shape phase help set conditions for successful theater operations. They are designed to dissuade or deter adversaries, assure friends, and set conditions for contingency plans. Shape activities are generally conducted as part of military engagement and security cooperation. Joint and multinational operations and various interagency activities occur routinely during the shape phase. Shape activities are executed continuously with the intent of enhancing international legitimacy and gaining multinational cooperation by shaping perceptions and influencing adversaries' and allies' behavior, developing allied and friendly military capabilities for self-defense and multinational operations, improving information exchange and intelligence sharing, providing U.S. forces with peacetime and contingency access, and mitigating conditions that could lead to a crisis.

DETER

1-56. The intent of the deter phase is to prevent an adversary from undesirable actions through the posturing of friendly capabilities and demonstrating the will to use them. The deter phase is generally weighted toward security and preparatory actions to protect friendly forces and to indicate the intent to execute subsequent phases of a planned operation. A number of flexible deterrent options (FDOs) or flexible response options (FROs) could be implemented during this phase. Once a crisis is defined, these actions may include mobilization, tailoring of forces, and initial deployment into a theater. Other actions may include establishing friendly A2 and AD capabilities and developing C2, intelligence, force protection, and logistic capabilities to support the JFC's concept of operations. Many actions in the deter phase build on activities from the previous phase and are conducted as part of security cooperation. They can also be part of stand-alone operations.

SEIZE INITIATIVE

1-57. In the seize initiative phase, JFCs seize the initiative through decisive use of joint force capabilities. In combat, this involves conducting reconnaissance, maintaining security, performing defensive and offensive tasks at the earliest possible time, forcing the enemy to culminate offensively, and setting the conditions for decisive operations. Rapid application of joint combat power may be required to enter a theater (through joint forcible entry) or to delay, impede, or halt an enemy's initial aggression and to deny an enemy its initial objectives. Operations to gain access to theater infrastructure and expand friendly freedom of action continue during this phase, while the JFC seeks to degrade enemy capabilities with the intent of resolving the crisis at the earliest opportunity.

DOMINATE

1-58. The dominate phase focuses on breaking an enemy's will to resist or, in noncombat situations, to control an OE. Success in the dominate phase depends on overmatching enemy capabilities at the right time and place. Operations can range from large-scale combat to various stability activities, depending on the

STABILIZE

1-59. The stabilize phase is typically characterized by a shift in focus from sustained combat operations to stability activities. These operations help reestablish a safe and secure environment and provide essential government services, emergency infrastructure reconstruction, and humanitarian relief. The intent in this phase is to help restore local political, economic, and infrastructure stability. Civilian officials may lead operations during part or all of this phase, but typically the JFC will provide significant supporting capabilities and activities. Until legitimate local entities are functioning, the joint force may be required to perform essential civil administration and integrate the efforts of other supporting interorganizational partners. The JFC assesses the impact of operations in this phase on the ability to transfer authority for remaining requirements to a legitimate civil entity, which marks the end of the phase.

ENABLE CIVIL AUTHORITY

1-60. The enable civil authority phase is primarily characterized by joint force support to legitimate civil governance. Commanders provide this support by agreement with the appropriate civil authority. In some cases, and especially for operations within the United States, commanders provide this support under direction of the civil authority. The purpose is to help the civil authority regain its ability to govern and administer to the services and other needs of the population. The military end state is achieved during this phase, signaling the end of the joint operation. Combatant command involvement with other nations and other government agencies beyond the termination of the joint operation may be required to achieve the national strategic end state.

THE ARMY IN JOINT OPERATIONS

1-61. The Army's primary mission is to organize, train, and equip its forces to conduct prompt and sustained land combat to defeat enemy ground forces and seize, occupy, and defend land areas. The Army accomplishes its mission by supporting the joint force in four strategic roles: shape operational environments, prevent conflict, conduct large-scale ground combat, and consolidate gains. The strategic roles clarify the enduring reasons for which the U.S. Army is organized, trained, and equipped. (*Note.* Strategic roles are not tasks assigned to subordinate units.) Figure 1-4 shows the Army's strategic roles in a general relationship to the joint phasing model. (See ADP 1 for a discussion of the roles of the Army and its core competencies.)

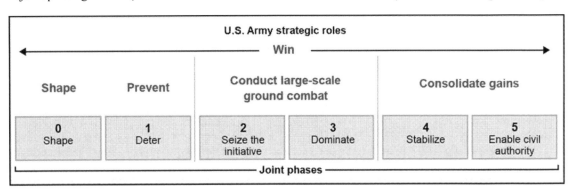

Figure 1-4. Army strategic roles and their relationships to joint phases

SHAPE OPERATIONAL ENVIRONMENTS

> *We in America have learned bitter lessons from two World Wars: It is better to be here ready to protect the peace, than to take blind shelter across the sea, rushing to respond only after freedom is lost. We've learned that isolationism never was and never will be an acceptable response to tyrannical governments with an expansionist intent.*
>
> President Ronald Reagan

1-62. Army operations to shape bring together all the activities intended to promote regional stability and to set conditions for a favorable outcome in the event of a military confrontation. Army operations to shape help dissuade adversary activities designed to achieve regional goals short of military conflict. As part of operations to shape, the Army provides trained and ready forces to geographic combatant commanders (GCCs) in support of their theater campaign plan. The theater army and subordinate Army forces assist the GCC in building partner capacity and capability and promoting stability across the AOR. Army operations to shape are continuous throughout a GCC's AOR and occur before, during, and after a joint operation within a specific operational area.

1-63. Shaping activities include security cooperation and forward presence to promote U.S. interests, developing allied and friendly military capabilities for self-defense and multinational operations, and providing U.S. forces with peacetime and contingency access to a host nation. Regionally aligned and engaged Army forces are essential to achieving objectives to strengthen the global network of multinational partners and preventing conflict. The Army garrisons forces and pre-positions equipment in areas to allow national leaders to respond quickly to contingencies. Operational readiness, training, and planning for potential operations by Army forces at home station are also a part of operations to shape. Army operations to shape correspond to continuous shaping activities within the joint phasing model. (Chapter 3 describes how Army forces, as part of a joint team, shape operational environments.)

PREVENT CONFLICT

To be prepared for war is one of the most effectual means of preserving peace.

President George Washington

1-64. Army operations to prevent include all activities to deter an adversary's undesirable actions. These operations are an extension of operations to shape designed to prevent adversary opportunities to further exploit positions of relative advantage by raising the potential costs to adversaries of continuing activities that threaten U.S. interests. Prevent activities are generally weighted toward actions to protect friendly forces, assets, and partners, and to indicate U.S. intent to execute subsequent phases of a planned operation. As part of a joint force, Army forces may have a significant role in the execution of directed FDOs. Additionally, Army prevent activities may include mobilization, force tailoring, and other predeployment activities; initial deployment into a theater to include echeloning command posts; employment of intelligence collection assets; and development of intelligence, communications, sustainment, and protection infrastructure to support the JFC's concept of operations. Army operations to prevent correspond to the deter phase in a joint operation. (Chapter 4 describes how Army forces, as part of a joint team, prevent conflict.)

CONDUCT LARGE-SCALE GROUND COMBAT

War's very object is victory—not prolonged indecision.

General Douglas MacArthur

1-65. During large-scale combat operations, Army forces focus on the defeat and destruction of enemy ground forces as part of the joint team. Army forces close with and destroy enemy forces in any terrain, exploit success, and break their opponent's will to resist. Army forces attack, defend, conduct stability tasks, and consolidate gains to attain national objectives. Divisions and corps are the formations central to the conduct of large-scale combat operations, organized, trained and equipped to enable subordinate organizations. The ability to prevail in ground combat is a decisive factor in breaking an enemy's will to continue a conflict. Conflict resolution requires the Army to conduct sustained operations with unified action partners as long as necessary to achieve national objectives. Conducting large-scale ground combat operations corresponds to seize the initiative and dominate phases of a joint operations. (Chapters 5, 6, and 7 describe how Army forces, as part of a joint force, conduct large-scale ground combat.)

CONSOLIDATE GAINS

Loss of hope, rather than loss of life, is the factor that really decides wars, battles, and even small combats.

Sir Basil Liddell Hart

Chapter 1

1-66. Army operations to consolidate gains include activities to make enduring any temporary operational success and to set the conditions for a sustainable environment, allowing for a transition of control to legitimate civil authorities. Consolidation of gains is an integral and continuous part of armed conflict, and it is necessary for achieving success across the range of military operations. Army forces deliberately plan to consolidate gains during all phases of an operation. Early and effective consolidation activities are a form of exploitation conducted while other operations are ongoing, and they enable the achievement of lasting favorable outcomes in the shortest time span. Army forces conduct these activities with unified action partners. In some instances, Army forces will be in charge of integrating forces and synchronizing activities to consolidate gains. In other situations, Army forces will be in support. Army forces may conduct stability tasks for a sustained period of time over large land areas. While Army forces consolidate gains throughout an operation, consolidating gains becomes the focus of Army forces after large-scale combat operations have concluded. Army operations to consolidate gains correspond with stabilize and enable civil authority phases of a joint operation. (Chapter 8 describes how Army forces, as part of a joint team, consolidate gains.)

WIN

1-67. Winning is the achievement of the purpose of an operation and the fulfillment of its objectives. The Army wins when it successfully performs its roles as part of the joint force during operations. It wins when it effectively shapes an OE for combatant commanders, and when it responds rapidly with enough combat power to prevent war through deterrence during crisis. When required to fight, the Army's ability to prevail in ground combat at any scale becomes a decisive factor in breaking the enemy's will to continue fighting. The Army wins when an enemy is defeated to such a degree that it can no longer effectively resist, and it agrees to cease hostilities on U.S. terms. To ensure that the military results of combat are not temporary, the Army follows through with its unique scope and scale of capabilities to consolidate gains and win enduring outcomes favorable to U.S. interests.

THE ARMY'S OPERATIONAL CONCEPT

The community of danger makes allies well disposed towards one another.

Xenophon

1-68. An *operation* is a sequence of tactical actions with a common purpose or unifying theme (JP 1). Army forces, as part of the joint and multinational force, contribute to the joint mission through the conduct of unified land operations. *Unified land operations* are simultaneous offensive, defensive, and stability or defense support of civil authorities' tasks to seize, retain, and exploit the initiative to shape the operational environment, prevent conflict, consolidate gains, and win our Nation's wars as part of unified action (ADRP 3-0). The goal of unified land operations is to apply land power as part of unified action to defeat an enemy on land and establish conditions that achieve the JFC's commander's end state.

DECISIVE ACTION

1-69. At the heart of the Army's operational concept is decisive action. *Decisive action* is the continuous, simultaneous combinations of offensive, defensive, and stability or defense support of civil authorities tasks (ADRP 3-0). During large-scale combat operations, commanders describe the combinations of offensive, defensive, and stability tasks in the concept of operations. As a single, unifying idea, decisive action provides direction for an entire operation. Based on a specific idea of how to accomplish the mission, commanders and staffs refine the concept of operations during planning and determine the proper allocation of resources and tasks. They adjust the allocation of resources and tasks to specific units throughout an operation, as subordinates develop the situation or conditions change.

1-70. The simultaneity of the offense, defense, and stability tasks is not absolute at every level. The higher the echelon, the greater the possibility of simultaneous offense, defense, and stability tasks. At lower echelons, all of a unit's combat power may be required to execute a specific task. As an example, a division may conduct offensive, defensive, and stability tasks simultaneously in some form. Subordinate brigades perform some combination of offensive, defensive, and stability tasks, but they are unlikely able to perform all three simultaneously during large-scale combat operations.

Overview of Army Operations

1-71. Offensive tasks are conducted to defeat and destroy enemy forces and seize terrain, resources, and population centers. Offensive tasks impose the commander's will on the enemy. The offense is the most direct and sure means of seizing and exploiting the initiative to gain physical and cognitive advantages over an enemy. In the offense, the decisive operation is a sudden, shattering action that capitalizes on speed, surprise, and shock effect to achieve the operation's purpose. If that operation does not destroy or defeat the enemy, operations continue until enemy forces disintegrate or retreat so they no longer pose a threat. Executing offensive tasks compels an enemy to react, creating or revealing additional weaknesses that an attacking force can exploit. (See ADRP 3-90 for a detailed discussion of offensive tasks.)

1-72. Defensive tasks are conducted to defeat an enemy attack, gain time, economize forces, and develop conditions favorable for offensive or stability tasks. Normally, the defense alone cannot achieve a decisive victory. However, it can set conditions for a counteroffensive or counterattack that enables Army forces to regain and exploit the initiative. Defensive tasks are a counter to enemy offensive actions. They defeat attacks, destroying as much of an attacking enemy as possible. They also preserve and maintain control over land, resources, and populations. The purpose of defensive tasks is to retain key terrain, guard populations, protect lines of communications, and protect critical capabilities against enemy attacks and counterattacks. Commanders can conduct defensive tasks to gain time and economize forces, so offensive tasks can be executed elsewhere. (See ADRP 3-90 for a discussion of defensive tasks.)

1-73. *Stability tasks* are tasks conducted as part of operations outside the United States in coordination with other instruments of national power to maintain or reestablish a safe and secure environment and provide essential governmental services, emergency infrastructure reconstruction, and humanitarian relief (ADP 3-07). These tasks support governance, whether it is imposed by a host nation, an interim government, or a military government. Stability tasks involve both coercive and constructive actions. They help to establish or maintain a safe and secure environment and facilitate reconciliation among local or regional adversaries. (See ADRP 3-07 for a discussion of stability.)

1-74. Conducting decisive action involves more than simultaneous execution of all its tasks. It requires commanders and staffs to consider their units' capabilities and capacities relative to each task. Commanders consider their missions, decide which tactics, techniques, and procedures to use, and balance the tasks of decisive action while preparing their commander's intent and concept of operations. They determine which tasks their force can accomplish simultaneously, if phasing is required, what additional resources it may need, and how to transition from one task to another.

SEIZE, RETAIN, AND EXPLOIT THE INITIATIVE

It is even better to act quickly and err than to hesitate until the time of action is past.

Carl von Clausewitz

1-75. Successful unified land operations require Army forces to seize, retain, and exploit the initiative by forcing an enemy to respond to friendly action. By presenting multiple dilemmas to an enemy, commanders force the enemy to react continuously until the enemy is finally driven into untenable positions. Seizing the initiative pressures enemy commanders into abandoning their preferred options and making mistakes. Enemy mistakes allow friendly forces to seize opportunities and create new avenues for exploitation. Throughout operations, commanders focus combat power to defeat enemy forces, protect populations and infrastructure, and consolidate gains to retain the initiative within an operation's overall purpose.

1-76. Commanders seize the initiative by acting. Without action, seizing the initiative is impossible. Faced with an uncertain situation, there is a natural tendency to hesitate and gather more information to reduce uncertainty. Waiting and gathering information might reduce uncertainty, but does not eliminate it. Waiting may even increase uncertainty while providing an enemy with time to seize the initiative. It is far better to manage uncertainty by acting and developing the situation instead of waiting. Exploiting the initiative requires positive action.

MULTI-DOMAIN BATTLE

1-77. The Army conducts operations across multiple domains and the information environment. All Army operations are multi-domain operations, and all battles are multi-domain battles. Examples of Army

Chapter 1

multi-domain operations include airborne and air assault operations, air and missile defense, fires, aviation, cyberspace electromagnetic activities, information operations, space operations, military deception (MILDEC), and information collection. Large-scale combat operations such as these entail significant operational risk, synchronization, capabilities convergence, and high operational tempo. Key considerations for operating in multiple domains include—

- Mission command.
- Reconnaissance in depth.
- Mobility.
- Cross-domain fires.
- Tempo and convergence of effects.
- Protection.
- Sustainment.
- Information operations.

1-78. Army forces may be required to conduct operations across multiple domains to gain freedom of action for other members of the joint force. Examples of these operations include neutralizing enemy integrated air defenses or long-range surface-to-surface fires systems, denying enemy access to an AO, disrupting enemy C2, protecting friendly networks, conducting tactical deception, or disrupting an enemy's ability to conduct information warfare.

1-79. Enemies are likely to initiate hostilities against a friendly force from initial positions of relative advantage. These include physical, temporal, and cognitive positions and cultural, informational, and other human factors peculiar to the land domain. Once an enemy is in a position of disadvantage, the joint force must rapidly exploit its advantage with force-oriented operations to destroy key enemy capabilities. Well synchronized, high-tempo offensive action, often in the form of ground maneuver, is required to defeat enemies with significant long-range fires and air defense capabilities. Understanding the importance of planning and executing area security and stability tasks is also necessary to consolidate gains across multiple domains in ways that support the purpose of friendly tactical operations to achieve strategic goals.

POSITIONS OF RELATIVE ADVANTAGE

In war trivial causes produce momentous events.

Julius Caesar

1-80. A *position of relative advantage* is a location or the establishment of a favorable condition within the area of operations that provides the commander with temporary freedom of action to enhance combat power over an enemy or influence the enemy to accept risk and move to a position of disadvantage (ADRP 3-0). Positions of relative advantage occur in all domains, providing opportunities for units to exploit. Commanders maintain momentum through exploitation of opportunities to consolidate gains, and they continually assess and reassess friendly and enemy effects for future opportunities. A key aspect in achieving a position of advantage is *maneuver*, the employment of forces in the operational area through movement in combination with fires to achieve a position of advantage in respect to the enemy (JP 3-0).

1-81. Positions of relative advantage are usually temporary and require initiative to exploit. While friendly forces are seeking positions of advantage, enemy forces are doing the same. There are multiple forms of positional advantage that provide opportunities to exploit. Some are considerations that should be understood when formulating tactical and operational concepts, while others are goals that can be worked towards as a means of destroying or defeating the enemy and achieving the overall purpose of the operation. Examples of positional advantage include—

- Physical and geographical (including strategic positioning, sanctuary, and control of key terrain).
- Combat power and warfighting function overmatch (including range, lethality, precision, and mass).
- Relationships and influence (including allies, interoperability, access, and indigenous forces).
- Legitimacy, ideas, and popular perception (including what is good versus bad, accepted versus opposed, and a believable narrative).

Overview of Army Operations

- Time (including speed of recognition, speed of decision making, speed of action, and operational tempo).
- Freedom of action (including secure lines of communications, standoff, depth, access to cyberspace, maritime and air enablers, and friendly A2 and AD measures).
- Moral (including alignment of words and deeds, just and unjust, and international support).
- Will (including doing what must be done, continuing as long as it takes, and maintaining support from domestic leaders).

1-82. Relative positional advantage is something to gain, protect, and exploit across all domains. Combining positional advantages across multiple domains during each phase of operations provides opportunities for exploitation through maneuver. Physical or geographic positions of relative advantage are often identified first as decisive points and then depicted in operational graphics as objectives. The greater the number of positions of advantage a commander can generate, the increased number of dilemmas that commander can present to an enemy. The combination of positional advantages change over time relative to changes in the OE, and this change includes how the enemy reacts to friendly forces' activities. It is the exploitation of positions of advantage through maneuver which deters, defeats, or destroys an enemy. Leaders at every echelon are expected to display the initiative necessary to assume prudent risk while taking timely advantage of opportunities that present themselves under ambiguous, chaotic conditions. It is not always possible to understand those opportunities before they arise, so it is important that units have a command climate that rewards those who make decisions and act boldly in the absence of orders.

MISSION COMMAND

Never tell people how to do things. Tell them what to do and they will surprise you with their ingenuity.

General George S. Patton, Jr.

1-83. Mission command is the exercise of authority and direction by the commander using mission orders to enable disciplined initiative within the commander's intent to empower subordinates in the conduct of unified land operations. Mission command is a principle of unified land operations that enables commanders to blend the art of command and the science of control while integrating the warfighting functions during operations.

1-84. Mission command requires an environment of mutual trust and shared understanding among commanders, staffs, and subordinates. It requires a command climate in which commanders encourage subordinates to accept prudent risk and exercise disciplined initiative to seize opportunities and counter threats within the commander's intent. Using mission orders, commanders focus their orders on the purpose of an operation rather than on the details of how to perform assigned tasks. Doing this minimizes detailed control and allows subordinates the greatest possible freedom of action. Finally, when delegating authority to subordinates, commanders set conditions for success by allocating adequate resources to subordinates based on assigned tasks.

1-85. Through mission command, commanders integrate and synchronize operations. Commanders understand that they do not operate independently but as part of a larger force united by a common operational purpose. They integrate and synchronize their actions with the rest of the force to achieve the overall objective of the operation. Commanders create and sustain situational understanding through collaboration dialogue within their organization and with unified action partners to facilitate unity of effort. They provide a clear commander's intent and use mission orders to assign tasks, allocate resources, and issue broad guidance. (See ADRP 6-0 for doctrine on mission command).

OPERATIONAL ART

"[T]here is a wide difference between right and wrong disposition of troops, just as stones, bricks, timber and tiles flung together anyhow are useless, whereas when the materials that neither rot nor decay, that is, the stones and tiles, are placed at the bottom and the top, and the bricks and timber are put together in the middle, as in building, the result is something of great value, a house, in fact.

Xenophon

1-86. Army commanders and staffs employ operational art to determine what tactics best achieve a strategic purpose. *Operational art* is the cognitive approach by commanders and staffs—supported by their skill, knowledge, experience, creativity, and judgment—to develop strategies, campaigns, and operations to organize and employ military forces by integrating ends, ways, and means (JP 3-0). Through operational art, commanders and staffs combine art and science to develop plans and orders that describe how (ways) the force employs its capabilities (means) to achieve the desired end state (ends) while considering risk. This requires commanders to answer the following questions:

- What conditions, when established, constitute the desired end state (ends)?
- How will the force achieve these desired conditions (ways)?
- What sequence of actions helps attain these conditions (ways)?
- What resources are required to accomplish that sequence of actions (means)?
- What risks are associated with that sequence of actions and how can they be mitigated (risks)?

1-87. Operational art is critical to leaders being able to organize the systemic defeat of an opposing force, first conceptually in their minds, and then translating their conceptual solutions into concrete execution. Systems thinking informs a situational template that puts a threat into perspective. Systemic defeats depend on knowing what functions are vital to an enemy's mission at any given time, and knowing how that function can be disabled. Identifying the logic inherent to a particular conflict, which spans domains, the human dimension, and the information environment, is both possible and essential. An understanding of the system by which an enemy fights to achieve its purpose requires an understanding of the unique logic of each case.

ELEMENTS OF OPERATIONAL ART

1-88. In applying operational art, commanders and their staffs use a set of intellectual tools to help them understand their OE, visualize their operational approach, and describe the operation's end state. Collectively, this set of tools is known as the elements of operational art. These tools help commanders understand, visualize, and describe operations and help them formulate their commander's intent and planning guidance. Commanders selectively use these tools in any operation. However, their application is broadest in the context of large-scale combat operations.

Elements of Operational Art
- End state and conditions
- Center of gravity
- Decisive points and spaces
- Lines of operations and lines of effort
- Operational reach
- Culmination
- Basing
- Tempo
- Phasing and transitions
- Risk

1-89. Not all elements of operational art apply at all levels of warfare. For example, a company commander may be concerned about the tempo of an upcoming operation, but is probably not concerned with an enemy's center of gravity. On the other hand, a corps commander may consider all elements of operational art in developing a plan in support of the JFC's campaign plan. The application of specific elements of operational art is situation and echelon dependent.

1-90. Commanders use the elements of operational art to help them form their *commander's visualization*—the mental process of developing situational understanding, determining a desired end state, and envisioning an operational approach by which the force will achieve that end state (ADRP 5-0). In building their visualization, commanders first seek to understand the OE. Next, commanders envision a set of desired future conditions that represents the operation's end state. Commanders complete their visualization by conceptualizing an *operational approach*—a broad description of the mission, operational concepts, tasks, and actions required to accomplish the mission (JP 5-0). (See ADRP 3-0 for a discussion of each element of operational art.)

RISK

Never forget that no military leader has ever become great without audacity.

Carl von Clausewitz

1-91. Risk, uncertainty, and chance are inherent in all military operations. Operational art balances risk and opportunity to create and maintain the conditions necessary to seize, retain, and exploit the initiative and achieve decisive results. During execution, opportunity is fleeting. The surest means to create opportunity is to accept risk while minimizing hazards to friendly forces. Commanders and staffs at all echelons evaluate enemy and friendly actions to determine where commanders can assume risk in one or more domains and the information environment. When assuming risk, commanders should allocate minimal combat power to secondary efforts.

1-92. Corps and division commanders weigh the main effort and shape their OEs as required to increase their responsiveness to operations as brigade combat teams (BCTs) engage in their close areas. These commanders mitigate risk by accelerating or decelerating the tempo of operations based on the situation. Accelerating the tempo of operations may be necessary to enable the attacking divisions to rapidly close with the enemy to mitigate the impact of any enemy superior fires overmatch. Commanders may choose to slow the tempo to prevent the creation of assailable flanks, if the enemy has greater superiority in maneuver forces. Situational understanding is essential to managing risk. Seeing, understanding, and responding to windows of vulnerability or opportunity within each domain and the information environment can reduce risk to the friendly force and enhance success in chaotic and high-tempo operations. (See appendix B for a listing of risk considerations).

DEFEAT AND STABILITY MECHANISMS

1-93. When developing an operational approach, commanders consider how to employ a combination of defeat mechanisms and stability mechanisms. Defeat mechanisms are dominated by offensive and defensive tasks, while stability mechanisms are dominant in stability tasks that establish and maintain security and facilitate consolidating gains in an AO.

Defeat Mechanisms

1-94. A *defeat mechanism* is a method through which friendly forces accomplish their missions against enemy opposition (ADRP 3-0). Commanders describe a defeat mechanism as the physical, temporal, or psychological effects it produces. Operational art formulates the most effective, efficient way to defeat enemy aims. Physically defeating an enemy deprives enemy forces of the ability to achieve those aims. Temporally defeating an enemy anticipates enemy reactions and counters them before they can become effective. Psychologically defeating an enemy deprives that enemy of the will to continue the conflict. The defeat mechanisms are—

- Destroy.
- Dislocate.
- Disintegrate.
- Isolate.

1-95. *Destroy* is a tactical mission task that physically renders an enemy force combat-ineffective until it is reconstituted. Alternatively, to destroy a combat system is to damage it so badly that it cannot perform any function or be restored to a usable condition without being entirely rebuilt (FM 3-90-1). An enemy cannot restore a destroyed force to a usable condition without entirely rebuilding it. When commanders seek to destroy enemy forces, they apply combat power on an enemy capability so that it can no longer perform any function. When commanders seek to destroy an enemy's narrative, commanders employ all means to affect the information environment to disprove, discredit, or make irrelevant the enemy narrative, or to deny its delivery. Destroying enemy forces' informational capabilities completely may be impossible, but destroying enough key nodes on their network puts them at a significant tactical disadvantage. From a temporal perspective, the massed destruction of enemy capabilities leaves an enemy no time to adjust plans or reposition forces quickly enough to recover.

1-96. *Dislocate* is to employ forces to obtain significant positional advantage, rendering the enemy's dispositions less valuable, perhaps even irrelevant (ADRP 3-0). Commanders often achieve dislocation by maneuvering forces into locations where an enemy does not expect them. Surprise can unhinge an enemy's operational approach and disrupt an enemy's ability to cognitively adapt to multiple simultaneous dilemmas. Operations conducted rapidly and simultaneously throughout the depth of an enemy's echelons prevent that enemy's ability to effectively reposition by depriving the enemy of time and space. Deception and disruption of enemy networks can have the same effect.

1-97. *Disintegrate* means to disrupt the enemy's command and control system, degrading its ability to conduct operations while leading to a rapid collapse of the enemy's capabilities or will to fight (ADRP 3-0). Commanders often achieve disintegration by specifically targeting an enemy's command structure and communications systems. Doing so requires aligning effects in time and space, across all available domains, to separate an enemy's leaders from its led.

1-98. *Isolate* is a tactical mission task that requires a unit to seal off—both physically and psychologically—an enemy from sources of support, deny the enemy freedom of movement, and prevent the isolated enemy force from having contact with other enemy forces (FM 3-90-1). When commanders isolate, they deny an enemy or adversary access to capabilities that enable an enemy unit to maneuver in time and space at will. Isolation that is both physical and virtual generates greater cognitive effects than one or the other by itself.

Stability Mechanisms

1-99. A *stability mechanism* is the primary method through which friendly forces affect civilians in order to attain conditions that support establishing a lasting, stable peace (ADRP 3-0). As with defeat mechanisms, combinations of stability mechanisms produce complementary and reinforcing effects that accomplish the mission more effectively and efficiently than single mechanisms do alone.

1-100. The four stability mechanisms are compel, control, influence, and support. Compel means to use, or threaten to use, lethal force to establish control and dominance, effect behavioral change, or enforce compliance with mandates, agreements, or civil authority. Control involves imposing civil order. Influence means to alter the opinions, attitudes, and ultimately the behavior of foreign friendly, neutral, adversary, and enemy targets and audiences through messages, presence, and actions. Support is to establish, reinforce, or set the conditions necessary for the instruments of national power to function effectively.

STRATEGIC AND OPERATIONAL REACH

> *The more I see of war, the more I realize how it all depends on administration and transportation...It takes little skill or imagination to see where you would like your army to be and when; it takes much knowledge and hard work to know where you can place your forces and whether you can maintain them there.*
>
> Field Marshal Archibald Wavell

1-101. Strategic and operational reach enable Army forces to deploy rapidly, fight upon arrival, and conduct extended campaigns as part of a joint and multinational force. Doing so requires proficiency at force projection, protection, and sustainment. Soldiers require an expeditionary mindset to prepare them for short notice deployments into uncertain, often austere, and lethal environments.

1-102. Strategic reach provides the capability to operate against threats operating anywhere in the world. The distance across which the United States can project decisive military power is its strategic reach. It combines joint military capabilities—air, land, maritime, space, special operations, cyberspace, and information—with those of the other instruments of national power (diplomatic, economic, and information). Army forces contribute to strategic reach by securing and operating bases far from the United States. However, Army forces depend on joint capabilities to deploy and sustain them across intercontinental distances. In some cases, Army forces use strategic lift to deploy directly to an operational area. In many instances, land operations combine direct deployment with movements from intermediate staging bases located outside an operational area. Access to bases and support is influenced by diplomatic and economic factors.

1-103. *Operational reach* is the distance and duration across which a force can successfully employ military capabilities (JP 3-0). Extending operational reach is a major concern for commanders throughout the duration of an operation. Commanders and staffs increase operational reach through deliberate, focused planning, well in advance of operations if possible, and the appropriate sustainment to facilitate endurance. Operational reach is a function of intelligence, protection, sustainment, endurance, and relative combat power. The limit of a unit's operational reach is its culminating point.

1-104. Endurance refers to the ability to employ combat power for protracted periods, in any location across the globe. It stems from the ability to create, protect, and sustain a force, regardless of the distance from its base and the austerity of the environment. Endurance involves anticipating requirements and making the most effective, efficient use of available resources. Their endurance gives Army forces their campaign quality. Endurance contributes to Army forces' ability to make permanent the transitory effects of other capabilities.

1-105. Momentum comes from retaining the initiative and executing high-tempo operations that overwhelm enemy resistance. Commanders control momentum by maintaining focus and pressure. They set a tempo that prevents friendly exhaustion and maintains sustainment. A sustainable tempo extends operational reach. Commanders maintain momentum by anticipating and transitioning rapidly between any combination of offensive, defensive, and stability tasks. Sometimes commanders push their force to its culminating point to take maximum advantage of an opportunity. For example, exploitations and pursuits may involve pushing all available forces to the limits of their endurance.

1-106. Protection contributes to operational reach. Commanders anticipate how enemy actions and environmental factors might disrupt operations and then determine the protection capabilities required to maintain sufficient reach. Protection closely relates to endurance and momentum, and it contributes to the commander's ability to extend operations in time and space. Commanders and staffs consider operational reach to ensure Army forces accomplish their missions before culminating. Commanders continually strive to extend operational reach.

1-107. Force posture (including forces, footprints, and agreements) affects operational reach, and it is an essential maneuver-related consideration during theater strategy development and adaptive planning. Force posture is the starting position from which planners determine additional contingency-basing requirements to support specific contingency plans and crisis action responses. These requirements directly support the development of operational lines of communications and lines of operations, and they affect the combat power and other capabilities that a joint force can generate. In particular, the arrangement and positioning of temporary contingency bases support the ability of a joint force to project power by shielding its components from enemy action and improving sortie or resupply rates. Political and diplomatic considerations affect basing decisions. U.S. force-basing options span a range from permanently based forces to temporary sea basing that accelerates the deployment and employment of capabilities independent of infrastructure ashore.

OPERATIONAL MOVEMENT AND MANEUVER

To remain separated for as long as possible while operating, and to be concentrated in good time for the decisive battle, that is the task of the leader of large masses of troops.

Field Marshal Helmuth von Moltke the Elder

1-108. Operational movement and maneuver combines global force projection with maneuver against an operationally significant objective, as shown in figure 1-5 on page 1-24. This example requires strategic reach that deploys maneuverable land power to an operational area in a position of advantage. Operational movement and maneuver requires enough operational reach to execute operations decisively without an operational pause. Successful operational movement and maneuver allows a unit to secure and defend a lodgment; develop support infrastructure and base camps; and receive, stage, and build up forces. Success demands full integration of all available means in a multi-domain approach. Thus, successful operational movement and maneuver combines force projection with land maneuver to operational depth in an integrated, continuous operation.

Chapter 1

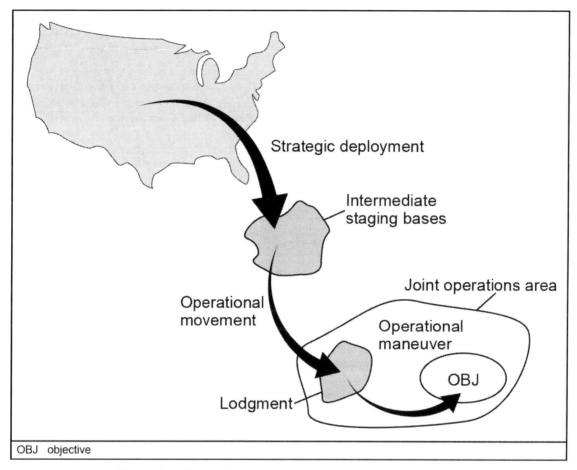

Figure 1-5. Operational maneuver from strategic distance

1-109. The most difficult form of operational maneuver from strategic distance projects forces directly from the United States into an operational area. Examples of this operational maneuver from strategic distance include the 1942 invasion of North Africa and the 1992 intervention in Somalia, when forces were projected from the United States and deployed directly into an operational area. In many cases, operational maneuver from strategic distance requires intermediate staging bases. From these bases, operational maneuver develops using intratheater lift and Army maneuver capabilities.

EXPEDITIONARY CAPABILITY

1-110. Swift campaigns, however desirable, are the exception. When objectives involve destroying large enemy forces, dominating terrain, and controlling populations, success usually requires employing land power for significant periods. Therefore, the Army requires an expeditionary capability and the endurance to prevail in protracted conflict against determined enemies.

1-111. Expeditionary capabilities begin with a mindset that pervades the force. Forward deployed units, forward positioned capabilities, and force projection—from anywhere in the world—contribute to the Army's expeditionary capabilities. Expeditionary capability is more than the ability to deploy quickly. It requires deploying the right mix of Army forces to the right place at the right time. Expeditionary capability provides JFCs with the ability to deter an adversary or take decisive action if deterrence fails.

1-112. Commanders consider their forces from operational and tactical perspectives, and they ensure deployed Army forces possess sufficient combat power to succeed. Expeditionary warfare balances the ability to mass lethal effects against the requirements to deploy, support, and sustain units. Commanders assemble force packages that maximize the lethality of initial-entry forces. Commanders tailor follow-on forces to increase both the lethality and operational reach of the entire force.

1-113. Deploying commanders integrate protection capabilities to ensure mission accomplishment and increase the survivability of deployed Army forces. As with the other attributes, lift constraints and time available may complicate integration of protection capabilities. In many operations, rapid offensive action to seize the initiative may better protect forces than extensive defensive positions around lodgments.

FORCE PROJECTION

1-114. Force projection is the military component of power projection. It is central to the National Military Strategy. Speed is paramount; force projection is a race between friendly forces and enemy or adversary forces. The side that most rapidly builds combat power can seize the initiative. Thus, it is not the velocity of individual stages or transportation means that are decisive; it is a combat-ready force deployed to an operational area before an enemy is prepared to act or before the situation deteriorates further.

1-115. Commanders visualize force projection as one seamless operation. Deployment speed sets the initial tempo of military activity in an operational area. Commanders understand how speed, sequence, and mix of deploying forces affect their employment options. Commanders prioritize the mix of forces on the time phased force and deployment list to project forces into an operational area where and when required. Singular focus on the land component, to the exclusion of complementary joint capabilities, can result in incorrect force sequencing.

1-116. Force projection encompasses five processes: mobilization, deployment, employment, sustainment, and redeployment. These processes occur in a continuous, overlapping, and repeating sequence throughout an operation. Force projection operations are inherently joint. They require detailed planning and synchronization. Sound, informed decisions made early about force projection may determine a campaign's success. (See chapter 4 for a discussion of force projection.)

ENTRY OPERATIONS

1-117. Attaining operational reach often requires gaining and maintaining operational access in the face of enemy A2 and AD capabilities and actions. Commanders conduct forcible entry operations to seize and hold a military lodgments in the face of armed opposition. Once an assault force seizes a lodgment, it normally defends to retain it, while the JFC rapidly deploys additional combat power by air and sea. When conditions are favorable, JFCs may combine a forcible entry with other offensive operations in a *coup de main*. This action can achieve the strategic objectives in a simultaneous major operation. For example, the 1989 invasion of Panama demonstrated operational maneuver from strategic distance in a *coup de main*.

1-118. A forcible entry operation can be by parachute, air, or amphibious assault. The Army's parachute assault and air assault forces provide a formidable forcible entry capability. Marine Corps forces specialize in amphibious assault; they also conduct air assaults as part of amphibious operations. Special operations forces (SOF) play an important role in forcible entry; they conduct shaping operations in support of conventional forces while executing their own missions. These capabilities permit JFCs to overwhelm enemy A2 measures and quickly insert combat power. The entry force either resolves the situation or secures a lodgment for delivery of larger forces by aircraft or ships. The three forms of forcible entry produce complementary and reinforcing effects that help JFCs seize the initiative early in a campaign.

1-119. Forcible entry operations are inherently complex and always joint. Often only hours separate the alert from the deployment. The demands of simultaneous deployment and employment create a distinct set of dynamics. Operations are carefully planned and rehearsed in training areas and marshalling areas. Personnel and equipment are configured for employment upon arrival without reception, staging, onward movement, and integration (RSOI). (See JP 3-18 for doctrine on joint forcible entry operations.)

OPERATIONAL FRAMEWORK

1-120. An *operational framework* is a cognitive tool used to assist commanders and staffs in clearly visualizing and describing the application of combat power in time, space, purpose, and resources in the concept of operations (ADP 1-01). The operational framework provides an organizing construct for

Chapter 1

visualizing and describing operations by echelon in time and space within the context of an AO, area of influence, and area of interest. It provides a logical architecture for determining the responsibilities, permissions, and restrictions for subordinate echelons, and by doing so enables freedom of action and unity of effort. When used in conjunction with effective operational graphics, it provides commanders the ability to provide intent, develop shared visualization, and ultimately create the shared understanding necessary for the exercise of initiative at every echelon.

1-121. The operational framework has four components. First, commanders are assigned an AO for the conduct of operations, from which, in turn, they assign AOs to subordinate units based on their visualization of the operation. Units should be assigned AOs commensurate with their ability to influence what happens within them. Second, within their assigned AO, commanders designate deep, close, support, and consolidation areas to describe the physical arrangement of forces in time, space, and purpose. Third, commanders establish decisive, shaping, and sustaining operations to further articulate an operation in terms of purpose. Finally, commanders designate the main and supporting efforts to designate the shifting and prioritization of resources.

OPERATIONAL FRAMEWORK CONSIDERATIONS

1-122. When establishing their operational framework, commanders and staffs consider the physical, temporal, virtual, and cognitive aspects of their own AO, their higher echelon's AO, and subordinate AOs. The physical, temporal, virtual, and cognitive aspects of an operational framework vary in terms of focus and priority depending upon the echelon, force capabilities, and the OE.

Physical Considerations

1-123. Physical considerations include geography, terrain, infrastructure, populations, distance, weapons ranges and effects, and known enemy locations. They also include other related factors that influence the use of friendly or enemy capabilities, such as climate and weather. When considering physical aspects, commanders and staffs look beyond the land domain. They look into each domain for relevant physical aspects and pay particular attention to the physical dimension of the information environment.

Temporal Considerations

1-124. Temporal considerations relate to those things related to time, including when capabilities can be used, how long they take to generate and employ, and how long they must be used to achieve desired effects. Temporal considerations largely inform commanders and staffs about when to start necessary movement, activities, or requests for effects at each echelon. Temporal considerations cross the physical domains and the dimensions of the information environment. For example, understanding the cognitive dimension of the information environment helps commanders determine operational tempo to prevent an enemy from making timely decisions.

Virtual Considerations

1-125. Virtual considerations are those pertaining to activities, capabilities, and effects relevant to the layers of cyberspace. When not addressed, virtual capabilities provide a form of sanctuary for adversaries and enemies. Commanders assign responsibilities, priorities, and desired effects across the operational framework. They consider what entities are present, what must be protected, what should be attacked, and the effects they want to generate. Friendly examples include the availability and methods of employment for offensive and defensive cyber capabilities, as well as other capabilities that can be used to target a threat. Threat examples include the identification of virtual systems, entities, formations and persons for targeting or countering in order to enable friendly freedom of action. Virtual entities or activities can include banking, virtual organizations, and recruiting that generate effects in the physical world.

Cognitive Considerations

1-126. Cognitive considerations relate to people and how they behave. They include unit morale and cohesiveness, as well as perspectives and decision making. Cognitive considerations should account for both the current situation and desired outcomes to ensure tasks, purpose, and end state are aligned. Commanders

Overview of Army Operations

consider the personal capabilities of their subordinates and adversaries and the attitudes of civilian populations relevant to operations. Physical, virtual, and information related capabilities all influence friendly, adversary, and enemy behavior. Cognitive considerations relate to decision making, both friendly and enemy, and the perceptions and behavior of populations and the enemy.

JOINT OPERATIONAL AREAS

1-127. Army forces are assigned AOs within a joint organizational construct. As such, it is important that Army commanders and staffs understand the various options JFCs have in organizing operational areas. Joint operational areas include—

- AORs.
- Theater of war.
- Theater of operations.
- Joint operations area (JOA).
- Joint special operations area (JSOA).
- Joint security area (JSA).
- Amphibious operational area.
- AO.

Figure 1-6 on page 1-28 depicts a notional combatant commander's AOR with subordinate operational areas.

Area of Responsibility

1-128. An AOR is established by the Unified Command Plan that defines geographic responsibilities for a GCC. AOR is synonymous with the term theater and should not be confused with other operational area terms such as theater of operation or AO. Only a GCC is assigned an AOR. Within an AOR, the combatant commander exercises combatant command (command authority) (COCOM) over assigned forces. All U.S. forces within an AOR (assigned, attached, operational control, or in transit through the region) fall under the control of that geographic combatant command for as long as they remain in the AOR.

1-129. GCCs conduct operations in their assigned AORs. When warranted, the President, Secretary of Defense, or GCCs may designate a theater of war or a theater of operations for each operation. GCCs can elect to control operations directly in these operational areas, or may establish subordinate joint forces for the purpose, while remaining focused on the broader AOR.

Theater of War

1-130. A theater of war is a geographical area established for the conduct of major operations and campaigns involving combat. A theater of war is established primarily when there is a formal declaration of war or it is necessary to encompass more than one theater of operations (or a JOA and a separate theater of operations) within a single boundary for the purposes of C2, sustainment, protection, or mutual support. A theater of war does not normally encompass a GCC's entire AOR, but may cross the boundaries of two or more AORs.

Theater of Operations

1-131. A theater of operations is an operational area defined by the GCC for the conduct or support of specific military operations. A theater of operations is established primarily when the scope of the operation in time, space, purpose, and employed forces exceeds what a JOA can normally accommodate. More than one joint force headquarters can exist in a theater of operations. A GCC may establish one or more theaters of operations. Different theaters will normally be focused on different missions. A theater of operations typically is smaller than a theater of war, but is large enough to allow for operations in depth and over extended periods of time. Theaters of operations are normally associated with major operations and campaigns and may cross the boundary of two AORs.

Joint Operations Area

1-132. For operations somewhat limited in scope and duration, or for specialized activities, the commander can establish a JOA. A JOA is an area of land, sea, and airspace, defined by a GCC or subordinate unified commander, in which a JFC (normally a joint task force [JTF]) conducts military operations to accomplish a specific mission. JOAs are particularly useful when operations are limited in scope and geographic area or when operations are to be conducted on the boundaries between theaters.

Joint Special Operations Area

1-133. A JSOA is an area of land, sea, and airspace assigned by a JFC to the commander of SOF to conduct special operations activities. It may be limited in size to accommodate a discreet direct action mission or may be extensive enough to allow a continuing broad range of unconventional warfare operations. A JSOA is defined by a JFC who has geographic responsibilities. JFCs may use a JSOA to delineate and facilitate simultaneous conventional and special operations. The joint force special operations component commander is the supported commander within the JSOA.

Joint Security Area

1-134. A JSA is a specific surface area, designated by the JFC as critical that facilitates protection of joint bases and supports various aspects of joint operations such as lines of communication, force projection, movement control, sustainment, C2, airbases and airfields, seaports, and other activities. JSAs are not necessarily contiguous with areas actively engaged in combat. JSAs may include intermediate support bases and other support facilities intermixed with combat elements. (See JP 3-10 for additional guidance on JSAs.)

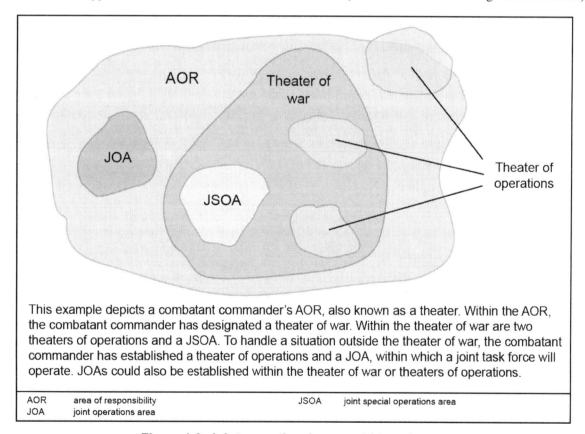

This example depicts a combatant commander's AOR, also known as a theater. Within the AOR, the combatant commander has designated a theater of war. Within the theater of war are two theaters of operations and a JSOA. To handle a situation outside the theater of war, the combatant commander has established a theater of operations and a JOA, within which a joint task force will operate. JOAs could also be established within the theater of war or theaters of operations.

| AOR | area of responsibility | JSOA | joint special operations area |
| JOA | joint operations area | | |

Figure 1-6. Joint operational areas within a theater

Overview of Army Operations

Amphibious Area of Operations

1-135. An amphibious area of operations is a geographic area within which is located the objective(s) to be secured by the amphibious force. This area must be of sufficient size to ensure accomplishment of the amphibious force's mission and must provide sufficient area for conducting necessary sea, air, and land operations.

LAND FORCE AREA OF OPERATIONS

1-136. An *area of operations* is an operational area defined by a commander for land and maritime forces that should be large enough to accomplish their missions and protect their forces (JP 3-0). AOs do not typically encompass the entire operational area of the JFC, but should be large enough for component commanders to accomplish their missions (including a designated amount of airspace) and protect their forces. Component commanders with AOs typically designate subordinate AOs within which their subordinate forces operate. These commanders employ the full range of joint and Service control measures and graphics as coordinated with other component commanders and their representatives to delineate responsibilities, deconflict operations, and achieve unity of effort.

1-137. The land component or ARFOR commander assigns subordinate AOs to maneuver units. Unit responsibilities within an assigned AO include—

- Terrain management.
- Information collection.
- Civil military operations.
- Movement control.
- Clearance of fires.
- Security.
- Personnel recovery.
- Airspace control.
- Minimum essential stability tasks.

Within an AO, commanders use control measures to assign responsibilities, coordinate fires and maneuver, and organize operations. To facilitate this integration and synchronization, commanders designate targeting priority, effects, and timing within their AOs.

1-138. Understanding the relationship between an AO, area of influence, and area of interest assists commanders in developing their operational framework. It also requires physically, temporally, cognitively and virtually understanding operations in depth. The proper application of the operational framework enables simultaneous operations and converging of effects against an enemy. An *area of influence* is a geographical area wherein a commander is directly capable of influencing operations by maneuver or fire support systems normally under the commander's command or control (JP 3-0). Understanding an area of influence helps commanders and staffs plan branches to the current operation in which the force uses capabilities outside the AO. An AO should not be substantially larger than a unit's area of influence. Ideally, an area of influence would encompass the entire AO. An AO that is too large for a unit to control can allow sanctuaries for enemy forces, creates friendly vulnerabilities and positions of advantage for the enemy to exploit and may limit joint flexibility. An *area of interest* is that area of concern to the commander, including the area of influence, areas adjacent thereto, and extending into enemy territory (JP 3-0). This area also includes areas occupied by enemy forces who could jeopardize the accomplishment of the mission. An area of interest for stability tasks may be much larger than that area associated with the offense and defense. Cognitive and virtual aspects of an area of interest are often broader than physical aspects.

1-139. Commanders assign subordinates responsibility for particular areas in order to create freedom of action, generate rapid tempo, and best use available combat power. Some capabilities, like cyberspace and information operations, can potentially generate effects far outside assigned AOs for tactical units. As such, retaining the capabilities and the authority for their employment at corps or higher echelons frees division and brigade leaders to focus on the extremely demanding lethal and physical aspects of close and deep operations. However, providing capability that cannot be used effectively or encumbers a particular echelon is wasteful of combat power.

1-140. When assigning AOs, each headquarters ensures its subordinate headquarters' capabilities align with their span of control and missions. The way in which headquarters graphically portray the operational framework and allocate resources to accomplish missions in time and space drives the pace of operations. Higher echelon headquarters create the conditions for subordinate echelons to succeed. Commanders and staffs consider the reliable availability of networks, effective span of leader control, the temporal focus at different echelons, the impact of lethality and stress on decision making, identification of targets, approval levels and timing constraints for effects delivery, sustainment requirements, physical and electro-magnetic signatures, airspace control, and clearance of fires. These considerations inform commander decisions about where capabilities are located in formations and who has ultimate responsibility for their most effective employment.

Close, Deep, Support, and Consolidation Areas

> *And to control many is the same as to control few. This is a matter of formations and signals.*
>
> Sun Tzu

1-141. Commanders designate close, deep, support, and consolidation areas to describe the physical arrangement of forces in time, space, and focus. Commanders will always designate a close area and a support area. They designate a deep area and consolidation area as required. The echelon above corps (normally the land component commander) designates the corps' AO. In addition to flank and rear boundaries, a corps forward boundary (a phase line) could be used to depict the geographic extent of the corps' responsibilities. This phase line is adjusted as required.

1-142. Figure 1-7 depicts a corps AO (organized into close, deep, support, and consolidation areas) within a theater of operations. Within the theater of operations, the corps is supported from the JSA. The joint force and corps also receives support from outside the theater of operations from the strategic support area. The strategic support area describes the area extending from a theater of war or theater of operations to a continental United States (CONUS) base or another combatant's AOR, that contains those organizations, lines of communication, and other agencies required to support forces in operations. The strategic support area includes the air and seaports supporting the flow of forces and sustainment into the theater.

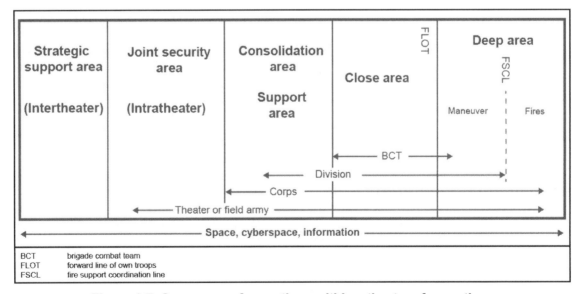

Figure 1-7. Corps area of operations within a theater of operations

1-143. The corps headquarters geographically divides its AO into subareas in which deep, close, support, and consolidate gains operations are conducted. The use of unit boundaries delineates responsibilities of subordinate units (the corps, divisions, and separate brigades) facilitates control, and enables freedom of action. The corps headquarters plans for and adjusts subordinate unit boundaries (including forward, rear, and lateral boundaries) and fire support coordination measures based on changes in the situation.

1-144. The placement of the fire support coordination line (FSCL) is a key consideration for deconflicting operations in the deep area. The *fire support coordination line* is a fire support coordination measure established by the land or amphibious force commander to support common objectives within an area of operation; beyond which all fires must be coordinated with affected commanders prior to engagement, and short of the line, all fires must be coordinated with the establishing commander prior to engagement. (JP 3-09). The area between the front line of own troops (FLOT) and the FSCL is typically the area over which friendly ground forces intend to maneuver in the near future. The corps coordinates for and controls fires in this area. Beyond the FSCL, the corps nominates targets to the land component commander and coordinates for joint interdiction. *Interdiction* is an action to divert, disrupt, delay, or destroy the enemy's military surface capability before it can be used effectively against friendly forces, or to otherwise achieve objectives (JP 3-03).

1-145. Subordinate unit AOs may be contiguous or noncontiguous. When AOs are contiguous, a boundary separates them (see figure 1-8 on page 1-32). When AOs are noncontiguous, they do not share a boundary (see figure 1-9 on page 1-33); the concept of operations links the elements of the force. The higher echelon headquarters is responsible for the area between noncontiguous AOs.

Close Area

1-146. The *close area* is the portion of a commander's area of operations assigned to subordinate maneuver forces (ADRP 3-0). Operations in the close area are operations that are within a subordinate commander's AO. Commanders plan to conduct decisive operations using maneuver in the close area, and they position most of the maneuver force within it. Within the close area, one unit may conduct the decisive operation while others conduct shaping operations (not to be confused with the Army operations to shape discussed in chapter 2). A close operation requires speed and mobility to rapidly concentrate overwhelming combat power at the right time and place and to exploit success.

1-147. The corps close operations consist of the current battles and engagements of its major maneuver units, together with the sustainment and protection activities supporting those units. The corps close operations include the close, deep, support, and consolidation operations of its committed divisions and separate maneuver brigades. Corps headquarters focus on assigning tasks and resourcing divisions with capabilities that help identify windows of opportunity, particularly in information collection, cyberspace operations, and electronic warfare. The corps headquarters reinforces divisions with supporting capabilities, including fires and aviation. In temporal terms, current planning ensures the success of the divisions by positively influencing conditions in the corps deep area while preparing to exploit success with branch plans and sequels for divisions to execute. Corps headquarters focus on what division headquarters cannot.

1-148. A division's close operations consist of the current battles and engagements of its maneuver brigades. The division headquarters focuses on information collection, sustainment, and planning that enable freedom of action for the brigades in close operations. The division headquarters also provides or coordinates capabilities in support of its BCTs such as field artillery, air and missile defense, aviation support, electronic warfare, mobility, and joint fires to either shape close operations through operations in the division deep area or reinforce BCTs in close operations. In temporal terms, current planning is focused on those things happening in the deep area that can enable success in the close area (across the warfighting functions). Division commanders commit their reserves to exploit opportunities or to counter threats in the close area. The division headquarters seeks to facilitate brigade positions of advantage for exploitation.

1-149. The division close area is primarily where brigades operate. Brigades focus on reconnaissance and security, defending areas, and securing or seizing objectives. The key to successful close operations is planning that facilitates rapid decision making to exploit opportunities, mass indirect and direct fires, properly use terrain, minimize visual and EMS signatures, disperse and rapidly maneuver, protect networks, and sustain brigades. Weapon ranges, both direct and indirect, and the mobility of formations define the characteristics of operations in the close area.

Chapter 1

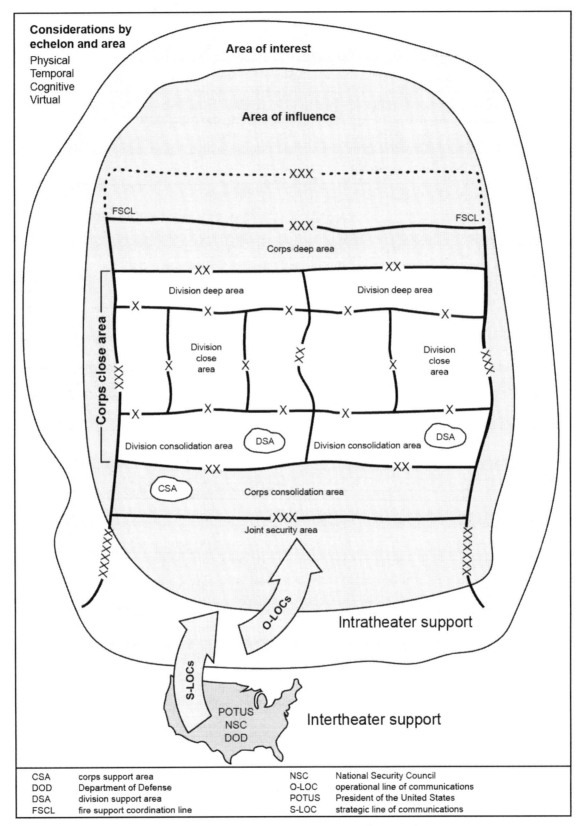

Figure 1-8. Contiguous corps area of operations

Overview of Army Operations

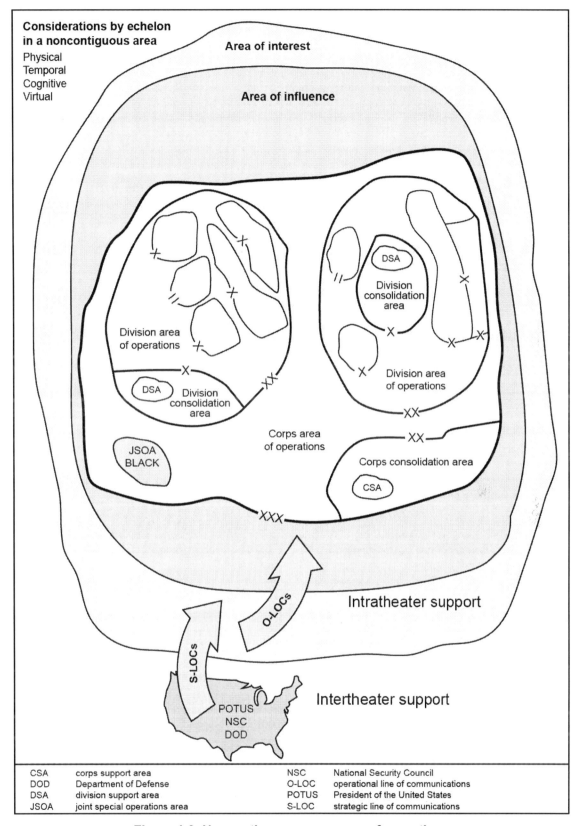

Figure 1-9. Noncontiguous corps area of operations

Chapter 1

Deep Area

1-150. A *deep area* is the portion of the commander's area of operations that is not assigned to subordinate units (ADRP 3-0). Operations in the deep area involve efforts to prevent uncommitted or out of contact enemy maneuver forces from being committed in a coherent manner or preventing enemy enabling capabilities, such as fires and air defense, from creating effects in the close area. A commander's deep area generally extends beyond subordinate unit boundaries out to the limits of the commander's designated AO. The purpose of operations in the deep area is to set the condition for success in the close area or to set the conditions for future operations. Operations in the deep area might disrupt the movement of operational reserves, for example, or prevent an enemy from employing long-range cannon, rocket, or missile fires. Planning for operations in the deep area includes considerations for information collection, airspace control, joint fires, obstacle emplacement, maneuver (air and ground), special operations, and information operations.

1-151. Enemy forces in the deep area are not out of contact in a multi-domain environment. Deep operations seeking relative advantage in cyberspace, space, and the information environment occur continuously across the range of military operations. Their purpose is to keep units in close operations unencumbered and to enable freedom of action when they are committed. Joint and army capabilities shape enemy forces and the OE before they can employ lethal or non-lethal means against friendly forces.

1-152. Corps headquarters play a significant role in physical and temporal deep area operations. Temporally, corps planners must project into the future and decide what conditions can be created and exploited to defeat the enemy and accomplish the corps mission. Corps deep operations are those activities which are directed against enemy forces not currently engaged in the close operation, but capable of engaging or inflicting damage in future close operations. Information collection, fires, EW, cyberspace operations, and tactical deception (TAC-D) focus on high-pay off targets in the corps deep area. Many of these capabilities are not resident in the corps. As such, the corps headquarters coordinates for these through planning and the targeting process for joint support (including joint fires; EW; cyberspace operations; and intelligence, surveillance, and reconnaissance).

1-153. Division deep operations, reinforced by corps capabilities, focus on information collection, fires, and maneuver on enemy organizations and capabilities beyond the range of the BCTs engaged in close operations. They identify opportunities for BCTs to exploit, disrupt enemy C2, and mass effects against key enemy capabilities, such as long range fires, to enable rapid movement. They assist brigades in identifying gaps and seams, and they coordinate the anticipatory sustainment necessary for maneuver in the deep area as conditions rapidly change.

1-154. The theater army has a deep area from a temporal and planning perspective, as it determines the desired cyberspace, space, and information capabilities for large-scale combat operations against specific adversaries during operations to shape. It provides guidance for mapping of EMS terrain and threat cyberspace capabilities, and it provides the intelligence support necessary to develop long-term situational awareness, including the collection and analysis of indications and warnings. The theater army headquarters also conducts planning for the consolidation of gains and what the posture of Army units in the theater should look like when operations to consolidate gains conclude.

Support Area

1-155. The *support area* is the portion of the commander's area of operations that is designated to facilitate the positioning, employment, and protection of base sustainment assets required to sustain, enable, and control operations (ADRP 3-0). It is where most of the echelon's sustaining operations occur. Within a division or corps support area, a designated unit, such as maneuver enhancement brigade, provides area security, terrain management, movement control, mobility support, and clearance of fires. This allows sustainment units to focus on their primary functions.

1-156. The corps headquarters is likely to position assets in the division support area to facilitate division operations and enable freedom of action. It controls movement short of division boundaries and allocates resources to support the concept of the operation. Planning in the support area largely influences current and future operations in the deep, close, and consolidation areas.

1-157. The division headquarters orchestrates the sustainment and protection tasks essential to ensuring freedom of action in the division close and deep areas. It is likely to have mission command responsibility for maneuver enhancement and sustainment brigades to execute those tasks. The division orchestrates movement and terrain management, protection of sustainment assets, and planning to support continuous operations.

Consolidation Area

1-158. The *consolidation area* is the portion of the commander's area of operations that is designated to facilitate the security and stability tasks necessary for freedom of action in the close area and to support the continuous consolidation of gains (ADRP 3-0). Corps and division commanders may establish a consolidation area, particularly in the offense as the friendly force gains territory, to exploit tactical success while enabling freedom of action for forces operating in the other areas. When designated, a consolidation area refers to an AO assigned to an organization which extends from its higher headquarters boundary to the boundary of forces in close operations where forces have established a level of control and large-scale combat operations have ceased.

1-159. The consolidation area does not necessarily need to surround, nor contain, the support area base clusters, but typically it does. It requires a purposefully task-organized, combined arms unit to conduct area security and stability tasks and employ and clear fires. This unencumbers units conducting close operations and enables the higher echelon headquarters to focus on close operations, deep operations, and future planning. The additional combat power (BCTs or divisions) necessary to execute consolidation of gains is additional to the required combat power needed for close and deep operations. The theater army and GCC must include the expected force requirements to consolidate gains during operation plan (OPLAN) refinement and operations to shape and prevent to ensure these additional forces are included in the required forces to successfully conduct large-scale combat operations. In large-scale combat operations, a maneuver enhancement brigade is assigned to the support area, and it provides support to forces in contact with the enemy. Thus, a division or corps headquarters would receive an additional subordinate unit (BCT or division, respectively) responsibility for the consolidation area. For example, a division headquarters would receive an additional BCT, and assign that BCT an AO that corresponds with the division's consolidation area.

1-160. The division and corps consolidation areas generally have different characteristics based on the situation. For a division, the BCT assigned responsibility for the consolidation area will initially focus primarily on security tasks that help maintain the tempo of operations in other areas, and it is likely to conduct offensive tasks to defeat or destroy enemy remnants in order to protect friendly forces positioned in or moving through the area. The division consolidation area grows as the BCTs in close operations advance. When division boundaries shift, as is likely during the offense, the corps consolidation area will grow, and the balance of security and stability tasks may shift towards more of a stability focus, as conditions allow. The division responsible for the corps consolidation area conducts tasks designed to set conditions for the handover of terrain to host-nation forces or legitimate civilian authorities.

1-161. The theater army plans and coordinates the capabilities necessary to exploit military success through the consolidation of gains across the AO. The planning process begins as early as possible; in some areas the threat is clearly understood even during operations to shape and plans can be relatively well developed. In others, planning must run in parallel with that planning necessary during operations to prevent and large-scale combat operations.

Decisive, Shaping, and Sustaining Operations

1-162. Decisive, shaping, and sustaining operations lend themselves to a broad conceptual orientation. The *decisive operation* is the operation that directly accomplishes the mission (ADRP 3-0). It determines the outcome of a major operation, battle, or engagement. The decisive operation is the focal point around which commanders design an entire operation. Multiple subordinate units may be engaged in the same decisive operation. Decisive operations lead directly to the accomplishment of a commander's intent. Commanders typically identify a single decisive operation, but more than one subordinate unit may play a role in a decisive operation.

1-163. A *shaping operation* is an operation that establishes conditions for the decisive operation through effects on the enemy, other actors, and the terrain (ADRP 3-0). Information operations, for example, may integrate Soldier and leader engagement tasks into an operation to reduce tensions between Army units and different ethnic groups through direct contact between Army leaders and local leaders. In combat,

Chapter 1

synchronizing the effects of close air support, attack helicopters, artillery fires, and obstacles to delay or disrupt repositioning forces are also shaping operations. Shaping operations may occur throughout an AO and involve any combination of forces and capabilities. Shaping operations set conditions for the success of the decisive operation. Commanders may designate more than one shaping operation.

1-164. A *sustaining operation* is an operation at any echelon that enables the decisive operation or shaping operation by generating and maintaining combat power (ADRP 3-0). Sustaining operations differ from decisive and shaping operations in that they focus internally (on friendly forces) rather than externally (on enemy forces or the environment). While sustaining operations are inseparable from decisive and shaping operations, they are not usually decisive themselves. Sustaining operations occur throughout an AO, not just within a support area. Failure to sustain may result in mission failure. Sustaining operations determine how quickly Army forces reconstitute and how far Army forces can exploit success.

Main and Supporting Efforts

1-165. Commanders designate main and supporting efforts to establish clear priorities of support and resources among subordinate units. The *main effort* is a designated subordinate unit whose mission at a given point in time is most critical to overall mission success (ADRP 3-0). It is usually weighted with the preponderance of combat power. Typically, commanders shift the main effort one or more times during execution. Designating a main effort temporarily prioritizes resource allocation. When commanders designate a unit as the main effort, it receives priority of support and resources in order to maximize combat power. Commanders establish clear priorities of support, and they shift resources and priorities to the main effort as circumstances and the commander's intent require.

1-166. Commanders may designate a unit conducting a shaping operation as the main effort until the decisive operation commences. However, the unit with primary responsibility for the decisive operation then becomes the main effort upon the execution of the decisive operation.

1-167. A *supporting effort* is a designated subordinate unit with a mission that supports the success of the main effort (ADRP 3-0). Commander's resource supporting efforts with the minimum assets necessary to accomplish the mission. Forces often realize success of the main effort through success of supporting efforts.

SEQUENCING OPERATIONS

1-168. Part of the art of planning is determining the sequence of actions that best accomplishes the mission. Ideally, commanders plan to accomplish a mission with simultaneous actions throughout an AO. However, operational reach, resource constraints, and the size of the friendly force may limit a commander's ability to do this. In these cases, commanders phase operations. Phasing is a way to view and conduct operations in manageable parts.

1-169. A *phase* is a planning and execution tool used to divide an operation in duration or activity (ADRP 3-0). A change in phase usually involves a change of mission, task organization, or rules of engagement. Phasing helps in planning and controlling operations during execution. Phasing may be indicated by time, distance, terrain, or event. Within a phase, a large portion of the force executes similar or mutually supporting activities. Achieving a specified condition or set of conditions typically marks the end of a phase.

1-170. Army commanders recommend to the JFC the best sequence of operations that achieve a tempo of operations to reach the desired objective. Commanders consider a variety of factors, including geography, strategic lift, command structure, logistics, enemy reinforcement, and the information environment. However, sequencing decisions for force-projection operations of ground forces is complicated by rapidly changing enemy situations. The sequence that commanders choose, therefore, should not hinder future options but should be flexible enough to accommodate change.

1-171. The sequence of large-scale combat operations (or the sequence of battles within large-scale combat operations) relates directly to the commander's decision on phasing. A phase represents a period during which a large number of forces are involved in similar activities (deployment, for example). A transition to another phase—such as a shift from deployment to the defense-indicates a shift in emphasis. For example, Phase I, the defense, could lead to Phase II, the counteroffensive, followed by a third phase that orients on consolidation and post-conflict activities. For example, World War II's Operation Overlord contained six distinct phases: buildup, rehearsals, embarkation, assault, buildup, and breakout.

1-172. Sustainment is crucial to phasing. Operational planners consider establishing logistics bases, opening and maintaining lines of communications, establishing intermediate logistics bases to support new phases, defining priorities for services and support, and securing sustainment nodes. Sustainment, then, is key to sequencing the major operations of the campaign.

1-173. Transitions mark a change of focus between phases or between the ongoing operation and execution of a branch or sequel. Shifting priorities between offensive, defensive, and stability tasks also involves a transition. Transitions require planning and preparation well before their execution so that a force can maintain the momentum and tempo of operations. A force is vulnerable during transitions, and commanders establish clear conditions for their execution. Commanders identify potential transitions during planning and account for them throughout execution. Commanders should appreciate the time required to both plan for and execute transitions. Assessment ensures that commanders measure progress toward such transitions and take appropriate actions to prepare for and execute them.

1-174. During planning, commanders establish conditions for transitioning into each phase. They adjust their phases to take advantage of opportunities presented by the enemy or to react to an unexpected setback. Actions by the enemy also determine conditions for phases. Changes in phases at any level can lead to a period of vulnerability for the force. At this point, missions and task organizations often change. Therefore, the careful planning of branches and sequels can reduce the risk associated with transition between phases.

1-175. Branches are contingency plans—options built into the basic plan—for changing the disposition, orientation, or direction of movement, and also for accepting or declining battle. They give commanders flexibility by anticipating enemy reactions that could alter the basic plan.

1-176. Sequels are subsequent operations based on the possible outcomes of the current operation: victory, defeat, or stalemate. A counteroffensive, for example, would be a logical sequel to a defense. An operation to consolidate gains would be the logical sequel to the offense. Executing a sequel will normally mean beginning another phase of the campaign. This is a continuous process during an operation; a commander should never be without options.

1-177. The dynamic nature of large-scale combat operations may demand frequent task-organization changes. It is unlikely that a single task organization suffices from the line of departure through the consolidation of gains. Commanders and staffs anticipate task organization requirements and integrate these anticipated changes into the operations process as branches or sequels, or during periods requiring a tactical pause. (See appendix A for command and support relationships and developing a task organization.)

CONTROL MEASURES

1-178. Commanders, assisted by their staffs, exercise control through control measures. A *control measure* is a means of regulating forces or warfighting functions (ADRP 6-0). Control measures are established under a commander's authority; however, commanders may authorize staff officers and subordinate leaders to establish them. Commanders may use control measures for several purposes: for example, to assign responsibilities, require synchronization between forces, impose restrictions, or establish guidelines to regulate freedom of action. Control measures are essential for coordinating subordinates' actions. They can be permissive or restrictive. Permissive control measures allow specific actions to occur; restrictive control measures limit the conduct of certain actions.

1-179. Control measures help commander's direct actions by establishing responsibilities and limits that prevent subordinate units' actions from impeding one another. They foster coordination and cooperation between forces without unnecessarily restricting freedom of action. Good control measures foster freedom of action, sound decision making, and individual initiative.

1-180. Some control measures are graphic. A *graphic control measure* is a symbol used on maps and displays to regulate forces and warfighting functions (ADRP 6-0). Graphic control measures are always prescriptive. They include symbols for boundaries, fire support coordination measures, airspace coordinating measures, air defense areas, and minefields. Commanders establish them to regulate maneuver, movement, airspace use, fires, and other aspects of operations. In general, all graphic control measures should relate to easily identifiable natural or man-made terrain features. The most important control measure is the boundary. Boundaries define the AO assigned to a commander. Commanders have full freedom of action to conduct operations within the boundaries of their AO unless the operations order establishes constraints. (See ADRP 1-02 for doctrine on Army terms and symbols.)

1-181. Commanders use the minimum number of control measures necessary to control their forces. Commanders tailor their use of control measures to conform to their higher commander's intent. They also consider the mission, terrain, and amount of authority delegated to their subordinates. Effectively employing control measures requires commanders and staffs to understand their purposes and ramifications, including the permissions or limitations imposed on their subordinates' freedom of action and initiative. Each measure should have a specific purpose: to mass the effects of combat power, synchronize subordinate forces' operations, or minimize the possibility of fratricide. Chapters 4 through 8 address commonly used graphic control measures used during all operations.

PATHS TO VICTORY

1-182. The Army is a globally engaged, regionally responsive force providing a full range of capabilities to combatant commanders. As part of a joint interdependent team, Army forces combine offensive, defensive, and stability tasks to seize, retain, and exploit the initiative and consolidate gains. Army forces shape operational environments, prevent conflict, fight and win in large-scale ground combat, and consolidate gains. Operations to shape, prevent, defend, attack, and consolidate gains summarize the Army's roles as part of a joint force.

1-183. During operations to shape, Army forces assist GCCs in shaping their regions through numerous cooperative actions with partner nations. Army forces alter conditions that, if left unchanged, can precipitate international crisis or war. The equipment, training, and financial assistance the United States provides to partner nations improves their ability to secure themselves. This assistance often improves access to key regions. Security cooperation also communicates the U.S. position to adversaries in those regions. If necessary, combat-ready Army units can deploy to threatened areas, reinforce host-nation forces, complement U.S. air and sea power, and unmistakably communicate American intent to partner and adversary alike. These are tangible effects of the Army's role in operations to shape. Other benefits are less tangible; these benefits are realized through face-to-face training involving Soldiers and military partners. Working together develops trust between military forces. The positive impression Army forces make upon multinational militaries, local leaders, and other government agencies produces lasting benefits.

1-184. During operations to prevent, Army forces set conditions to deny adversaries the ability to achieve objectives at acceptable military and political costs. Prevent activities include actions to protect and secure friendly forces, assets, and partners, and prevent activities indicate U.S. intent to execute subsequent phases of planned operations, if required. As an extension of operations to shape, Army forces conduct security cooperation tasks designed to thwart the short-term success of an adversary, such as improving the readiness and effectiveness of conventional military forces. Army forces may also assist partners with unconventional capabilities that enable protracted resistance against a more powerful neighbor, should they face occupation after the defeat of their conventional military forces. Credible capability and capacity for long-term irregular warfare can deter adversaries sensitive to the economic and military costs of long-term conflict. Building irregular capability and capacity in partner nations is likely to be more effective in situations where resource constraints and geography make other options unrealistic.

1-185. Prevention requires a credible force. Friends and adversaries must believe that the Army is credible in order to prevent conflicts. Credibility equates to capability, and capability is built upon combat-ready forces that are forward positioned or can be tailored and deployed rapidly. Partner nations under external threat need to understand that introducing U.S. forces alters the regional military balance in their favor and bolsters their capability to resist aggression. Credible Army forces, prepared to win during large-scale combat operations, reduce the risk of miscalculation by an adversary. When combined with the capabilities of the joint force, Army forces are a powerful deterrent.

1-186. During large-scale combat operations, Army forces attack and defend against enemy forces. JFCs require Army units skilled in the use of combined arms and able to employ capabilities across multiple domains in complementary ways. If an enemy cannot be defeated from a distance using Army and joint capabilities, then Army units close with and destroy that enemy. Enemies whose operational approach employs a system consisting of an integrated fires complex and integrated air defense require a level of centralized control and intelligence, surveillance, and reconnaissance capability that is vulnerable to deception, disruption, and isolation. Offensive tasks that effectively isolate parts of the system, destroy key components of it so as to enable rapid tactical maneuver at operational depth, and exploit disruptions to the system before enemies can establish or re-establish the ability to mass effects will dislocate and ultimately collapse the enemy's ability to resist. Establishing and retaining mobility generates the positions of relative advantage on the ground for maneuver forces to exploit.

1-187. Tactical success wins battles, but it is not enough to win wars. Enduring security and political outcomes require ending enemies' abilities to resist by following through on battlefield successes. Army forces play a vital role in the consolidation of gains. Rapidly consolidating gains is a form of exploitation that transitions military positions of advantage into enduring operational and strategic outcomes. Ensuring that enemies cannot transition a conventional military defeat into a protracted conflict that negates initial successes is foundational to victory. Every part of an enemy's ability to resist must be accounted for and addressed during planning. Execution of consolidate gains tasks may occur sequentially with brigades, but simultaneously at the division, corps, and army level. Building irreversible momentum towards the desired end state is a continuous process, and commanders should pursue momentum towards the end state relentlessly.

This page intentionally left blank.

Chapter 2
Army Echelons, Capabilities, and Training

[T]he real object of having an army is to provide for war.

Elihu Root

This chapter is divided into three sections. Section 1 provides and overview of Army echelons, capabilities, and training. Section II provides a general discussion of Army forces in a theater. Section III discusses Army capabilities by warfighting function. Section IV discusses how Army forces train for large-scale combat operations.

SECTION I – OVERVIEW OF ARMY ECHELONS, CAPABILITIES, AND TRAINING

2-1. As the Nation's decisive land force, the Army provides a mix of headquarters, units, and capabilities to geographic combatant commanders (GCCs) in support of a theater campaign plan (TCP) and specific joint operations. In order to effectively command these organizations, the Army provides an echeloned array of higher headquarters designed toward a specific function or mission. Winning in large-scale ground combat requires Army forces that can integrate landpower in a multi-domain approach to defeat enemy forces and control terrain. The theater army, corps, and division headquarters give the combatant commander several options necessary for the employment of landpower. (See paragraphs 2-4 through 2-103 for a discussion of Army echelons.)

2-2. The ability of Army forces to shape operational environments, prevent conflict, defeat enemy forces in large-scale ground combat, and consolidate gains relates to the quantity of combat power they can continuously generate and apply. Combat power includes all capabilities provided by unified action partners that are integrated, synchronized, and converged with the commander's objectives to achieve unity of effort in sustained operations. The purpose of combat power is to accomplish missions. (See paragraphs 2-104 through 2-266 for a discussion of Army capabilities).

2-3. The combatant commander has ever-changing needs for trained and ready Army forces while executing the TCP. Shaping operational environments (OEs) and preventing conflict requires flexible and credible U.S. military power to dissuade potential adversaries from threatening vital American security interests. During large-scale combat operations, joint force commanders require Army units that can defeat an enemy and consolidate gains to ensure enduring outcomes. (See paragraphs 2-267 through 2-320 for a discussion on training for large-scale combat operations.)

SECTION II – ARMY ECHELONS

The land domain is the area of the Earth's surface ending at the high water mark and overlapping with the maritime domain in the landward segment of the littorals.

JP 3-31, *Command and Control for Joint Land Operations*, 2014

THEATER ARMY

2-4. The theater army is the senior Army headquarters in an area of responsibility (AOR), and it consists of the commander, staff, and all Army forces assigned to a combatant command. Each theater army has operational and administrative responsibilities. Its operational responsibilities include command of forces, direction of operations, and control of assigned areas of operations (AOs). Its administrative responsibilities

Chapter 2

encompass the Service-specific requirements for equipping, sustaining, training, unit readiness, discipline, and personnel matters. As required, the theater army provides Army support to other services and common user logistics.

2-5. The theater army always maintains an AOR-wide focus, providing support to Army and joint forces across the AOR, in accordance with the GCC's priorities of support. For example, the theater Army continues shape and prevent activities in various operational areas at the same time it is supporting large-scale combat operations.

OPERATIONAL RESPONSIBILITIES

2-6. Through assignment or allocation of Army forces to a GCC by the Secretary of Defense, the theater army may exercise operational control (OPCON) (or other delegated command authority) of Army forces until the combatant commander attaches those forces to another subordinate Service or joint command. OPCON provides authority to organize and employ commands and forces as the commander considers necessary to accomplish missions. It does not include authoritative direction for logistics or matters of administration, discipline, internal organization, or unit training. (See appendix A for a discussion of command and support relationships).

2-7. The theater Army serves as the Army Service component command (ASCC) of the geographic combatant command. The *Army Service component command* is the command responsible for recommendations to the joint force commander on the allocation and employment of Army forces within a combatant command (JP 3-31). As an ASCC, the theater army executes several functions in support of the GCC. These functions include—

- Executing the combatant commander's daily operational requirements.
- Setting the theater.
- Setting the joint operations area (JOA).
- Serving as a joint task force (JTF) or joint force land component for crisis response and limited contingency operations.

2-8. Key tasks associated with the theater army's roles include—

- Serving as the primary interface between the Department of the Army, Army commands, and other ASCCs.
- Developing Army plans to support the TCP plan within an AOR.
- Tailoring Army forces for employment in an AOR.
- Controlling reception, staging, onward movement, and integration (RSOI) for Army forces in an AOR.
- Exercising OPCON of deployed Army forces not subordinated to a joint force commander (JFC).
- Exercising administrative control (ADCON) of all Army forces operating within the AOR.
- Providing support as directed by the combatant commander to unified action partners.
- Exercising OPCON of all joint forces attached to it as either a joint force land component headquarters or JTF headquarters, as required by the combatant commander.

(See ATP 3-93 for more information on the theater army.)

ADMINISTRATIVE RESPONSIBILITIES

2-9. The ASCC completes its administrative responsibilities through ADCON—the direction or exercise of authority over subordinate or other organizations in respect to administration and support. ADCON includes organization of Service forces, control of resources and equipment, personnel management, unit logistics, individual and unit training, readiness, mobilization, demobilization, discipline, and other matters not included in the operational missions of the subordinate or other organizations. It is a Service authority, not a joint authority. It is exercised under the authority of and is delegated by the Secretary of the Army. Regardless of whether Army forces are OPCON to the theater army or not, the theater army commander retains responsibility for ADCON of all Army forces. The ASCC normally has responsibility for—

Army Echelons, Capabilities, and Training

- Training units.
- Supplying.
- Administering (including the morale and welfare of personnel)
- Maintaining.

ARMY SUPPORT TO OTHER SERVICES AND COMMON-USER LOGISTICS

2-10. The Army provides certain support to other Services across all phases of a joint operation through several types of authorities. Collectively known as Army support to other Services (ASOS), these responsibilities include all executive agent responsibilities assigned to Department of the Army by the Secretary or Deputy Secretary of Defense.

2-11. An executive agent is the head of a Department of Defense (DOD) component that has been assigned specific responsibilities, functions, and authorities to provide defined levels of support for operational missions or administrative or other designated activities that involve two or more of the DOD components. Whether the term executive agent is used or not, the theater army, on behalf of the Army, is responsible for support functions in all theaters as designated by the GCC or higher. The supporting requirements the theater army provides as part of ASOS can include—

- Missile defense.
- Fire support.
- Base defense.
- Transportation.
- Fuel distribution.
- General engineering.
- Intra-theater medical evacuation.
- Logistics management.
- Communications.
- Chemical, biological, radiological, and nuclear (CBRN) defense.
- Explosive ordnance disposal.

2-12. In addition to ASOS requirements, a GCC may designate a Service (usually the dominant user or most capable service) to serve as the lead Service and provide common-user logistics for the entire theater, areas within a theater, or specific joint operations. *Common-user logistics* are materiel or service support shared with or provided by two or more Services, Department of Defense agencies, or multinational partners to another Service, Department of Defense agency, non-Department of Defense agency, and/or multinational partner in an operation (JP 4-09). The GCC frequently tasks the Army component of a joint force to provide sustainment support to other Service components. Additionally, the GCC may task the Army component of a joint force to provide specific support to multinational commands or other agencies.

THEATER ARMY ORGANIZATION

2-13. The size and composition of forces available to a theater Army vary based on the combatant commander's continuing requirements for Army support. During operations to shape and operations to prevent, the theater army may require more sustainment or civil affairs (CA) units. During the conduct of large-scale combat operations, the theater army may be task organized with an Army Air and Missile Defense Command and Army CBRN units to support theater force protection operations. During operations to consolidate gains, the theater army may require more military police units and a theater aviation brigade.

2-14. Prior to the outbreak of large-scale ground combat, theater armies usually have access to five enabling capabilities (sustainment, signal, medical, military intelligence, and CA). Figure 2-1 on page 2-4 illustrates an example of forces that may be available to a theater army to provide these capabilities. These forces are either allocated or assigned to the combatant commander, who establishes command and support relationships with the theater army as required. Not every theater army will have the forces shown. In some cases, a brigade is task organized to an Army command (or direct reporting unit), and aligned to the theater army. In other cases, the theater army has a brigade instead of a full command.

Chapter 2

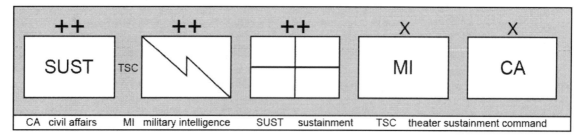

Figure 2-1. Theater enabler organizations

THEATER SUSTAINMENT COMMAND

2-15. The theater sustainment command (TSC) is the senior Army sustainment headquarters supporting an AOR. It commands task-organized, subordinate organizations executing theater opening, theater distribution and sustainment operations, and theater closing in support of theater army objectives. The TSC executes distribution and materiel management responsibilities through a distribution management center. The TSC may be augmented with appropriate personnel and equipment to perform as a joint command for logistics.

2-16. The TSC has AOR-wide responsibilities and manages distribution in subordinate operational areas as directed. The TSC sets the theater conditions for successful sustainment operations. The TSC may maintain oversight of sustainment operations or, more likely, execute this oversight through employment of a subordinate expeditionary sustainment command (ESC). The ESC will employ sustainment brigades to execute theater opening, theater sustaining, and theater distribution operations. Generally, when two or more sustainment brigades deploy to a JOA, the Army tailors the TSC with an ESC. (See ATP 4-94 for doctrine on the TSC.)

Expeditionary Sustainment Command

2-17. For joint operations requiring substantial commitment of Army forces, at least one ESC is attached to the TSC. The ESC commands attached sustainment units in a land AO defined by the JFC. The ESC concentrates on synchronizing operational-level sustainment operations to meet the day-to-day and projected operational requirements of the supported force. The ESC supports the deployed force within a JOA while the TSC maintains an AOR-wide focus.

2-18. The ESC plans and executes sustainment, distribution, theater opening, and RSOI for Army forces. The ESC may serve as the basis for an expeditionary joint command for logistics when directed by the combatant commander or designated coalition or JTF commander.

2-19. When an ESC serves as a joint command for logistics, the headquarters should be augmented by personnel and equipment from other Services. Additional ESCs may be attached to the TSC as the theater of operations expands or if the combatant commander establishes a joint security area or intermediate staging base.

2-20. Depending on the situation, ESCs may have a command or support relationship other than with the TSC. When the TSC is deployed, the ESC normally has a command relationship (attached) with the TSC and a support relationship with supported units. When the TSC is not deployed or if leaders want to achieve specific effects, the ESC may have a command relationship, such as attached, OPCON, or under tactical control (TACON) of another headquarters. For example, an ESC which began an operation subordinate to the TSC may subsequently have a command relationship with a corps headquarters or JTF.

Sustainment Brigades

2-21. The theater army commander and TSC commander task-organize sustainment brigades for a joint operation or campaign. Each sustainment brigade is a multifunctional sustainment organization with a flexible, headquarters capable of accomplishing multiple sustaining missions. The number of combat sustainment support battalions (CSSBs) and functional logistics battalions attached to the sustainment brigade varies with the task organization established by the ESC commander. The specific task organization of each sustainment brigade will vary based on its mission.

2-22. The sustainment brigades normally remain attached to the TSC or ESC. However, the sustainment brigade may have a command or support relationship with the maneuver headquarters.

2-23. Sustainment brigades are task organized based on deliberate analysis by the sustainment brigade's higher headquarters based on the mission, type, and size of the supported formation. Subordinate units of the sustainment brigade may include CSSBs, functional logistics battalions, and functional logistics companies, platoons, and detachments. Selected CSSBs may also be organized to provide specific types of support to brigade combat teams (BCTs) and to other support brigades lacking full internal sustainment capability. The sustainment brigade usually has human resources and financial management units attached. Under normal circumstances, the sustainment brigade will not have medical organizations attached.

> **Sustainment: OPERATIONS DESERT SHIELD and DESERT STORM**
>
> The logistics support for OPERATIONS DESERT SHIELD and DESERT STORM demonstrates the impact of developing a theater plan for supporting deployed forces. The replacement operations management system processed and deployed 18,000 filler personnel. Of the 139,000 reserve personnel mobilized, 124,500 of them were in 1,033 units and 14,900 were from the individual ready reserve. While 40 percent of the Army's combat service support assets were deployed to the theater, 60 percent of those assets came from the Reserve Components.
>
> The modern division consumes as much as a World War II field army. During OPERATION DESERT SHIELD, the defensive phase of the Gulf War, each division required 345,000 gallons of diesel fuel, 50,000 gallons of aviation fuel, 213,000 gallons of water, and 208 40-foot tractor-trailers of other supplies each day, ranging from barrier material to ammunition. During OPERATION DESERT STORM, a 100-hour offensive, a single division consumed 2.4 million gallons of fuel transported on 475 5,000-gallon tankers.

THEATER-LEVEL SIGNAL SUPPORT

2-24. The joint force depends upon an integrated communications architecture that connects strategic, operational, and tactical commanders across the globe. The *Department of Defense information network* is the set of information capabilities, and associated processes for collecting, processing, storing, disseminating, and managing information on-demand to warfighters, policy makers, and support personnel, whether interconnected or stand-alone, including owned and leased communications and computing systems and services, software (including applications), data, security services, other associated services, and national security systems (JP 6-0). Operation and defense of DODINs is largely a matter of overarching common processes, standards, and protocols integrated by USCYBERCOM. ARCYBER oversees network operations and defense for the Army. The Army connects to DODINs through the U.S. Army Network Enterprise Technology Command and its subordinate signal commands and brigades, including—
- Signal command (theater).
- Theater strategic signal brigade.
- Theater tactical signal brigade.

(See FM 6-02 for doctrine on signal support to operations.)

MEDICAL COMMAND (DEPLOYMENT SUPPORT)

2-25. The medical command (deployment support) is the senior medical command within a theater. The medical command (deployment support) commands medical units that provide health care in support of deployed forces. The medical command (deployment support) is a regionally focused command and provides subordinate medical organizations to operate under the medical brigade (support) or multifunctional medical battalion (medical battalion [multifunctional]). (See FM 4-02 for doctrine on the Army health system.)

MILITARY INTELLIGENCE BRIGADE-THEATER

2-26. A military intelligence brigade-theater (MIB-T) is assigned to the combatant command and may be attached, OPCON, or TACON to the theater Army by the combatant commander. The MIB-T provides regionally focused collection and analysis in support of theater army daily operations requirements and specific joint operations in the AOR. In particular, the theater army headquarters relies heavily on the MIB-T for threat characteristics, intelligence estimates, threat and civil considerations, data files and databases, and all-source intelligence products. These products support theater army planning requirements, including development of Army plans supporting the theater campaign plan and maintenance of operation plans (OPLANs) and contingency plans. The theater army headquarters depends on the MIB-T for intelligence operations and analytic support.

2-27. The MIB-T's regional focus enhances its capabilities to develop and exploit language skills and cultural insights specific to the AOR. The MIB-T's regional focus also provides the benefits of continuity and cultural context to its analytic intelligence products. The MIB-T can collect, analyze, and track the threat characteristics and doctrine of partner nations, enemies, and adversaries over many years. These abilities allow the MIB-T to create and maintain a valuable data base of intelligence regarding regional military forces, persons of interest, and evolving doctrine and capabilities of regional military forces.

CIVIL AFFAIRS COMMAND

2-28. Each theater army receives support from an apportioned Reserve Component civil affairs command (CACOM). The CACOMs provide theater-level planning, coordination, policies, and programs in support of a GCC's regional civil-military operations strategy and stabilization, reconstruction, and development efforts through planning teams. CACOMs provide support to theater armies, JTFs, land component commands, or the senior Army forces headquarters. Each CACOM has the capability to establish civil-military operations centers to integrate, coordinate and synchronize civil-military operations in support of civil administration or transitional military authority.

2-29. The CACOM's subordinate brigades provide support to corps with planning teams that conduct liaison, coordination, education and training, and assessment functions. The brigades establish civil-military operations centers to coordinate and interface with U.S. forces and indigenous populations and institutions, humanitarian organizations, intergovernmental organizations, nongovernmental organizations, multinational forces, host-nation government agencies, civilian agencies of the U.S. Government, and other unified action partners. Civil military operations centers may facilitate continuous coordination among the key participants conducting CA operations from local levels to international levels within a given AO, and they develop, manage, and analyze the civil inputs to the common operational picture (COP).

2-30. CA battalions are task-organized to support Army corps or divisions through planning teams providing liaison, coordination, education and training, and area assessment functions. (See FM 3-57 for doctrine on CA operations.)

THE EXPANDED THEATER

2-31. For large-scale combat operations or protracted joint operations, the theater army may be reinforced by an array of Army capabilities deployed from the United States and supporting theater armies. Other Army functional or multifunctional headquarters and units may be made available to the theater army based on requirements of the AOR, including forward stationing, base operations, security force assistance missions, theater security cooperation activities, or ongoing military operations. These Army functional or multifunctional units may have either a command or a support relationship with the theater army. In some cases, the Department of the Army tasks certain functional or multifunctional battalions to support more than one theater army. Figure 2-2 provides an example of a theater army task-organized for large-scale combat operations.

Army Echelons, Capabilities, and Training

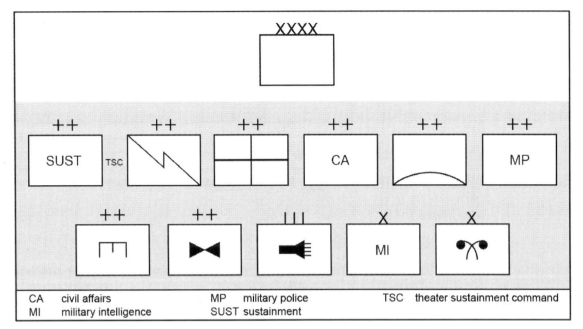

Figure 2-2. Example of a theater army organization for large-scale combat operations

ARMY AIR AND MISSILE DEFENSE COMMAND

2-32. When required by the GCC, an Army air and missile defense command (AAMDC) may be attached or OPCON to the theater army. Air defense artillery (ADA) units in the AOR may have a command or support relationship with the AAMDC.

2-33. The JFC establishes air and missile defense (AMD) priorities, allocates forces, and apportions air power. The JFC typically assigns overall responsibility for counter-air and interdiction missions to the joint force air component commander (JFACC). Normally, the JFACC also serves as both the area air defense commander (AADC) and the airspace control authority. The AADC is responsible for defensive counter-air operations. The AADC coordinates with joint and multinational partners to develop procedures for the theater air defense plan. The AAMDC supports the AADC for AMD throughout the AO.

2-34. The AAMDC commander has several responsibilities. These include command of all subordinate Army AMD units; Army AMD coordinator for the land component and ARFOR; and deputy AADC (if designated). The AAMDC commander has overall responsibility for planning Army AMD operations in support of the JFC. The AAMDC task-organizes and assigns missions to subordinate brigades. The AAMDC has dedicated liaison teams able to deploy to a major theater. It provides elements to the joint force air component, joint force land component, and joint special operations task force to integrate AMD operations.

2-35. The threat from hostile ballistic missiles, aircraft, cruise missiles, and unmanned aircraft systems (UASs) may originate from outside the JOA. The AAMDC commander distributes AMD assets to protect not only the JOA, but also other high-value assets across the AOR. If the AAMDC is not located in the JOA, the ADA brigade commander may serve as the deputy AADC. That brigade takes responsibility for AMD planning and for providing liaison to the joint force land component headquarters and AADC or JFACC. (See JP 3-01 and FM 3-01 for doctrine on theater AMD.)

MILITARY POLICE COMMAND

2-36. The theater army normally receives one military police (MP) command when more than one MP brigade is required. The MP command reinforces and augments tactical-level MP efforts and develops theater detention areas or dislocated civilian operations bases. The commander of the MP command also serves as the commander of detainee operations. The MP brigade is one of the Army's functional brigades. It can command up to five mission-tailored MP battalions; integrate capabilities from all three MP disciplines

Chapter 2

(police operations, detention operations, and security and mobility support); and integrate police intelligence operations. It can also command other non-MP units (focused on performing area support, detention, or dislocated civilian tasks) and synchronize MP support across multiple organizations that control an AO in support of stability tasks. A functional MP brigade is not designed to control terrain; significant augmentation and task organization would be required to assign an MP brigade an AO.

THEATER ENGINEER COMMAND

2-37. The theater engineer command (TEC) provides theater-wide engineer support as well as engineer support to forces deployed within a JOA. The TEC supervises geospatial support, construction, real property maintenance activities, sustainment of lines of communications, engineer logistics management, and base camp development. The command supports Department of State or host-nation efforts to restore essential services, and it aids in infrastructure development. The TEC is typically task organized with subordinate engineer brigades.

2-38. The TEC typically serves as the senior engineer headquarters for the theater army and all assigned or attached engineer brigades and other engineer units. When directed, it may also command engineers from other Services and multinational forces and provide oversight of contracted construction engineers. The TEC provides peacetime training and support of military engagement for its supported combatant commanders. The command also coordinates closely with the DOD construction agent in the JOA. (See DODD 4270.5 for more information about responsible areas for senior contract construction agents. See ATP 3-34.23 for doctrine on theater-level engineer capabilities.)

THEATER AVIATION COMMAND

2-39. The theater aviation command provides air traffic service, airfield management, aeromedical evacuation, theater aviation support, and coordination of aviation staging and onward movement in support of corps, Army, or joint operations in theater.

2-40. The theater aviation brigade (general support) supports the theater with additional general support aviation battalions to perform assault, heavy lift, aeromedical evacuation, and air movement. Each theater aviation brigade can conduct assault or general support aviation tasks in support of the theater army and its subordinate commands. Unlike combat aviation brigades (CABs), a theater aviation brigade lacks attack and reconnaissance battalions. The theater aviation brigade has a mix of lift helicopters and fixed-wing aircraft. The brigade can conduct air assault, air movement, and sustaining operations. It will normally not have attached UASs. The theater aviation brigade reinforces CABs with additional assault, general support, heavy lift, and aeromedical evacuation. If properly task organized, with additional mission command and staff and maintenance assets, the theater aviation brigade can conduct other traditional CAB missions. If task organized with a theater fixed-wing battalion, the theater aviation brigade flies fixed-wing sorties in support of the ARFOR, theater army, and joint force land component. (See FM 3-04 for more discussion of the theater aviation brigade and theater fixed wing battalion.)

2-41. A theater aviation sustainment maintenance group may be attached to a theater army to provide depot level maintenance support. The theater aviation sustainment maintenance group may be subsequently attached to the TSC. (See TC 3-04.7 for more information on theater aviation sustainment maintenance.)

2-42. Army air traffic service units at the theater level consist of the theater airfield operations group with its subordinate airfield operations battalions. These units are normally attached to the theater aviation command; they establish and operate airfields as needed in the AOR. Advanced operations bases can operate fully instrumented airfields with airport surveillance radar approach, precision approach radar, and controlling airspace necessary to support airfield operations. (See FM 3-04.120 for more information on Army air traffic service units.)

PSYCHOLOGICAL OPERATIONS GROUP

2-43. Psychological operations groups from the Army Reserve support conventional Army forces within an AOR. These groups provide the formations that support Army corps, divisions, and brigades with military information support operations (MISO). A group forming a military information support task force normally operates in support of a corps, but it may provide direct support to a division. These groups' capabilities are tactical in nature, and they lack the level of analysis, production, and dissemination required at the operational and strategic levels. They receive augmentation, including language and cultural expertise, regional analysis, and mass communications delivery capabilities. Depending upon its missions, a military information support task force receives additional augmentation from the supported maneuver unit and the Army Reserve strategic dissemination company. As the Army Reserve provides the only conventional force psychological operations capability, early mobilization and integration into predeployment training and the force deployment process is critical.

2-44. A tailored joint military information support task force supports the combatant commander, JTF commander, theater special operations commander, and joint special operations task force commander. The military information support task force plans, develops, and (when directed) executes MISO. Joint military information support task forces function as the central coordination point for all MISO activities executed in an AOR. Joint military information support task forces contribute to the planning and execution of MISO to achieve the joint commander's overall objectives. (See FM 3-53 for more information on MISO).

CHEMICAL, BIOLOGICAL, RADIOLOGICAL, AND NUCLEAR BRIGADE

2-45. Army CBRN brigades offer a range of capabilities to the theater. Units from these brigades can be task organized to support the combatant commander, subordinate JFCs, Army force commanders, and functional components faced with CBRN threats or hazards. The CBRN task organization within a theater depends on the weapons of mass destruction (WMD) threat. (See ATP 3-11.36 for more information on CBRN units.)

OTHER ORGANIZATIONS FOR THEATER SUPPORT

2-46. Several other organizations provide support to the theater army. These organizations include the explosive ordnance disposal (EOD) group, theater information operations group, regional support groups, battlefield coordination detachment, Army special operations forces (SOF), and security force assistance brigades.

EXPLOSIVE ORDNANCE DISPOSAL GROUP

2-47. A theater army supporting large-scale combat operations is allocated an EOD group. An EOD group commander can exercise mission command for two to six EOD battalions. An EOD battalion conducts staff planning and staff control of EOD assets within a division AO. EOD groups and battalions position their EOD companies at locations where they can best provide support throughout an operational area. The EOD group headquarters commands all Army EOD assets and operations in a theater and can serve as the basis for a counter-improvised explosive device task force. The group may also form the core of a specialized combined JTF with the mission of providing various protection and exploitation enablers such as counter-improvised explosive device, exploitation, or countering weapons of mass destruction (CWMD) task forces. The group can also provide enabling support, analysis, and support to targeting efforts, theater exploitation, and CWMD. The senior EOD commander normally functions as the EOD special staff officer for the senior deployed Army headquarters. (See ATP 4-32.1 and ATP 4-32.3 for doctrine on explosive ordnance disposal.)

THEATER INFORMATION OPERATIONS GROUPS

2-48. The Army provides information operations support to the theater army through elements dedicated to helping theater organizations analyze and operate within an increasingly complex information environment. These units augment theater forces with deployable, mission-tailored, support teams and continental U.S. based operational planning support, intelligence analysis, and technical assistance. These support organizations include the 1st Information Operations Command (Land) from the Regular Army and theater information operations groups from the Reserve Component. The 1st Information Operations Command (Land) is assigned to the United States Army Intelligence and Security Command and OPCON to ARCYBER. These organizations provide the following support—

- Field support teams provide information operations subject matter expertise to supported commands. These teams help those commands with the planning, execution, and assessment of information operations.
- Vulnerability assessment teams help the supported commands identify information operations and cyberspace vulnerabilities within their operational procedures, policies, practices, and training. These teams also collaborate with the supported commands as they work to resolve identified vulnerabilities.
- Operations security (OPSEC) support teams assist supported commands in assessing and developing unit operations security programs.
- Reach-back elements provide information operations and cyberspace operational planning support, intelligence analysis, and technical assistance for deployed forces requesting support.
- Cyberspace opposing forces provide a non-cooperative cyberspace threat during major exercises and training center rotations to help fully challenge the ability of deploying units to operate in a hostile cyberspace threat environment.

REGIONAL SUPPORT GROUPS

2-49. A theater army may receive a regional support group to provide contingency and expeditionary base operations support. These groups have responsibilities for managing facilities, providing administrative and logistics support of Soldier services, and ensuring the security of personnel and facilities on a base camp.

BATTLEFIELD COORDINATION DETACHMENT

2-50. A battlefield coordination detachment (BCD) is a specialized, regionally focused Army element that serves as the senior Army operational commander's liaison with the air component. A BCD is co-located with the joint air operations center (JAOC), combined air operations center, or the Air Force air operations center. The BCD is the Army's interface for systems connectivity to the JAOC and for personnel integration with their JAOC counterparts. BCD tasks include facilitating the exchange of current intelligence and operational data, processing air support requests, monitoring and interpreting the land battle situation, coordinating AMD, coordinating airlift, and integrating airspace requirements. (See ATP 3-09.13 for doctrine on the BCD.)

2-51. In large-scale combat operations, the BCD supports the joint force land component. Army corps relay requirements and requests to the land component, who, in turn, relays land component requirements and requests for joint force air component support through the BCD. The BCD represents the joint force land component commander throughout the joint air tasking cycle in the JAOC.

ARMY SPECIAL OPERATIONS FORCES

2-52. The theater special operations command is the subordinate special operations command through which the GCC normally exercises OPCON of all Army special operations forces (ARSOF) within the AOR. (See FM 3-05.) The commander of the theater special operations command serves as the primary advisor to the combatant commander for applying regionally aligned ARSOF. As directed by the GCC, the theater army provides support to deployed SOF. The special operations commander coordinates with the theater army for sustainment requirements. The ADCON of ARSOF and logistics support of SOF-unique items will normally remain in special operations channels.

SECURITY FORCE ASSISTANCE BRIGADE

2-53. Security force assistance brigades provide theater army commanders the capability to support theater security cooperation activities and build partner-nation security force capacity. Each security force assistance brigade is organized with a headquarters and headquarters company, two advisory maneuver battalions (either an infantry or combined arms battalion), one advisory cavalry squadron, one advisory field artillery battalion, one advisory engineer battalion (with embedded signal and military intelligence companies), and an advisory brigade support battalion to focus primarily on tactical and operational advising. Security force assistance brigades have the capability to conduct tactical advisory missions.

CORPS

2-54. Large-scale combat operations may require a corps headquarters to function as a tactical land headquarters under a joint or multinational land component command. A corps is normally the senior Army headquarters deployed to a JOA. It commands Army and multinational forces in campaigns and major operations.

2-55. A corps headquarters is organized, trained, and equipped to control the operations of two to five divisions, together with supporting theater-level organizations. The distinguishing differences between corps and division operations are their scope and scale. During large-scale combat operations, a corps operates as a formation, not just as a headquarters. Normally, a corps exercises OPCON over two or more U.S. Army divisions and a variety of supporting brigades, it exercises TACON over various multinational units and U.S. Marine Corps units, and it is supported by various theater sustainment organizations. The corps has both operational and administrative responsibilities.

CORPS OPERATIONAL RESPONSIBILITIES

2-56. Corps conduct offensive, defensive, and stability tasks through a series of coordinated and integrated division and separate brigade operations. These operations achieve positions of relative advantage across multiple domains in order to destroy or defeat an enemy and achieve the overall purpose of the operations. Commanders direct decisive action tasks to create and exploit positions of relative advantage by using the appropriate combination of defeat and stability mechanisms that best accomplish the mission. (See chapter 1 for a discussion of defeat and stability mechanisms.)

2-57. The corps commander synchronizes the employment of joint capabilities in conjunction with Army decisive action. Corps operations shape an OE and set the conditions for tactical actions by the division and lower echelons. In large-scale combat operations, the corps task-organizes and maneuvers divisions to destroy enemy land forces, seize key terrain and critical infrastructure, and dominate the land portion of the JOA. Corps tasks associated with the conduct of large-scale combat operations include—

- Conduct shaping operations within the corps AO.
- Task-organize and employ divisions and brigades.
- Integrate and synchronize operations of divisions and brigades.
- Mass effects at decisive points.
- Allocate resources and set priorities.
- Leverage joint capabilities.

(See ATP 3-92 for information on corps operations.)

2-58. A corps receives capabilities and units from the theater army to conduct operations. In addition to its divisions, the corps may directly control BCTs and several different types of multifunctional and functional brigades. There is no standard configuration for a corps, but a corps will generally require a maneuver enhancement brigade (MEB), a CAB, an ESC, a field artillery brigade, and a MIB-T in order to conduct large-scale combat operations. Other units may provide direct or general support. Figure 2-3 on page 2-12 shows an example corps task organization.

2-59. Based on the assigned tasks of the divisions and the allocation of brigades, the corps commander determines the appropriate command and support relationships for subordinate divisions and brigades. (See tables A-2 and A-3 in appendix A for listings of these command and support relationships.) A corps may retain some brigades in reserve or for consolidation of gains activities.

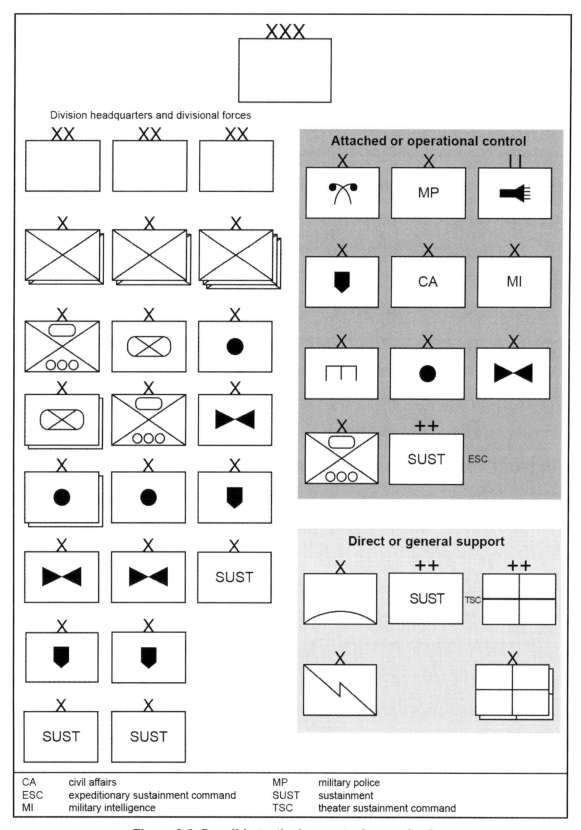

Figure 2-3. Possible tactical corps task organization

Corps Administrative Responsibilities

2-60. All JTFs that include Army forces have an ARFOR. The *ARFOR* is the Army component and senior Army headquarters of all Army forces assigned or attached to a combatant command, subordinate joint force command, joint functional command, or multinational command (FM 3-94). When a corps is an ARFOR, the ARFOR consists of the corps commander, the corps headquarters, and all the Army forces attached to the JTF.

2-61. As an ARFOR, the corps provides administrative and logistics support to all Army forces assigned to these organizations as specified by the theater army. The theater army commander specifies the ADCON responsibilities of the ARFOR, with the theater army retaining control of RSOI, logistics support of the deployed force, personnel support, and medical support. Administrative responsibilities retained by the corps include internal administration and discipline, training within the JOA, and Service-specific reporting. (See FM 3-94, chapter 2, for further details.)

DIVISIONS

2-62. Divisions are the tactical units of execution for a corps. A division's primary role is as a tactical headquarters commanding brigades in decisive action. A division combines offensive, defensive, and either stability or defense support of civil authorities tasks in an AO assigned by its higher headquarters, normally a corps. It task-organizes its subordinate forces to accomplish its mission. During large-scale combat operations, a division operates as a formation and not only as a headquarters. The corps commander determines the number and types of BCTs necessary for the divisions to accomplish their respective missions. Divisions have operational and administrative responsibilities.

Division Operational Responsibilities

2-63. A division headquarters is organized, trained, and equipped to command the operations of two to five BCTs. Divisions are typically task-organized with a combination of armored, infantry, and Stryker BCTs. These BCTs are dependent on the enabling capabilities at division and corps level. Division tasks associated with the conduct of large-scale combat operations include—
- Conduct shaping operations within the division AO.
- Task-organize and employ BCTs and multifunctional and functional brigades.
- Integrate and synchronize operations of BCTs and multifunctional and functional brigades.
- Mass effects at decisive points (focus BCTs, multifunctional brigades, functional brigades, and joint capabilities)
- Allocate resources and set priorities.
- Leverage joint capabilities.

(See ATP 3-91 for more information on division operations.)

2-64. A division receives capabilities and units from its corps to conduct operations. In addition to the BCTs, a division may directly control several different types of multifunctional and functional brigades. There is no standard configuration for a division, but a division will require a CAB, a MIB-T, division artillery (DIVARTY), a MEB, brigade engineer battalions, and a sustainment brigade to provide the base capabilities necessary for the conduct of large-scale combat operations. Other units may provide direct or general support.

2-65. The size, composition, and capabilities of the forces task-organized under the division may vary between divisions involved in the same campaign, and they may change from one operational phase to another. Operations primarily focused on destruction of a conventional enemy military force (the conduct of offense and defense tasks) require a different mix of forces and capabilities from those required for an operation primarily focused on the protection of civil populations (the conduct of stability tasks). Figure 2-4 on page 2-14 shows an example division task organization.

2-66. Each division has an organic DIVARTY. The DIVARTY is a brigade-level command that also fills the role of the force field artillery headquarters for the division. The division commander specifies the commensurate responsibilities of the force field artillery headquarters and the duration of those responsibilities. These responsibilities may range from simple mentoring and technical oversight of BCT

Chapter 2

organic field artillery battalions to making recommendations on task organizing all field artillery units, including organic, attached, or placed under the OPCON of a division. This includes the field artillery battalions organic to the BCTs. The DIVARTY commander is the fire support coordinator for the division and is the primary advisor to the division commander for the fires warfighting function. (See FM 3-09 for more discussion on the DIVARTY.)

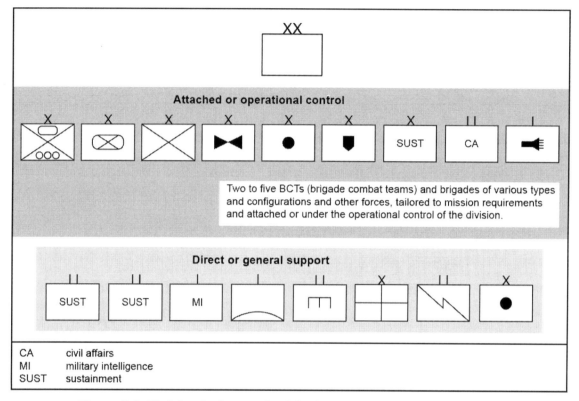

Figure 2-4. Division task organized for large-scale combat operations

DIVISION ADMINISTRATIVE RESPONSIBILITIES

2-67. A corps retains ARFOR responsibilities through the theater army back to the Department of the Army, unless the corps commander shifts Service responsibilities to another headquarters. A corps commander may designate a subordinate Army division commander as the deputy ARFOR commander for performing those duties. When a division receives attachments, that division assumes administrative control of those units, including logistics, medical and administration.

BRIGADE COMBAT TEAMS

Success in war can be achieved only by all branches and arms of the service mutually helping and supporting one another in the common effort to attain the desired end.

Major General Leonard Wood

2-68. Modular brigade combat teams are combat proven organizations with inherent capabilities that are well suited for the range of military operations in a multi-domain environment. A BCT is the Army's primary combined arms, close-combat force. BCTs maneuver against, close with, and destroy the enemy. BCTs seize and retain key terrain, exert constant pressure, and break the enemy's will to fight. They are the principal ground maneuver units of a division. BCTs have organic combined arms capabilities, including battalion sized maneuver, field artillery, reconnaissance, and sustainment units. Each BCT has organic medical support. Division maneuver combines joint capabilities with the organic capabilities of BCTs to provide mutual support that enables BCTs to conduct operations within contiguous or noncontiguous AOs.

Army Echelons, Capabilities, and Training

2-69. There are three types of BCTs: armored, infantry, and Stryker. BCTs normally operate as part of a division. However, they can operate outside of a division headquarters structure directly for a corps or higher echelon commander. In this case, that higher headquarters assigns the BCT its mission, AO, and supporting elements. That higher echelon headquarters orchestrates the BCT's actions with other elements of the larger force. Figure 2-5 shows the organization of an armored BCT.

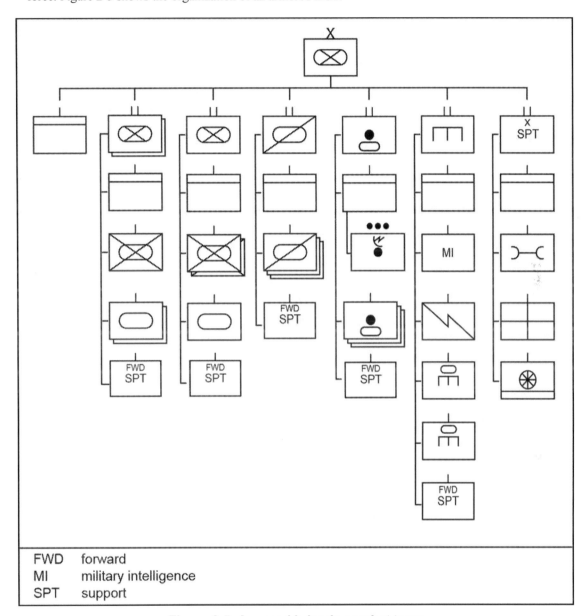

Figure 2-5. Armored brigade combat team

2-70. A BCT has organic capabilities across the warfighting function (mission command, movement and maneuver, intelligence, fires, sustainment, and protection warfighting functions). These capabilities are scalable to meet mission requirements. These organic capabilities include maneuver, field artillery, intelligence, signal, engineer, CBRN, and sustainment capabilities.

2-71. A higher echelon commander can task organize a BCT by adding or subtracting smaller units and capabilities through the use of command and support relationships. Higher echelon commanders may augment BCTs with additional combat power for specific missions. Augmentation might include aviation, armor, infantry, field artillery, AMD, MP, CA, military information support elements, engineers, CBRN, and

EOD. Some of this augmentation may come from other BCTs. Multifunctional and functional brigades, such as a field artillery brigade, a MEB, a CAB, an engineer brigade, and a sustainment brigade can support BCT operations. (See FM 3-96 for additional information on BCT operations.)

MULTIFUNCTIONAL AND FUNCTIONAL BRIGADES

2-72. Theater armies, corps, and divisions may be task-organized with an assortment of multifunctional and functional units to support their operations. These brigades add capabilities such as attack and reconnaissance aviation, fires, contracting support, or sustainment. The theater army may tailor subordinate corps and division headquarters with combinations of multifunctional brigades.

MULTIFUNCTIONAL BRIGADES

2-73. Multifunctional brigades provide a variety of functions in support of operations. Normally, they are attached to a corps or division, but they may be under the command of a joint or multinational headquarters. Multifunctional support brigades include the CAB, the expeditionary CAB, the field artillery brigade, and the MEB.

Combat Aviation Brigade

2-74. A CAB is organized and equipped to synchronize the operations of multiple aviation battalions simultaneously to support corps or division operations. The commander can reorganize a CAB in response to the situation. It can provide tailored support to adjacent supported maneuver commanders at the BCT echelon and below. The CAB commander is normally the senior Army aviation officer in the corps or division structure, and the CAB commander advises adjacent and higher echelon commanders on manned and unmanned aviation system employment.

2-75. The CAB provides a corps or division commander with a maneuver advantage that can overcome the constraints of limiting terrain and extended distances. Attack, reconnaissance, utility, and cargo aircraft may maneuver independently under corps or division control in the echelon deep area or within an assigned AO. Alternatively the CAB's attack, reconnaissance, utility, and cargo assets may be under OPCON, TACON, general support, or direct support to another brigade as situationally appropriate. Furthermore, a CAB may receive OPCON of ground maneuver forces to conduct security or reconnaissance of the corps' or division's flanks or front or to accomplish other economy of force missions. The use of aviation assets requires additional detailed planning and synchronization using specific airspace control processes to maximize results. (FM 3-04 describes the framework and imperatives of air-ground operations.)

Expeditionary Combat Aviation Brigade

2-76. The expeditionary CAB is a multifunctional unit that is designed to air assault maneuver forces; position personnel, supplies, and equipment; evacuate casualties; conduct personnel recovery; and provide mission command. When task-organized with an attack reconnaissance battalion or attack reconnaissance squadron, expeditionary CABs also provide accurate and timely information collection; provide reaction time and maneuver space; and destroy, defeat, disrupt, or delay enemy forces.

Field Artillery Brigade

2-77. A field artillery brigade's primary task is conducting corps-level strike operations. It is capable of employing Army fires and incorporating electronic warfare (EW). In addition, a brigade can request joint fires and coordinate with airspace control elements. The field artillery brigade can detect and attack targets using a mix of its organic target acquisition and fires capabilities, a supported division's information collection capabilities, and access to higher echelon headquarters information collection capabilities provided by the intelligence enterprise.

2-78. Field artillery brigades are typically the force field artillery headquarters for the formation to which they are aligned. The field artillery brigade is capable of providing and coordinating joint lethal and nonlethal effects. The field artillery brigade has the necessary fire support and targeting structure to execute the entire decide, detect, deliver, and assess process. Normally, an additional field artillery brigade will be OPCON to the corps to serve as the counter-fire headquarters. (See FM 3-09 and ATP 3-09.24 for additional information on field artillery brigade operations.)

Maneuver Enhancement Brigade

2-79. The MEB is a multifunctional headquarters designed to perform support area and maneuver support operations for the echelon it supports. Higher echelon commanders base the MEB's task-organization on identified mission requirements for the echelon it is supporting. The MEB can perform MP, engineer, and CBRN missions simultaneously in addition to all of the doctrinal responsibilities associated with being assigned an AO. In addition, the MEB has the following responsibilities within an assigned support area:
- Support to base camp and base cluster defense.
- Liaison and coordination.
- Infrastructure development.
- Host-nation support integrations.
- Area damage control.

2-80. The MEB supports an Army, joint, interagency, or multinational headquarters. The MEB headquarters is staffed and optimized to conduct combined arms operations integrating a wide range of maneuver support related technical branches and combat forces. The MEB organizes, provides, or employs battalion task force and company team combined arms technical experts to conduct maneuver support tasks across all operational environments. The MEB may include a mix of CBRN, CA, engineer, EOD, MP, and potentially ADA units in addition to a tactical combat force. The number and type of organizations placed under this brigade depends on the mission, threat, and number and type of battalions or companies operating in the brigade's AO. The MEB provides staff planning for and control of the units required to conduct decisive action in the echelon support area, support area operations, and maneuver support operations.

2-81. A corps or division commander can task-organize parts of a MEB to a BCT or other brigade for a specific mission, or a MEB may complement or reinforce a BCT with forces under the MEB's control that are performing selected missions or tasks within the BCT AO. This may include engineers, MP, CBRN, or other units for area protection, CBRN support to a field artillery unit, the CAB, and the supporting sustainment brigade. Although a MEB may frequently attach and detach more units than other support brigades, it must also continually provide integrated and synchronized services like the other support brigades. (See FM 3-81 for additional information on MEB operations.)

Sustainment Brigade

2-82. A sustainment brigade is capable of providing general support to one or more divisions, BCTs, multifunctional and functional brigades, ancillary units, and unified action partners. This is in addition to supporting the corps headquarters and headquarters battalion and other units operating in its assigned area. The sustainment brigade's attached units will normally be comprised of CSSBs. The Army designed the sustainment brigade to command up to seven functional and multifunctional logistics battalions. The brigade focuses on management and distribution of supplies, field services, human resources support, execution of financial management support, and allocation of field echelon maintenance in an assigned area. This support extends the operational reach of supported maneuver commanders. The operation and the size of the supported force will dictate the quantities of sustainment brigades employed and the quantity and type of units attached to them. Each of these units must be clearly identified in the corps or division task organization to allow the sustainment brigade commander and the supported unit commanders to clearly understand the sustainment resources available to support their operations.

2-83. A theater army normally attaches sustainment brigades to an ESC with a general or area support relationship to a corps or division headquarters. However, during high tempo large-scale combat operations, a sustainment brigade may be placed OPCON to a corps or division based on the mission and operational variables. (See ATP 4-93 for additional information on the sustainment brigade.)

Chapter 2

FUNCTIONAL BRIGADES

2-84. A functional brigade is a brigade or group that provides a single function or capability. These brigades can provide support for a theater, corps, or division, depending upon how each is tailored. Functional brigade organization varies extensively. Functional brigades include the AMD brigade, the CA brigade, the engineer brigade, the expeditionary-military intelligence brigade (E-MIB), the signal brigade, the MP brigade, and the theater tactical signal brigade (TTSB). Paragraphs 2-84 through 2-103 discuss the roles and organizational structures of these functional brigades. (See table 2-1.)

Table 2-1. Additional functional brigades

Brigade Type	Function	Reference
Field support brigade	Integrates Army Materiel Command acquisition, logistics, and technology capabilities (excluding theater support contracting and logistics civil augmentation program) in support of the operational and tactical level commanders.	ATP 4-91
Chemical, biological, radiological, and nuclear defense brigade	Provides command and control of two to six chemical battalions and other assigned or attached separate companies.	FM 3-11
Contract support brigade	Provides primary theater strategic and operational level contracting support planning and advising.	ATP 4-92
Medical brigade	Provides mission command and coordinates operations of all medical and other assigned or attached medical units.	FM 4-02
Petroleum brigade	Conducts command, control, staff planning, coordination and supervision of bulk petroleum as well as supervision of potable water production, storage, and distribution for the theater.	ATP 4-43
Aerial intelligence brigade	Provides mission command for the Army's echelons above brigade manned and unmanned aerial intelligence, surveillance, and reconnaissance assets and units.	ATP 3-55.3

Air Defense Artillery Brigade

2-85. ADA brigades are structured to perform several functions supporting the AAMDCs and designated GCC organizations supporting AMD integration and operations. ADA brigade functions include mission command activities, integration, planning, and liaison with joint, higher echelon units, and subordinate battalions. ADA brigades are the force providers for the AAMDCs, meeting the commander's AMD objectives. ADA brigades, both active and reserve component, must be prepared to integrate a mix of active and reserve component forces. ADA brigades are aligned under the AAMDCs and deployed to control the fires of subordinate units. Each brigade consists of a headquarters, a brigade staff, and its subordinate battalions.

2-86. ADA brigade tasks include protection of operational level bases and base camps, military or political headquarters, and ports of debarkation. ADA brigades deploy early to protect aerial ports of debarkation, seaports of debarkation, early arriving forces, and critical supplies according to the JFC's defended asset list. As a lodgment expands, ADA forces may reposition to better protect critical assets, communications, transportation, and maneuver forces. ADA brigades and any available joint and multinational air defense forces combine to form an integrated AMD structure after completion of deployment operations. Echelon commanders, with staff support, designate their echelon priority assets. The JFC and joint staff take these echelon priority asset lists into consideration during the development of the JTF defended asset list. An ADA brigade may position its firing units and radars within a corps or division AO. Corps and divisions may integrate these air defense firing units and radars into their defense plans. Those firing units and radars conform to local security measures. Short-range air defense units will generally be placed in direct support or general support of divisions. (See FM 3-01 for additional information on Army ADA operations.)

Civil Affairs Brigade

2-87. The CA brigade provides a CA capability to joint force and land component commanders. The CA brigade mitigates or defeats threats to civil society and conducts actions normally performed by civil governments across the range of military operations. This occurs by engaging and influencing the civil population and authorities through the planning and conducting of CA operations or enabling civil-military operations to shape the civil environment and set the conditions for military operations.

2-88. The CA brigade headquarters provides a control structure and staff supervision of the operations of its assigned CA battalions or other attached units. CA force structure contains expertise in five functional specialty areas. These specialty areas include security, justice and reconciliation, humanitarian assistance and social well-being, governance and participation, and economic stabilization. Within each functional specialty area, technically qualified and experienced individuals, known as CA functional specialists, advise commanders and help or direct their civilian counterparts. (See FM 3-57 for additional information on CA.)

Engineer Brigade

2-89. Typically, a task-organized engineer brigade will be allocated to a corps headquarters. An engineer brigade can control up to five mission-tailored engineer battalions that are not organic to maneuver units. These battalions have capabilities from any of the three engineer disciplines, combat engineering, general engineering, and geospatial engineering, to enhance mobility, countermobility, protection, and sustainment.

2-90. An engineer brigade develops plans, procedures, and programs for engineer support (including requirements determination, operational mobility and countermobility, general engineering, power generation, area damage control, military construction, geospatial engineering, engineering design, construction materials, and real property maintenance activities). An engineer brigade integrates and synchronizes engineer capabilities across a corps AO and reinforces subordinate corps units in the execution of engineer tasks by allocating mission-tailored engineer forces.

2-91. Engineer tasks alter terrain to overcome obstacles (including gaps), create, maintain, and improve lines of communication, create fighting positions, improve protective positions, and build structures and facilities (including base camps, aerial ports, seaports, utilities, and buildings). An engineer brigade is also capable of rapid deployment in modular elements to support the needs of the operational commander. These elements are capable of providing a wide range of technical engineering expertise and support. (See ATP 3-34.23 for additional information on engineer operations.)

Expeditionary-Military Intelligence Brigade

2-92. The Army has three E-MIBs in its force structure. The Army designed them to augment the corps and division capability to process, exploit, and disseminate national and joint force signals intelligence and geospatial intelligence. E-MIBs also provide counterintelligence, human intelligence, and ground-based signals intelligence collection to corps and division headquarters. The E-MIB also supports site exploitation operations. The E-MIB does not conduct reconnaissance. (The corps and division commanders task available maneuver forces to conduct reconnaissance.) The corps commander retains control of the E-MIB or task organizes elements of the E-MIB to divisions as required.

2-93. The E-MIB staff works closely with the corps assistant chief of staff, intelligence's (G-2's) analysis and control element. The E-MIB commander and staff also assist the corps commander in information collection management. (See ATP 2-19.3 for more discussion of E-MIB capabilities and operations.)

Military Police Brigade

2-94. The Army allocates an MP brigade to a division when the magnitude of functional MP requirements exceeds the capability of the MEB to control MP activities. In these instances, MP brigade-level control capability is required to allocate, synchronize, control, and provide technical oversight for MP assets and to provide consistent application of MP capabilities across the division AO. The situation requires a brigade-size

Chapter 2

mission command capability if that situation needs more than two MP battalions' worth of capabilities within the division AO. Some functional MP elements remain under control of the MEB, even if a corps or higher echelon headquarters provides a functional MP brigade to the division. (See FM 3-39 for additional information on the MP brigade and MP operations.)

Theater Tactical Signal Brigade

2-95. A TTSB provides functional signal support for corps and division operations. TTSBs provision communications and information systems support to a theater army headquarters, their subordinate units, and as required, to joint, inter-organizational, and multinational partners throughout the area of responsibility. The TTSB and its subordinate units install, operate, maintain, and defend the Department of Defense information network-Army (DODIN-A). Each TTSB leverages the extension and reachback capabilities to provide joint communications and information systems services to the GCC and subordinate commanders to conduct mission command. (See FM 6-02 for more information on the TTSB.)

Cyberspace Support

2-96. There is one cyber protection brigade that provides worldwide defensive cyberspace support, including support for corps and below. The cyber protection brigade is subordinate to Network Enterprise Technology Command. The cyber protection brigade provides mission command and support to assigned cyber protection teams that conduct defensive cyberspace operations (DCO). Cyber protection teams conduct DCOs to enable freedom of maneuver in cyberspace for supported commanders. Cyber protection brigade mission support includes recruitment, planning, coordination, and training required to enable cyber protection teams to conduct DCOs and meet cybersecurity objectives to support the DOD, the Army, combatant commands, and interagency operations worldwide.

2-97. Corps, division, and brigade headquarters have the resources available to execute cyberspace electromagnetic activities. The available resources enable operating in cyberspace and the electromagnetic spectrum (EMS) and affecting enemy and adversary cyberspace and use of the EMS to provide freedom of maneuver. Staff sections and nonorganic support provide the mission support for operations at corps echelons and below.

2-98. Nonorganic assets enhance organic capabilities by providing additional personnel and equipment to meet mission requirements. Expeditionary signal, joint EW capabilities, cyberspace mission forces, and national agencies provide additional assets based on operational requirements, and they require coordination to ensure the appropriate equipment and personnel are provided. Units can request support for nonorganic capabilities to provide effects on, in, and through cyberspace and the EMS.

2-99. Important enablers that support the Army's defense-in-depth include the Army Cyber Operations and Integration Center and the regional cyber centers. The Army Cyber Operations and Integration Center is an operational element of the ARCYBER headquarters, and it is the top-level control center for all Army cyberspace activities. It provides situational awareness and Department of Defense information network (DODIN) operations reporting for the DODIN-A. The center coordinates with the regional cyber centers and provides operational and technical support as required.

2-100. The regional cyber center is the single point of contact for operational status, service provisioning, incident response, and all Army network services in its assigned theater. It coordinates directly with tactical units to provide DODIN-A services, support to DODIN operations, and (when required) defensive cyberspace operations to enable mission command and the warfighting functions. (See FM 3-12 for more information on the cyber protection brigade.)

Space Support

2-101. The space brigade is the primary Army space force provider. It is a multi-component organization comprised of Regular Army, Army Reserve, and Army National Guard Soldiers. The space brigade coordinates with combatant commanders, ASCCs, and space support elements to execute space operations; deploy combat ready Army space forces; perform theater space operations; and conduct space control planning, coordination, integration, and execution in support of the combatant commander's priorities.

2-102. Corps and division headquarters have organic space support elements (SSEs) to plan, prepare, execute, and assess planning; integrate and coordinate space capabilities; and support commanders in the exercise of mission command through space operations. The primary function of the SSE is to synchronize space activities throughout the operations process.

2-103. The SSE is responsible for maintaining situational awareness and updating the space portion of the COP. SSE members coordinate space operations objectives and tasks with their counterparts at higher and lower echelons. The SSE serves as the primary mission command element within the staffs for space operations. The SSE also works alternative compensatory control measures and special technical operations missions. (See FM 3-14 for more information on the space brigade.)

SECTION III – ARMY CAPABILITIES (COMBAT POWER)

No army can be efficient unless it be a unit for action; and the power must come from above, not below.

Lieutenant General William T. Sherman

2-104. Large-scale combat operations executed through simultaneous offensive, defensive, and stability tasks require continuously generating and applying combat power, often for extended periods. *Combat power* is the total means of destructive, constructive, and information capabilities that a military unit or formation can apply at a given time (ADRP 3-0). Combat power includes all capabilities provided by unified action partners that are integrated, synchronized, and converged with the commander's objectives to achieve unity of effort in sustained operations. The purpose of combat power is to accomplish missions.

2-105. Commanders conceptualize capabilities in terms of combat power. Combat power has eight elements: leadership, information, mission command, movement and maneuver, intelligence, fires, sustainment, and protection. The Army collectively describes the last six elements as the *warfighting functions*—a group of tasks and systems united by a common purpose that commanders use to accomplish missions and training objectives (ADRP 3-0). Commanders apply combat power through the warfighting functions using leadership and information as shown in figure 2-6 on page 2-22.

2-106. During operations, every unit, regardless of type, either generates or maintains combat power. All units contribute to operations. Commanders ensure Army forces have enough potential combat power to combine the elements of decisive action in ways appropriate to the situation. Ultimately, Army forces combine elements of combat power to defeat an enemy and prevail during large-scale combat operations.

2-107. Combat power is not a numerical value. It can be estimated but not quantified with precision. Combat power is always relative. It has meaning only in relation to conditions and enemy capabilities. It is relevant solely at the point in time and space where it is applied. In addition, how an enemy generates and applies combat power may fundamentally differ from that of Army forces. It is dangerous to assume that enemy capabilities are a mirror image of friendly capabilities.

2-108. Before an operation, combat power is unrealized potential. Through leadership, this potential is transformed into action. Commanders use information to integrate and enhance action. Information is also applied through the warfighting functions to shape an OE and complement action. Combat power becomes decisive when applied by skilled commanders leading well-trained Soldiers and units. Ultimately, commanders achieve success by applying superior combat power at the decisive place and time.

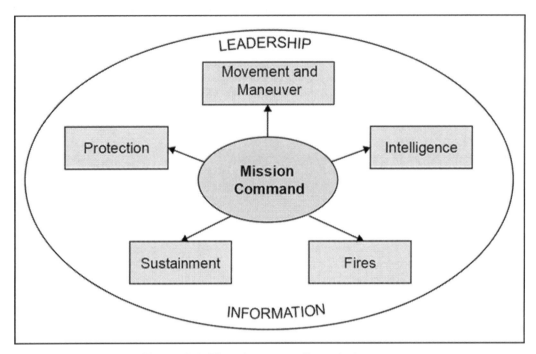

Figure 2-6. The elements of combat power

THE ELEMENTS OF COMBAT POWER

2-109. Combat power has eight elements: leadership, information, mission command, movement and maneuver, intelligence, fires, sustainment, and protection. These elements are discussed in paragraphs 2-110 through 2-266.

LEADERSHIP

I am doing all that I can and yet I feel that I am still leaving much undone.

Field Marshal Earl Kitchener

2-110. *Leadership* is the process of influencing people by providing purpose, direction and motivation to accomplish missions and improve the organization (ADP 6-22). Leadership is the multiplying and unifying element of combat power. Confident, competent, and informed leadership intensifies the effectiveness of all other elements of combat power by formulating sound operational ideas and assuring discipline and motivation in the force. Good leaders are the catalyst for success. Effective leadership can compensate for deficiencies in all the warfighting functions because it is the most dynamic element of combat power. The opposite is also true; poor leadership can negate advantages in warfighting capabilities. An Army leader, by virtue of assumed role or assigned responsibility, inspires and influences people to accomplish organizational goals. Army leaders motivate people to pursue actions, focus thinking, and shape decisions for the greater good of the organization. They instill in Soldiers the will to win.

2-111. Leaders influence not only Soldiers but other people as well. Leadership is crucial in dealing with civilians in any conflict. Face-to-face contact with people in an AO encourages cooperation between civilians and Soldiers. Army leaders work with members of other Services and civilian organizations. These leaders strive for the willing cooperation of unified action partners. The Army requires self-aware, adaptive leaders who can both defeat the enemy in large-scale combat operations and master the complexities of operations dominated by stability tasks.

2-112. Leadership ensures Soldiers understand the purpose of operations and use their full capabilities. In every operation, leaders clarify purpose and mission, direct operations, and set the example for courage and competence. They hold their Soldiers to the Army Values and ensure their Soldiers comply with the law of war.

INFORMATION

Unless you understand the actual circumstances of war, its nature and its relations to other things, you will not know the laws of war, or how to direct war, or be able to win victory.

Mao Tse Tung

2-113. Information is a powerful tool in an OE. In modern conflict, information has become as important as lethal action in determining the outcome of operations. Every engagement, battle, and major operation requires complementary information operations to both inform a global audience and to influence audiences within an operational area; information is a weapon against enemy command and control (C2), and it is a means to affect enemy morale. It is both destructive and constructive. Commanders use information to understand, visualize, describe, and direct the warfighting functions. Soldiers constantly use information to persuade and inform target audiences. They also depend on data and information to increase the effectiveness of the warfighting functions.

2-114. Since information shapes the perceptions of a civilian population, it also shapes much of an OE. All parties in a conflict use information to convey their message to various audiences. These include enemy forces, adversaries, and neutral and friendly populations. Information is critical in land combat where successful human interaction is a major factor in success. Information must be proactive as well as reactive. The enemy adeptly manipulates information and combines message and action effectively. Countering enemy messages with factual and effective friendly messages can be as important as the physical actions of Soldiers. The effects of each warfighting function should complement themes and messages, while themes and messages stay consistent with Soldiers' actions.

2-115. Information, as an element of combat power, includes both the combat information necessary to gain understanding and make decisions and the use of information to dominate the information environment. Information enables commanders at all levels to make informed decisions on how to best apply combat power. Ultimately, this creates opportunities to achieve definitive results. Knowledge management enables commanders to make informed, timely decisions despite the uncertainty of operations. Information management helps commanders make and disseminate effective decisions faster than the enemy can. Commands also use information operations to integrate information-related capabilities to inform and influence populations and gain a position of relative advantage over an adversary or enemy.

MISSION COMMAND

Lead me, follow me, or get out of my way.

General George S. Patton Jr.

2-116. The *mission command warfighting function* is the related tasks and systems that develop and integrate those activities enabling a commander to balance the art of command and the science of control in order to integrate the other warfighting functions (ADRP 3-0). The mission command warfighting function integrates the other warfighting functions (movement and maneuver, intelligence, fires, sustainment, and protection) into a coherent whole. By itself, the mission command warfighting function will not secure an objective, move a friendly force, or restore an essential service to a population. Instead, it provides purpose and direction to the other warfighting functions.

MISSION COMMAND WARFIGHTING FUNCTION TASKS

2-117. While staffs perform essential functions, commanders are ultimately responsible for accomplishing assigned missions. Throughout operations, commanders encourage disciplined initiative through a clear commander's intent while providing enough direction to integrate and synchronize the force at the decisive place and time. To this end, commanders perform three primary mission command warfighting function tasks. The commander tasks are—

- Drive the operations process through the activities of understanding, visualizing, describing, directing, leading, and assessing operations.
- Develop teams, both within their own organizations and with unified action partners.
- Inform and influence audiences, inside and outside their organizations.

2-118. Staffs support commanders in the exercise of mission command by performing four primary mission command warfighting function tasks. The staff tasks are—
- Conduct the operations process: plan, prepare, execute, and assess.
- Conduct knowledge management, information management, and foreign disclosure.
- Conduct information operations.
- Conduct cyberspace electromagnetic activities.

2-119. Six additional tasks reside within the mission command warfighting function. These tasks are—
- Conduct CA operations.
- Conduct military deception.
- Install, operate, and maintain the DODIN.
- Conduct airspace control.
- Conduct information protection.
- Plan and conduct space activities.

2-120. Paragraphs 2-121 through 2-151 highlight several of those tasks. (See ADRP 6-0 for a complete discussion of the mission command warfighting function.) Commanders need support to exercise mission command effectively. At every echelon of command, each commander establishes a *mission command system*—the arrangement of personnel, networks, information systems, processes and procedures, and facilities and equipment that enable commanders to conduct operations (ADP 6-0). Commanders organize the five components of their mission command system to support decision making and facilitate communication. The most important of these components is personnel.

Conduct the Operations Process

> *I have published under my name a good many operational orders and a good many directives...but there is one paragraph in the order than I have always written myself...the intention paragraph.*
>
> Field-Marshall Viscount William Slim

2-121. The Army's framework for exercising mission command is the *operations process*—the major mission command activities performed during operations: planning, preparing, executing, and continuously assessing the operation (ADP 5-0). The operations process is a commander-led activity, informed by the philosophy of mission command. Commanders, supported by their staffs, use the operations process to drive the conceptual and detailed planning necessary to understand, visualize, and describe their operational environment; make and articulate decisions; and direct, lead, and assess operations. Figure 2-7 shows the operations process with the commander at the center leading the process.

2-122. The operations process serves as an overarching model that commanders, staffs, and subordinate leaders use to integrate the warfighting functions across all domains and synchronize the force to accomplish missions. This includes integrating numerous processes such as the intelligence process, the military decision-making process, and targeting within the headquarters and with higher echelon, subordinate, supporting, supported, and adjacent units.

Army Echelons, Capabilities, and Training

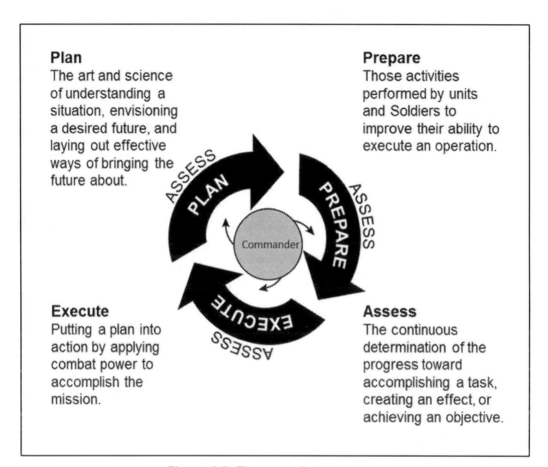

Figure 2-7. The operations process

2-123. The activities of the operations process (plan, prepare, execute, and assess) are not discrete; they overlap and recur as circumstances demand. Planning starts an iteration of the operations process. Upon completion of the initial order, planning continues as leaders revise the plan based on changing circumstances. Preparing begins during planning and continues through execution. Execution puts a plan into action by applying combat power to seize, retain, and exploit the initiative and consolidate gains. Assessing is continuous and influences the other three activities. The operations process, while simple in concept, is dynamic in execution-especially in fast-paced, large-scale combat operations. Commanders must organize and train their staffs and subordinates as an integrated team to simultaneously plan, prepare, execute, and assess operations.

2-124. Army operational planning requires the complete definition of the mission, expression of the commander's intent, completion of the commander and staff estimates, and development of a concept of operations. These form the basis of a plan or order and set the conditions for a successful battle. The initial plan establishes the commander's intent, the concept of operations, and the initial tasks for subordinate units. It allows the greatest possible operational and tactical freedom for subordinate leaders. It is flexible enough to permit leaders to seize opportunities consistent with the commander's intent, thus facilitating quick and accurate decision making during combat operations. The plan not only affects the current operation, but it also sets the stage for future operations. (See ADRP 5-0 for doctrine on the operations process.)

2-125. Both commanders and staffs have important roles within the operations process. The commander's role is to drive the operations process through the activities of understanding, visualizing, describing, directing, leading, and assessing operations, as shown in figure 2-8. Commanders balance their time between leading their staffs through the operations process and providing purpose, direction, and motivation to subordinate commanders and Soldiers.

2-126. The staff's role is to assist commanders with understanding situations, making and implementing decisions, controlling operations, and assessing progress. Staff members advise and make recommendations to the commander within their area of expertise based on their running estimates. They support the commander in communicating decisions and intentions through plans and orders. While commanders make key decisions throughout the operations process, they are not the only decision makers. Commanders delegate trained and trusted staff members with decision-making authority, freeing commanders from routine decisions. This enables commanders to focus on key aspects of an operation.

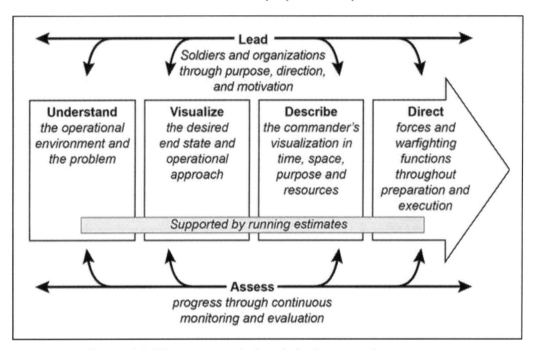

Figure 2-8. The commander's role in the operations process

Conduct Information Operations

> *There has never been a time in our history when there was so great a need for our citizens to be informed and to understand what is happening in the world. The cause of freedom is being challenged throughout the world today....Propaganda is one of the most powerful weapons the Communists have in this struggle. Deceit, distortion, and lies are systematically used by them as a matter of deliberate policy.*
>
> President Harry Truman

2-127. Commanders and staffs synchronize information-related capabilities through information operations. *Information operations* are the integrated employment, during military operations, of information-related capabilities in concert with other lines of operation to influence, disrupt, corrupt, or usurp the decision-making of adversaries and potential adversaries while protecting our own (JP 3-13). Information operations are commander centric and coordinated as part of planning and targeting. The effects of information operations can be simultaneously observed within an AO and worldwide.

2-128. Figure 2-9 depicts information operations authorities and capabilities in the context of operations to shape, prevent, win, and consolidate gains. As operations move right on the conflict continuum, authorities for the use of information-related capabilities increase. For example, during operations to shape, information operations focus on messaging to inform and influence allies, to deter adversaries, and to conduct defensive cyberspace operations. During large-scale combat operations, information-related capabilities become much more offensively oriented and focused on the support of maneuver commanders. National, joint, and Army capabilities such as offensive space control, offensive cyberspace operations, and EW become available to support ground forces engaging the enemy.

Army Echelons, Capabilities, and Training

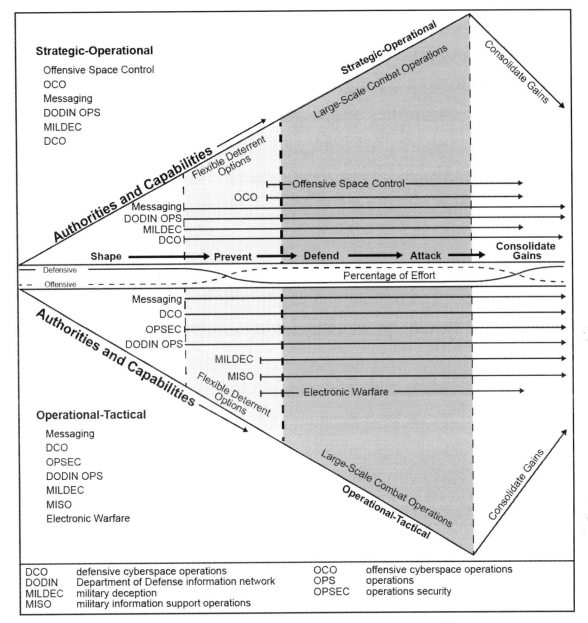

Figure 2-9. Dynamic continuum of information operations

2-129. Commanders and their units must coordinate what they do, say, and portray. The development of themes and messages in support of a military operation is fundamental to that process. A theme is a unifying or dominant idea or image that expresses the purpose for military action. Themes are tied to objectives, lines of effort, and end state conditions. Themes are overarching and apply to the capabilities of public affairs activities, MISO, and Soldier and leader engagements. A message is a verbal, written, or electronic communication that supports a theme focused on a specific actor and in support of a specific action. Commanders approve and employ themes and messages as part of planned activities designed to influence specific foreign audiences for various purposes that support current or planned operations.

2-130. Staffs synchronize information operations throughout the operations process. Information operations specialists must effectively synchronize various information-related capabilities to support the concept of operations. The use of messaging is a critical supporting effort during large-scale combat operations. As in previous wars and conflicts, the message legitimizing why U.S. forces are there and why they fight must be

Chapter 2

communicated and understood across all echelons and audiences. Commanders must communicate and enforce a positive narrative and be aware that an enemy may lead with information effects and only support them with physical effects.

2-131. While all unit operations, activities, and actions affect the information environment, information related capabilities requiring synchronization and coordination as part of information operations include—
- Military deception (MILDEC).
- MISO.
- Soldier and leader engagement, including police engagement.
- CA operations.
- Combat camera.
- OPSEC.
- Public affairs.
- Cyberspace Operations.
- EW.
- Space operations.
- Special technical operations.

2-132. Paragraphs 2-133 through 2-151 highlight several information-related capabilities. (See FM 3-13 for doctrine on information operations.)

Military Deception

> *In wartime, truth is so precious that she should always be attended by a bodyguard of lies.*
> Winston S. Churchill

2-133. *Military deception* is actions executed to deliberately mislead adversary military, paramilitary, or violent extremist organization decision makers, thereby causing the adversary to take specific actions or (inactions) that will contribute to the accomplishment of the friendly mission (JP 3-13.4). MILDEC operations are planned from the top down. Subordinate deception plans must support higher echelon plans.

2-134. Tactical deception (TAC-D) is a deception activity planned and executed by, and in support of, tactical-level commanders to cause adversaries to take actions or inactions favorable to the friendly tactical commanders' objectives. TAC-D is conducted to influence military operations in order to gain a tactical advantage over an adversary, mask vulnerabilities in friendly forces, or to enhance the defensive capabilities of friendly forces. TAC-D is unique to the tactical requirements of the local commander and not necessarily linked or subordinate to a greater joint MILDEC plan. In order to ensure that it does not compromise an existing or future MILDEC, TAC-D must be approved two echelons up, not to exceed the combatant command.

2-135. The corps echelon is where properly conceived and employed MILDEC is most likely to succeed and mislead enemy commanders as to the true disposition, capabilities, and intentions of friendly forces. Divisions and lower echelons often lack the density of forces and capabilities to successfully deceive a regional peer enemy. Tactical headquarters employ TAC-D measures assigned by a higher echelon headquarters or develop their own measures to support their own concept of operations. These measures must be nested. The tactical headquarters higher echelon headquarters must vet and approve all TAC-D operations. TAC-D requires synchronization across multiple domains and echelons. An enemy is not likely to be deceived when friendly actions do not match the available information. Tactical units assigned to conduct MILDEC activities are often not informed of their mission's true purpose.

2-136. It is extremely difficult for BCTs and other types of brigades to employ effective TAC-D because of the lack of resources available at the brigade echelon. Operations security—including effective camouflage, concealment, and the use of decoys—may only delay an enemy from identifying the objective of a BCT operation for short periods. OPSEC may assist in determining the timing for the movement of forces, managing the daily volume of electronic signatures, or the timing of sustainment operations. BCT TAC-D may create only a very short period of time to exploit a position of relative advantage. False insertions, feints, deception fires, smoke screens, decoys, EW and other tactics, techniques, and procedures

Army Echelons, Capabilities, and Training

can create confusion or uncertainty in the mind of an enemy commander which will facilitate friendly maneuver. When possible, TAC-D is used by corps and division support brigades in the close and deep area to add to the dilemmas presented to an opposing enemy commander.

2-137. Commanders should assume all ground forces are under continuous observation by enemies and adversaries employing their national space-based capabilities and operate accordingly. Commanders employ camouflage and deception measures to minimize unit exposure and identification for targeting by long-range precision fires or direct action by enemy conventional, special purpose, and irregular forces. The harder it is to identify friendly forces to target, the less likely they will be targeted. Rapidly maneuvering friendly forces also reduces the ability of an enemy to target those friendly forces. The combination of deception, good OPSEC, and enemy limitations are all related. Minimizing the appearance of actual lucrative targets while providing false targets should be a MILDEC goal. (See FM 6-0 for more information on MILDEC.)

Military Information Support Operations

> *The mind of the enemy and the will of his leaders is a target of far more importance than the bodies of his troops.*
>
> Brigadier General Samuel B. Griffith II

2-138. Psychological factors are an integral part of all operations. MISO include words and actions specifically planned to reduce the combat effectiveness of enemy armed forces and to influence hostile, neutral, and friendly groups to support friendly force operations. However, MISO may take time to achieve a desired change in enemy behavior, depending on the target and the objective. Commanders are responsible for the integration of MISO into military planning. Commanders need to evaluate their plans and operations in view of their psychological impact and alignment with the guidance provided from their higher echelon headquarters. Tactical headquarters at the BCT echelon and below are primarily concerned with MISO designed to produce short-term results to gain a position of relative advantage, such as hastening the surrender of surrounded and isolated enemy forces. Some tactical MISO objectives may achieve goals to consolidate gains. Corps and division headquarters allocate supporting psychological operations companies, platoons, and teams to subordinate units as required.

Messaging Against Iraqi Forces

Prewar intelligence for OPERATION IRAQI FREEDOM indicated that the Iraqi army might be susceptible to an aggressive campaign to promote capitulation or mass surrender. Unfortunately, the surrender leaflets did not work as well in OIF as they did in DESERT STORM. Of course, one major difference between the two wars was that during DESERT STORM, Iraqi soldiers suffered through an extensive bombing for a month before receiving ground forces. As a result of the bombing, those forces were far more receptive to surrender appeals. In DESERT STORM, Iraqi troops were not defending their homeland, and the motivation to stay and die in Kuwait was arguably much weaker.

On Point: The U.S. Army in OIF I

2-139. Corps and divisions facilitate MISO to create periods of relative advantage by providing enemy target audiences information designed for a specific effect that supports friendly operations. Within the constraints of policy, planners and operators creatively erode support for enemy operations, build distrust between adversarial actors, and inform civilian populations of designated evacuation routes or areas that provide sanctuary.

2-140. MISO at the tactical level focuses on isolating enemy forces at the division echelon and below. MISO units may also contribute to the deception plans of higher echelon headquarters, and they have the ability to plan and execute TAC-D. MISO planners weigh short-term effects with the impact certain operations may have when a friendly force operates as an occupying force. (See FM 3-53 for additional information on MISO.)

Chapter 2

Public Affairs

2-141. Army public affairs is public information, command information, and community engagement activities directed toward both external and internal populations with interest in the DOD. Public affairs is the commander's responsibility at all echelons of command. Public affairs officers assist commanders and serve on the brigade's special staff and the commander's personal staff (at division through theater army echelons).

2-142. The U.S. military has an obligation to communicate with its members and the U.S. public, and it is in the national interest to communicate with international populations. The proactive release of accurate information to domestic and international audiences puts joint operations in context, facilitates informed perceptions about military operations, counters adversarial propaganda, and helps achieve national, strategic, and operational objectives. Effective public affairs help—

- Build partnerships.
- Deter adversaries.
- Enhance allied support.
- Support future U.S. security interests.
- Inform expectations and options.
- Counter inaccurate information, deception, and propaganda.
- Reinforce military success.
- Support military objectives.
- Articulate military capabilities.
- Communicate U.S. actions and policies.
- Support strategic narratives, themes, and goals.

2-143. Public affairs is inherent in all military activities, and it is a key enabler for managing and delivering public information through public communication. In Army public affairs, public communication is the communication between the Army and international, national, and local populations through coordinated programs, plans, themes, and messages. It involves the receipt and exchange of ideas and opinions that contribute to shaping public understanding of, and discourse with, the Army. Public communication includes the release of official information through news releases, public service announcements, media engagements, and social networks. (See FM 3-61 for public affairs doctrine.)

2-144. The abundance of information sources, coupled with technology such as smart phones, digital cameras, video chat, and social media enterprises, allows information to move instantaneously around the globe. Public affairs personnel and units frequently review and analyze media reports at the international, national, and local levels. The results of these analyses are provided to the supported commander on a regular basis. It is imperative for public affairs personnel to rapidly develop themes and messages to ensure that facts, data, events, and utterances are put in context. Social media use should conform to all relevant DOD and Service guidance and take into account OPSEC, operational risk, and privacy. (See JP 3-61 for more discussion on social media.)

Cyberspace Electromagnetic Activities

2-145. *Cyberspace electromagnetic activities* is the process of planning, integrating, and synchronizing cyberspace and electronic warfare operations in support of unified land operations (ADRP 3-0). Incorporating cyberspace electromagnetic activities (CEMA) throughout all phases of an operation is key to obtaining and maintaining freedom of maneuver in cyberspace and the EMS while denying it to enemies and adversaries. CEMA synchronizes capabilities across domains and warfighting functions and maximizes complementary effects in and through cyberspace and the EMS. Intelligence, signal, information operations, cyberspace, space, and fires operations are critical to planning, synchronizing, and executing cyberspace and EW operations.

2-146. CEMA adhere to the joint principles of operations, and of these principles, mass, unity of effort, surprise, and security are the most relevant. Commanders and staffs ensure compliance with relevant authorities and associated legal frameworks before conducting cyberspace and EW operations. When conducting cyberspace operations, the Army acts according to command lines of authority. For example, the

Army provides forces for offensive cyberspace operations (OCO), but does not execute these operations except as part of the joint force and as approved by the JFC. OCO and defensive cyberspace response actions are subject to authorities that reside with the President and Secretary of Defense.

2-147. Only forces with proper legal and command authority can create offensive effects, including DCO response actions, in cyberspace or the EMS. The use of cyberspace capabilities requires detailed planning and submitting requests using the cyber effects request format. Corps, divisions, and BCT headquarters currently possess only limited capabilities to conduct cyberspace surveillance. They have no organic capabilities to conduct cyberspace reconnaissance or other offensive tasks. The majority of cyberspace tasks at the corps echelon and below focus on enabling friendly mission command and fires while protecting the DODIN. Adversaries and enemies continuously seek to gain advantage by penetrating friendly networks to disrupt operations. Army forces seek to protect the DODIN through performing defensively oriented tasks to ensure disciplined adherence to security protocols and limiting digital footprints to avoid detection. At all times, ground forces need to safeguard friendly systems with access to the network from capture and exploitation.

2-148. Deliberate planning is required to request specific cyberspace effects that enable or enhance combined arms operations. Commanders request effects using the terms deny, degrade, disrupt, destroy, and manipulate. The Army considers these as separate effects rather than a subset of deny. These terms are common for targeting guidance or to describe effects for information operations. These are desired effects that support operations and are achievable using cyberspace capabilities. Planners use these terms to describe and plan cyberspace and electronic warfare effects. The most common effects associated with cyberspace operations are deny, degrade, disrupt, destroy, manipulate, and deceive.

2-149. Effects in and through cyberspace may have the same consequences as other types of traditional effects. Effects during operations include lethal and nonlethal actions and may be direct or indirect. Direct effects are first order consequences and indirect effects are second, third, or higher order consequences. Similar characteristics of direct and indirect effects in cyberspace can be cumulative or cascading. These effects are planned and controlled in order to meet the commander's objectives. Cumulative refers to compounding effects and cascading refers to influencing other systems with a rippling effect. The desired effects in cyberspace can support operations as another means to shape the operational environment to provide an advantage.

2-150. Echelons that do not have organic capabilities or authorities for cyberspace or EW operations may integrate supporting effects from forces with those with capabilities to support operations. The approval authority to execute the cyberspace and EW operations is described in the operations orders.

2-151. Corps, division, and BCT commanders can request support for nonorganic capabilities to provide effects on, in, and through cyberspace and the EMS. Commanders need to understand what factors influence the effects, duration, and timeliness of delivery of cyberspace effects. Nonlethal cyberspace attacks are initially penetrations that allow the cyberspace reconnaissance and surveillance necessary to determine how best to exploit a position of advantage on a network or system. Exploitation of those positions of advantage can be in the form of destruction, disintegration, or isolation of that cyberspace network or system. The exact effect delivered depends on the degree of relative advantage and available technical capabilities. Army forces prepare and submit effect requests using the cyber effects request format. (See FM 3-12 for more information on requesting cyberspace effects.)

ADDITIONAL TASKS

2-152. In addition to command and staff tasks, the mission command warfighting function has several additional tasks. Paragraphs 2-153 through 2-164 discuss additional mission command warfighting function tasks.

Airspace Control

2-153. *Airspace control* includes the capabilities and procedures used to increase operational effectiveness by promoting the safe, efficient, and flexible use of airspace (JP 3-52). Theater armies, corps, divisions and BCTs require the capability to conduct airspace control. Airspace control increases combat effectiveness while placing minimum constraint upon airspace users. Airspace control relies upon airspace management capabilities provided by joint and Army airspace control elements and often host-nation air traffic control.

2-154. The goal of airspace control at the corps echelon to enable responsive fires and maneuver across the corps AO, and airspace control provides unique coordination challenges in planning and execution with unified action partners. This includes coordination with the land component's BCD as its representative within the air component's air operations center, with the United States Air Force (USAF) theater air-ground system unit in direct support to the corps headquarters and with other theater air-ground system elements executing air operations over the corps AO.

2-155. The corps initial airspace control focus is working with higher headquarters to build an airspace control system that supports corps operations. Airspace control procedures for the entire force are set out in the joint airspace control plan published by the joint airspace control authority, developed with input from the functional components. Corps airspace planners working with the joint force land component planners and the BCD, influence the development of the airspace control plan very early in its development to ensure that the eventual airspace control plan and its related airspace control order and special instructions support corps operations with minimal interference. Building the airspace control systems also includes working with USAF to resource required USAF elements. Corps and division airspace control capabilities are significantly enhanced with the addition of aligned direct support USAF tactical air control system elements, primarily the tactical air control party and the air support operations center.

2-156. The corps headquarters understands airspace control guidance published in the airspace control order, air tasking order, area air defense plan, and land component's plan. The corps headquarters takes into account placement of the fire support coordination line and its impact on the operations of other components. Failure to understand these requirements can cause delays in clearing airspace or impact the operations of another component. During planning, the corps headquarters plans airspace use for corps-controlled Army fires, such as Army Tactical Missile System fires and Army aviation strikes.

2-157. During execution, the corps headquarters decentralizes airspace control to subordinate elements within their respective AOs for the execution of operations. Enabling subordinate divisions with direct support air support operations centers allows the divisions to form a joint air-ground integration center (JAGIC) and control delegated division assigned airspace by the airspace control authority. (See ATP 3-91.1 for a discussion of JAGIC operations.) Figure 2-10 illustrates some common airspace coordinating measures. The corps authorizes direct liaison between subordinate elements and other theater air-ground system airspace control nodes provided by other Services to minimize risk and maximize effectiveness. However, the corps retains responsibility for policies and for the development of future airspace use even when authorizing direct liaison authorized to subordinate units. The corps airspace element, working with the corps field artillery brigade, enables responsive deep fires by rapidly coordinating airspace for the brigade's rockets and missiles. (See JP 3-52 for an explanation of joint force airspace control. See ATP 3-52.2 for multi-Service airspace control techniques and procedures.)

2-158. The corps headquarters provides airspace control resources to support multinational forces under the OPCON or TACON of the corps commander. Corps airspace control planners support multinational forces with the same resources support as the Army functional brigades working directly for the corps. Airspace management occurs at the lowest levels possible to enable clearance of fires and control low altitude above ground airspace congestion given the threats posed by the enemy's integrated air defense system. (See FM 3-52 for additional information on Army airspace control.)

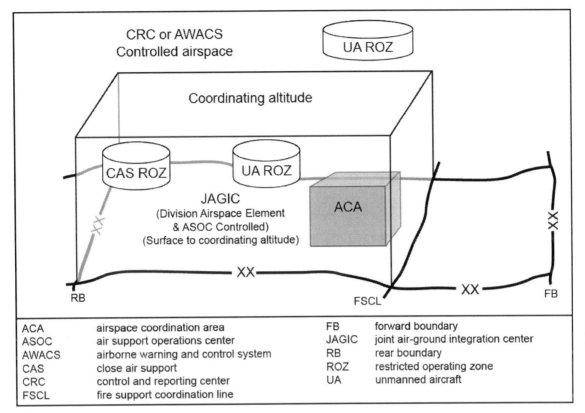

Figure 2-10. Example division airspace coordinating measures

Civil Affairs Operations

2-159. CA operations are actions planned, executed, and assessed by CA forces that enhance awareness of and manage the interaction with the civil component of the operational environment; identify and mitigate underlying causes of instability within civil society; or involve the application of functional specialty skills normally the responsibility of civil government.

2-160. CA operations consist of three competencies, CA activities, CA supported activities and military government operations. CA forces conduct CA operations to support the commander's concept of operations. These forces are the commander's primary asset to purposefully engage nonmilitary organizations, institutions and populations. CA capabilities establish, maintain, influence, or exploit relations between military forces and civil authorities (government and nongovernment) and the civilian populace in a friendly, neutral, or hostile AO to facilitate military operations and to consolidate operational objectives. CA forces may assist or perform activities and functions that are normally the responsibility of local government.

2-161. CA operations may occur before or during military operations or post conflict. They may also occur, if directed, in the absence of other military operations. A CA brigade will normally provide support to a corps. CA battalions and companies may be attached to or placed in support of subordinate divisions and BCTs respectively.

2-162. CA forces may be attached to a BCT during close operations to enable the transition to activities to consolidate gains. These assets may initially interact with host-nation civilian and military forces to enable the use of routes with host-nation support, to evacuate non-combatants, or to assess civil considerations during the conduct of operations to defeat an enemy which may damage infrastructure. In close operations, CA teams try to prevent civilians from entering into lethal areas. (See FM 3-57 for CA doctrine.)

Chapter 2

Plan and Conduct Space Activities

2-163. Space operations influence the conduct of all corps, division, and brigade operations. Space operations enable operations by providing positioning, navigation, timing, satellite communications, space-based intelligence, surveillance, reconnaissance, missile warning, and weather. Army units are consumers of information generated in the space domain and influence space operations by their demands for space-based capabilities. Space operations heavily impact daily operations in Army units. Deliberate planning and targeting processes must request space-based capabilities and effects through the JFC in a timely manner. Commanders who assume short notice responsiveness for space-based capabilities without understanding the limitations of those capabilities increase risk to the mission and their units.

2-164. Corps, division, and brigade commanders need to know the potential impact on operations if enemy action or natural phenomena interrupt the provision of those space-based services. Army forces must retain the ability to shoot, move, and communicate during large-scale combat operations when space-based capabilities are denied, degraded, or disrupted. Training and rehearsing combat skills and ensuring the availability of analog alternatives to space (or cyberspace) enabled systems is critical to successfully persisting in the chaos and friction of modern, large-scale combat operations. Units must train to operate with widespread denial, degradation, or disruption of friendly space capabilities. (See FM 3-14 for additional information on space activities.)

MISSION COMMAND SYSTEMS

2-165. Commanders organize their staffs and other components of the mission command system into command posts (CPs) to assist them in effectively conducting the operations process. A *command post* is a unit headquarters where the commander and staff perform their activities (FM 6-0).

Command Post Operations

2-166. CP personnel, information systems, and equipment must be able to support continuous operations while in communication with their higher echelon, subordinate, supporting, supported, and adjacent units. Commanders arrange personnel and equipment within the CP to facilitate internal planning, coordination, information sharing, and rapid decision making. They ensure staffs are trained on standard operating procedures and that the unit's battle rhythm effectively integrates and synchronizes the activities of the operations process within the headquarters and with external organizations. (See ATP 6-0.5 for doctrine on CP organization and operations.)

2-167. Commanders echelon their headquarters into CPs and assign responsibilities to each CP to assist them in controlling operations. While each CP is designed for a particular purpose (for example controlling current operations), activities common in all CPs include, but are not limited to—

- Maintaining running estimates in support of the commander's decision making.
- Controlling operations.
- Assessing operations.
- Developing and disseminating orders.
- Coordinating with higher, subordinate, and adjacent headquarters.
- Conducting knowledge management, information management, and foreign disclosure
- Conducting DODIN operations.
- Maintaining the COP.
- Performing CP administration, including sleep plans, security, and feeding schedules.

2-168. Table 2-2 lists the types of CPs typically employed by echelon and type of unit. Specific echelon and type of unit publications provide detailed information on CP organization and operations for specific units. For example, FM 3-96 provides doctrine for the organization and employment of the brigade main CP. (Organization varies extensively. See specific doctrine for each type of unit.)

Table 2-2. Command post by echelon and type of unit

Echelon or Type of Unit	Description	Command Posts
Theater army	A theater army headquarters is the Army Service component command assigned to a geographic combatant commander. It is organized, staffed, and equipped to perform three roles: • Theater army for a geographic combatant commander. • Joint task force headquarters (with augmentation) in crisis response and limited contingency operations. • Joint force land component headquarters (with augmentation) for crisis response and limited contingency operations.	Main command post (CP) Contingency CP Mobile command group
Field army	A field army headquarters is the Army component assigned to a subordinate unified command. The field army headquarters is staffed and equipped to perform three roles: • Army component and ARFOR for a subordinate unified commander. • Joint force land component headquarters (with augmentation) for large-scale combat operations. • Joint task force headquarters (with augmentation) for crisis response and limited contingency operations.	Main CP Operational CP Mobile command group
Corps	A corps headquarters is the Army's most versatile headquarters. The corps headquarters is staffed and equipped to— • Serve as the joint force land component commander (or multinational) headquarters (with augmentation) in crisis response and limited contingency operations. • Serve as a joint task force headquarters in a crisis response or limited contingency operation. • Serve as a tactical headquarters in large-scale combat operations.	Main CP Tactical CP Support area CP Early entry CP Mobile command group
Division	A division headquarters operates as a tactical headquarters under operational control of an Army corps or Marine expeditionary force headquarters. The division headquarters is staffed and equipped to— • Serve as a tactical headquarters in large-scale combat operations. • Serve as the joint force land component headquarters (or multinational) headquarters (with augmentation) in crisis response and limited contingency operations. • Serve as a joint task force headquarters in a crisis response or limited contingency operation.	Main CP Tactical CP Support area CP Early entry CP Mobile command group
Brigade combat team	The brigade combat team headquarters operates as a tactical headquarters normally under operational control of an Army division.	Main CP Tactical CP Mobile command group
Multifunctional brigade	A multifunctional brigade headquarters coordinates support for brigade combat teams and other forces.	Main CP Tactical CP
Functional brigades and battalions	Functional brigade and battalion headquarters coordinate a single function or capability.	Main CP Tactical CP
Combined arms and infantry battalions	Combined arms and infantry battalion headquarters operate as tactical headquarters assigned or attached to a brigade combat team.	Main CP Tactical CP Combat trains CP Field trains CP

Chapter 2

Main Command Post

2-169. A *main command post* is a facility containing the majority of the staff designed to control current operations, conduct detailed analysis, and plan future operations (FM 6-0). The main CP is the unit's principal CP serving as the primary location for plans, analysis, sustainment coordination, and assessment. It includes representatives of all staff sections and a full suite of information systems to plan, prepare, execute, and assess operations. The main CP is larger in size and in personnel and less mobile than the tactical CP. The chief of staff or executive officer provides staff supervision of the main CP. All units battalion echelon and above are resourced a main CP. Functions of the main CP include, but are not limited to—

- Controlling operations.
- Receiving reports for subordinate units and preparing reports required by higher echelon headquarters.
- Planning operations, including branches and sequels.
- Integrating intelligence into current operations and plans.
- Synchronizing the targeting process.
- Planning and synchronizing sustaining operations.
- Assessing the overall progress of operations.

Contingency Command Post

2-170. A contingency CP is a facility tailored from the theater army headquarters that enables a commander to conduct crisis response and limited contingency operations within an AOR. Employing the contingency CP for a mission involves a trade-off between the contingency command post's immediate response capability and its known limitations. These limitations include the scale, scope, complexity, intensity, and duration of operations that it can effectively command without significant augmentation. The contingency CP depends upon the main CP for long-range planning and special staff functional support. (See FM 3-94 for doctrine on theater army CP organization and operations.)

Operational Command Post

2-171. An operational CP is a facility containing a tailored portion of a field army headquarters used to control operations for a limited period or for a small-scale contingency. The operational CP provides a field army commander, or designated individual, the capability to form an ARFOR, land component, or JTF headquarters within a JOA. Depending on the situation, the operational CP staff may require additional augmentation, since its design provides minimal essential capabilities. The operational CP may require joint augmentation if it is designated a JTF headquarters or joint force land component headquarters. The operational CP personnel and equipment are deployable by fixed-wing aircraft from their garrison locations into a JOA. However, the operational CP has limited organic transportation once deployed into the JOA and typically occupies a semi-permanent fixed facility. The operational CP relies on the main CP for detailed planning, analysis, and special staff support.

Tactical Command Post

2-172. A *tactical command post* is a facility containing a tailored portion of a unit headquarters designed to control portions of an operation for a limited time (FM 6-0). The tactical CP maintains continuous communication with subordinates, higher echelon headquarters, other CPs, and supporting units. The tactical CP is fully mobile and includes only essential Soldiers and equipment. The tactical CP relies on the main CP for planning, detailed analysis, and coordination. A deputy commander or operations officer generally leads the tactical CP. Corps through battalion commanders employ a tactical CP as an extension of the main CP. Functions of a tactical CP include—

- Controlling the decisive operation or a specific shaping operation.
- Controlling a specific task within a larger operation such as a gap crossing, a passage of lines, a relief in place, or an air assault operation.
- Controlling the overall unit's operations for a limited time when the main CP is displacing or otherwise not available.

Army Echelons, Capabilities, and Training

- Performing short-range planning.
- Providing input to targeting and future operations planning.
- Providing a forward location for issuing orders and conducting rehearsals.
- Forming the headquarters of a task force with subordinate units task-organized under its control.

2-173. When a commander does not employ the tactical CP, the staff assigned to it reinforces the main CP. Unit standard operating procedures should address the specifics for this, including procedures to quickly detach the tactical CP from the main CP. Some multifunctional support brigades and functional brigades and battalions are not resourced with a tactical CP by table of organization and equipment; however, based on the situation, commanders can form a tactical CP from the personnel and equipment authorized from the main CP to assist them with mission command.

Support Area Command Post

2-174. Depending on the situation, including the threat, size of the support area, and number of units within the support and consolidation areas, division and corps commanders may form a support area command post (SACP) to assist in controlling operations. The SACP enables division and corps commanders to exercise mission command over disparate functionally focused elements operating within the support and consolidation areas that may exceed the effective span of control of the MEB or division and corps main CPs.

2-175. The SACP is not a separate section in the units table of organization and equipment. Commanders form a SACP from the equipment and personnel from the main and tactical CPs. The SACP normally co-locates with the MEB, which provides the SACP with signal connectivity, sustainment, security and workspace. Functions of the SACP include—

- Planning and directing sustainment.
- Terrain management.
- Movement control.
- Area security.

2-176. When augmented by the MEB staff, the SACP may also:

- Plan and control combined arms operations with units under division or corps control.
- Manage airspace
- Employ fires.

2-177. Normally, an assistant division commander for a division or a deputy corps commander for a corps leads the SACP. The specific functions and responsibilities assigned to the SACP will be assigned or designated by corps or division commanders to their deputy or assistant commanders through an order.

2-178. A properly resourced SACP assists corps and divisions commanders in shaping the support and consolidation areas that complement the corps or division's scheme of maneuver. This allows the main CP to focus on close and deep operations.

Early-Entry Command Post

2-179. While not a separate section of a unit's table of organization and equipment, commanders can establish an early-entry command post to assist them in controlling operations during the deployment phase of an operation. An *early-entry command post* is a lead element of a headquarters designed to control operations until the remaining portions of the headquarters are deployed and operational (FM 6-0).

2-180. Based on the situation, an early entry command post normally consists of personnel and equipment from the tactical CP with additional intelligence analysts, planners, and other staff officers from the main CP. The early-entry CP performs the functions of the main and tactical CPs until those CPs are deployed and operational. A deputy commander, assistant division commander, chief of staff, executive officer, or operations officer normally leads the early entry CP.

Chapter 2

Command Group

2-181. While not a CP, commanders form a command group to assist them in controlling operations when they are not located at a CP. A *command group* consists of the commander and selected staff members who assist the commander in controlling operations away from a command post (FM 6-0). Command group personnel include staff representation that can immediately affect current operations, such as maneuver, fires (including the air liaison officer), and intelligence. The mission dictates the command group's makeup. For example, during a deliberate breach, the command group may include an engineer officer, a fire support officer, and an air defense officer. When visiting a dislocated civilian collection point, the commander may take a translator, CA operations officer, a medical officer, an MP, and a chaplain.

2-182. Personnel and equipment to form the command group come from the main CP or tactical CP for brigade and lower echelon headquarters. Division and corps headquarters are equipped with a mobile command group on their tables of organization and equipment. The mobile command group serves as the commander's mobile CP. It consists of ground and air components equipped with information systems. These vehicles and aircraft allow commanders to move to critical locations to personally assess a situation, make decisions, and influence operations. The mobile command group's information systems and small staff allow commanders to do this while retaining communications with the entire force.

COMMAND POST ORGANIZATION AND EMPLOYMENT CONSIDERATIONS

2-183. CPs provide locations from which commanders, assisted by their staffs, exercise mission command Commanders organize their mission command system into CPs based on mission requirements and the situation. Planning considerations for CP organization and employment can be categorized as—
- Those contributing to effectiveness.
- Those contributing to survivability.

2-184. In many cases, these factors work against each other and therefore neither can be optimized. Tradeoffs are made to acceptably balance effectiveness and survivability.

Effectiveness

2-185. CP personnel, equipment, and facilities are arranged to facilitate coordination, exchange of information, and rapid decision making. A CP must effectively communicate with higher, subordinate, adjacent, supporting and supported units and have the ability to move as required. Considerations for CP effectiveness include design layout, standardization, continuity, and capacity.

2-186. Well-designed CPs integrate command and staff efforts. Within a CP, the location of CP cells and staff elements are arranged to facilitate internal communication and coordination. This arrangement may change over the course of operations as the situation changes. Other layout considerations include—
- The ease of information flow.
- User interface with communications systems.
- The positioning of information displays for ease of use.
- The integrating of complementary information on maps and displays.
- Adequate workspace for the staff and commander.
- The ease of displacement (setup, tear down, and movement).

2-187. Standardization increases efficiency of CP operations. Commanders develop detailed standard operating procedures for all aspects of CP operations, including CP layout, battle drills, meeting requirements, and reporting procedures. CP standard operating procedures are enforced and revised throughout training. Doing this makes many CP activities routine. Trained staffs are prepared to effectively execute drills and procedures in demanding stressful times during operations.

2-188. CPs must be staffed, equipped, and organized to control operations without interruptions. Commanders carefully consider the primary functions of each CP and CP cell and resource them accordingly in order to support continuous operations. To support continuous operations, unit standard operating procedures address shift plans, rest plans, and procedures for loss of communications with the commander, subordinates, or another CP. Maintaining continuity during displacement of a CP or catastrophic loss requires designating alternate CPs and passing control between CPs. Continuity of command requires commanders to designate seconds in command and inform them of all critical decisions. Primary staff officers should also designate alternates.

2-189. CPs should be staffed and organized to manage the information needed to operate effectively. The capacity to conduct (plan, prepare, execute, and continuously assess) operations concerns both staffing and information systems. So too, does the ability to manage relevant information. CP personnel must be trained and have the requisite tactical and technical proficiency.

Survivability

2-190. CP survivability is vital to mission success. Depending on the threat, CPs need to remain small and highly mobile, especially at lower echelons. CPs are easily acquired and targeted when concentrated. Methods to enhance CP survivability include dispersion; size; electronic signature; redundancy; mobility; camouflage; concealment; deception; use of decoys; continuity of operations; and primary, alternate, contingency, and emergency (PACE) planning. Additional measures include cover or shielding by terrain features or urban structures. (See ATP 3-37.34 for more information on CP survivability.)

2-191. Dispersing CPs enhances the survivability of the commander's mission command system. Commanders place minimum resources in the close area forward and keep more elaborate facilities in the support area. This makes it harder for enemies to find and attack them. It also decreases support and security requirements forward.

2-192. A CP's size affects its mobility and survivability. Large CPs can increase capacity and ease face-to-face coordination. Their size, however, makes them vulnerable to multiple acquisitions and attacks. Smaller CPs are easier to protect, but they may lack capacity to control operations effectively. The key to success is achieving the right balance.

2-193. Reducing CP size reduces CP signature and enhances mobility. However, some personnel and equipment redundancy is required for continuous operations. During operations, personnel and equipment are lost or fail under stress. Having the right amount of redundancy allows CPs to continue to operate effectively when this happens. CPs must deploy efficiently and move within an AO as the situation requires. CP mobility is important, especially at lower echelons during combat operations. Lower-echelon CPs and those employed forward in the close area may need to move quickly and often. Both small size and careful transportation planning facilitate rapid displacement of CPs.

CONTINUITY OF OPERATIONS PLANNING

> *Diverse are the situations under which an officer has to act on the basis of his own view of the situation. It would be wrong if he had to wait for orders at times when no orders can be given. But most productive are his actions when he acts within the framework of his senior commander's intent.*
>
> Field Marshal Helmuth von Moltke the Elder

2-194. Although Army forces have grown accustomed to communicating freely without fear of jamming or interception, U.S. enemies and adversaries are likely to use technological advances in cyberspace and vulnerabilities in the EMS to conduct cyberspace or EMS attacks. Commanders must be prepared to operate with degraded communications.

2-195. U.S. forces' increased reliance on reachback information and network capabilities creates vulnerabilities to attack from various sources. Employment of the mission command philosophy is essential to overcome the fog and friction that a decentralized, disaggregated, and degraded communications environment adds to the battlefield when other defensive counter-measures fail.

2-196. In a hostile EW environment where satellite links are denied or degraded, Army forces retain the ability to communicate on-the-move and at-the-halt over protected connections, both line of sight and beyond line of sight. Units at the battalion echelon and below must remain within line of site communications range. Establishing retransmission points, both ground and aerial, extends the communication range.

Chapter 2

2-197. Army units must develop, train, and implement techniques across all six warfighting functions that ensure continuity of operations and enable an accurate COP throughout a communications degraded environment. These methods include, but are not limited to—

- The ability to anticipate and recognize degraded and denied operations to ensure the timely employment of defensive countermeasures.
- Adjustments to the dispersion of units, graphic control measures, bandwidth management, operational tempo, and centralization or decentralization of critical assets as required.
- Redundancy in communication, targeting, and collection assets.
- Procedures to transfer data and information, both manually and verbally.
- The performance of all critical mission command and warfighting tasks through analog or manual operations.
- The establishment of "push versus pull" procedures to anticipate requirements when normal reporting is constrained.

Communications Planning Considerations

2-198. Planning for successful communications requires detailed planning by every staff member, not just the assistant chief of staff, signal (G-6) or battalion or brigade signal staff officer (S-6). Communications planners must understand the commander's concept of operations and intent, and they must have a clear picture of the overall communications architecture. The G-6 or S-6, in coordination with other members of the staff, initiates the signal planning process during the military decision-making process and considers the—

- Communications capability requirements for each warfighting function.
- Capabilities and limitations of all available communications systems.
- Potential joint, inter-organizational and multinational communications requirements.
- Detailed line of sight analysis.
- Redundancy in means to communicate.
- Integration of all available signal assets.
- Method of deployment (assets are sequenced to coincide with the arrival of forces).
- Locations of all command post nodes.
- The use of retransmission, digital network links, and node placement.
- Satellite communications requirements.
- Spectrum requirements for emitters, sensors, radars, or any other assets that rely on a frequency.
- Initial task organization and expected changes.
- Proper signal and communications security procedures.
- Conduct of communications rehearsals.

Primary, Alternate, Contingency, and Emergency Plans

2-199. A PACE plan is a key requirement for communications planning. A PACE plan establishes the primary, alternate, contingency, and emergency methods of communications for each warfighting function, typically from higher to lower echelons. Establishing a PACE plan requires care that an alternate or contingency method of communications does not rely solely on the primary method. For example, having voice over internet protocol as an alternate method of communications would be a poor choice if the primary is network data, because when a primary is down the alternate may be as well. The key to a good PACE plan is to establish redundancy so that communications are always available. Most units will have two PACE plans; one for communications to higher echelon headquarters and one for subordinate units. A PACE plan for a higher echelon headquarters will likely be established by the higher headquarters.

2-200. The PACE plan should be as simple as possible, yet it should maintain flexibility to provide communications support as reliably as possible during dynamic operations. If at all possible, PACE plans should revolve around warfighting functions; this assists units in delineating differences in reporting requirements for each warfighting function as each function reports, receives, and processes information differently than the other. There are four principal war fighting functions for the purposes of PACE planning:

movement and maneuver, intelligence, fires, and sustainment. The G-6 or S-6 does not dictate PACE plans for these warfighting functions, but assists the staff in developing them.

2-201. Units should identify appropriate PACE systems for each phase of the operation and publish them in the signal annex. An emergency means of communications does not always have to be equipment; it may be a procedure such as moving back to the last known effective communications point or linking up at a grid coordinate. The PACE concept has always been a valuable tool to ensure that there is a backup communications plan in place in case the primary plan fails.

MOVEMENT AND MANEUVER

Nine-tenths of tactics are certain, and taught in books: but the irrational tenth is like the kingfisher flashing across the pool, and that is the test of generals. It can only be ensured by instinct, sharpened by thought practicing the stroke so often that at the crisis it is as natural as a reflex.

T.E. Lawrence

2-202. The *movement and maneuver warfighting function* is the related tasks and systems that move and employ forces to achieve a position of relative advantage over the enemy and other threats (ADRP 3-0). Direct fire and close combat are inherent in maneuver. The movement and maneuver warfighting function includes tasks associated with force projection related to gaining a position of relative advantage over the enemy. Movement is necessary to disperse and displace the force as a whole or in part when maneuvering. Maneuver is the employment of forces in the operational area through movement in combination with fires to achieve a position of advantage with respect to the enemy. It works through movement and with fires to achieve a position of relative advantage over the enemy to accomplish the mission and consolidate gains. Commanders use maneuver for massing the effects of combat power to achieve surprise, shock, and momentum. Effective maneuver requires close coordination of movement with fires. The movement and maneuver warfighting function includes the following tasks:

- Move
- Maneuver.
- Employ direct fires.
- Occupy an area.
- Conduct mobility and countermobility.
- Conduct reconnaissance and surveillance.
- Employ battlefield obscuration.

2-203. Corps, division, and BCT commanders' schemes of maneuver seek to surprise the enemy. They do this by choosing unexpected directions, times, or types of movement and maneuver. Surprise delays enemy reactions, overloads and confuses enemy command and control systems, induces psychological shock in the enemy, and reduces the coherence of the enemy combined arms team. Commanders achieve tactical surprise by attacking or counterattacking in bad weather and over seemingly impassable terrain. They use camouflage and concealment to lure enemy forces into prepared engagement areas. They conduct feints and demonstrations to divert the enemy commander's attention from their main effort or decisive operation. They maintain a tempo of operations that allows them to operate within the enemy commander's decision cycle. They select portions of the enemy force for destruction leading to the enemy's defeat in detail. They employ sound OPSEC and MILDEC.

2-204. FM 3-90-1 and FM 3-90-2 provide the foundation for movement and maneuver during large-scale combat operations. Corps, division, and BCT commanders normally do not specify the form of maneuver to be adopted by subordinate units. However, the assignment of missions and tasks, AOs, and the allocation of forces may impose such limitations on a subordinate unit that its commander has little choice on the form of maneuver adopted. Tactical commands normally employ a combination of the six basic forms of maneuver—envelopment, flank attack, frontal attack, infiltration, penetration, and turning movement—in their performance of the four offensive tasks. The distinction in the form of maneuver adopted by a tactical echelon exists primarily in the intent of the echelon commander, since the subordinate elements may use other forms

Chapter 2

of maneuver. Chapter 5 addresses tactical enabling tasks. Chapters 6 and 7 address maneuver in the defense and the offense respectively.

INTELLIGENCE

In war nothing is more important to a commander than the facts concerning the strength, dispositions, and intentions of his opponent, and the proper interpretation of those facts.

President Dwight D. Eisenhower

2-205. The *intelligence warfighting function* is the related tasks and systems that facilitate understanding the enemy, terrain, weather, civil considerations, and other significant aspects of the operational environment (ADRP 3-0). The intelligence warfighting function tasks are—
- Intelligence support to force generation.
- Intelligence support to situational understanding.
- Conduct information collection.
- Intelligence support to targeting and information capabilities.

(See ADRP 2-0 for doctrine on the intelligence warfighting function.)

2-206. Intelligence is inherently a joint, interagency, intergovernmental, and multinational activity that is conducted across the intelligence enterprise. The intelligence enterprise provides a combination of ground, space, cyberspace, aerial, and seaborne collection capabilities and analytical elements that provide the most comprehensive intelligence possible. Army regional expertise and support begins within each theater through the theater army G-2 and supporting MIB-T down to the corps G-2s and E-MIBs and to each tactical echelon. (See ADRP 2-0 for doctrine on the intelligence warfighting function.)

2-207. The intelligence warfighting function encompasses more than the military intelligence branch:
- The commander drives the operations process and focuses the intelligence effort that supports it.
- Intelligence is commander-centric. The commander performs the central role within intelligence and enables the intelligence warfighting function.
- The entire staff is important to the intelligence warfighting function and each staff member contributes to intelligence in a different way.
- During combat operations, the G-2 and battalion or brigade intelligence staff officer (S-2), assistant chief of staff, operations (G-3) or battalion or brigade operations staff officer (S-3), G-6 or S-6, and fire support coordinator form the staff core that is essential to synchronize and integrate intelligence.
- Every Soldier contributes to information collection.

2-208. Intelligence drives operations and operations support intelligence; this relationship is continuous. The commander and staff need accurate, relevant, and predictive intelligence in order to understand threat centers of gravity, goals and objectives, and courses of action. Precise intelligence is also critical to target threat capabilities at the right time and place and to open windows of opportunity across domains during large-scale combat operations. Commanders and staffs must have detailed knowledge of threat strengths, weaknesses, organization, equipment, and tactics to plan for and execute friendly operations.

2-209. In order to produce intelligence, units must continuously execute aggressive information collection activities. *Information collection* is an activity that synchronizes and integrates the planning and employment of sensors and assets as well as the processing, exploitation, and dissemination systems in direct support of current and future operations (FM 3-55). Each unit carefully integrates and synchronizes its surveillance, reconnaissance, intelligence operations, and security operations as a layered information collection effort. The combination of combat information and intelligence produced as a result of the information collection effort and intelligence from other units and the intelligence enterprise support successful operations. (See FM 3-55 and ATP 3-55.4 for doctrine on information collection.)

Fighting for Intelligence

2-210. Information collection begins immediately following receipt of mission. Units must be prepared to fight for intelligence against a range of threats, enemy formations, and unknowns. These challenges include integrated air defense systems (IADSs) and long range fires, counter reconnaissance, cyberspace and EW operations, deception operations, and camouflage. It may be necessary for commanders to allocate maneuver, fires, and other capabilities to conduct combat operations to enable information collection.

2-211. Priority intelligence requirements, information requirements, and targeting requirements inform the integrated information collection plan. All units (maneuver, fires, maneuver support, and sustainment units) are part of the information collection effort. Commanders and staffs integrate and synchronize all activities that provide useful information as a part of the information collection effort, including Soldier and leader engagements, patrols, observation posts and listening posts, convoys, and checkpoints.

2-212. During planning, combat information and intelligence is especially useful in determining the viability of potential courses of action. For example, a commander who lacks the intelligence to know where most of the enemy's units and systems are located cannot conduct a deliberate attack. The unit must collect more information, conduct a reconnaissance in force, a more risky movement to contact, or a hasty attack.

2-213. During the execution phase of the operations process, a layered and continuous information collection effort ensures detection of any enemy formations, lethal fires capabilities, or specialized capabilities that provide the enemy advantage. In turn, this allows the commanders and staffs to adjust the scheme of maneuver and fires as the enemy situation develops.

The Commander's Role in Intelligence

2-214. Close interaction between the commander and G-2 or S-2 is essential as the staff supports unit planning and preparation through the integrating processes and continuing activities. The G-2 or S-2 supports the commander's ability to understand the operational environment and visualize operations by leading the intelligence preparation of the battlefield (IPB) process and portraying the enemy throughout the military decision-making process, developing the information collection plan, updating the intelligence running estimate, and developing intelligence products and reports. The commander's role is to direct the intelligence warfighting function through his relationship with the G-2 or S-2.

2-215. Commanders must stay constantly engaged with their G-2 or S-2. The following are examples of actively engaging the intelligence warfighting function:

- During planning, the commander—
    - Builds an effective staff team and fosters a collaborative environment that encourages critical thinking, candor, and cooperation which empowers the entire staff.
    - Accepts prudent risks and some operational uncertainty. Intelligence cannot eliminate uncertainty.
    - Prioritizes resources and capabilities to support information collection. The commander must resource the intelligence architecture adequately to provide adequate network capability (for example, bandwidth) and access. Network access and effective unit communications are especially critical.
    - Allocates adequate time for information collection or determines the appropriate balance between the time allotted for collection and operational necessity. It takes time and tactical patience to collect information and then develop effective intelligence products.
    - Provides the staff initial guidance and then later, intent and concept of the operation. It is important that the G-2 or S-2 understands the commander's guidance and both share the same perspective of the operation and the operational environment.
    - Must own the priority intelligence requirement. The commander personally engages in the development and approval of priority intelligence requirements that are clear, answerable, focused on a single question, and necessary to drive an operational decision.
    - Ensures that information collection and intelligence activities are fully integrated into plans and operations.

- Ensures the staff exploits information and intelligence from higher echelons, other units, and unified action partners. Intelligence sharing and an enterprise approach are important factors that facilitate cooperation and access to critical reinforcing intelligence capabilities during operations.
- Ensures the G-2 or S-2 leads the rest of the staff (who must actively contribute to the process) through IPB. IPB is a critical systematic and thorough analysis of the terrain, weather, threat, civil considerations, and other significant aspects of the operational environment. The commander must help shape the focus and scope of IPB to effectively drive the rest of the military decision-making process. (See ATP 2-01.3 for doctrine on IPB.)
- Ensures the G-2 or S-2 is focused on managing, directing, and coordinating the intelligence effort while the military intelligence commander is focused on commanding his unit and conducting intelligence operations.
- Ensures the G-2 or S-2 and G-3 or S-3 develops the information collection plan, which includes how the intelligence section operates, what requests for information are necessary, and how organic assets will answer the priority intelligence requirement. The collection plan should include contingency plans and meet the principles of cue, mix, and redundancy. (See ATP 2-01 for doctrine on planning requirements and assessing collection.)
- Ensures the G-6 or S-6 is integral to planning adequate network access during the operation as a critical enabler to operations and the intelligence warfighting function. Therefore, the G-2 or S-2 and G-6 or S-6 must coordinate closely throughout all planning activities.
- Ensures the G-2 or S-2, in coordination with the staff, especially the G-6 or S-6, develops a thorough intelligence architecture that describes how each intelligence discipline, complementary intelligence capability, processing, exploitation, and dissemination capability, and multinational partners support intelligence operations.
- Ensures the G-2 or S-2 is actively involved with all integrating processes, continuing activities, and the military decision-making process. For example, the G-2 or S-2 must actively participate in the targeting process with IPB products, especially named areas of interest (NAIs), time phase lines, event template matrix, high-value targets, and target areas of interest (TAIs) recommendations. The G-2 or S-2 thoroughly integrates battle damage assessment and targeting requirements with the other requirements in the collection plan.

- During preparation, the commander—
  - Ensures the G-2 or S-2 is actively involved in all rehearsals and portrays a sophisticated, capable, and realistic enemy.
  - Ensures the G-2 or S-2 and the G-3 or S-3 have issued all information collection orders in a timely manner, tracked the preparation of all information collection assets, coordinated airspace and other key control measures, and resynchronized ongoing collection and processing, exploitation, and dissemination requirements, as needed.
  - Ensures the G-2 or S-2 and G-6 or S-6 coordinate to ensure the network is operational and responsive.
  - Ensures the G-2 or S-2 has completed information sharing coordination with unified action partners and ongoing processes are in place for subsequent phases of the operation.
- During execution and assessment, the commander—
  - Ensures the G-2 or S-2 and G-3 or S-3 are able to assess ongoing operations against the IPB products and information collection plan. As the situation dictates, the S-2 or G-2 and S-3 or G-3 modify the information collection plan, prompt commander decisions, and recommend branches or sequels as necessary.
  - Ensures the G-2 or S-2 continually assesses the enemy and operational environment, answers priority intelligence requirements, and updates the running estimate.
  - Ensures the G-2 or S-2 and fire support coordinator are tracking battle damage assessment and providing operational context based on the battle damage assessment.
  - Ensures the G-2 or S-2 and G-3 or S-3 allow adequate time to plan for the next operation based on the results of the current operation.

2-216. Operations, targeting, intelligence, and communications are inextricably linked. Therefore, the commander must drive the operations process and enable the intelligence warfighting function. Commanders use their staff to synchronize intelligence with the other warfighting functions in order to visualize the operational environment and disrupt the threat simultaneously throughout an AO. Successful intelligence is a result of carefully developed requirements, staff integration and synchronization, continuous information collection, and when necessary, the willingness to fight for information.

FIRES

Hard pounding this, gentlemen: let's see who will pound the longest.
Duke of Wellington

2-217. The *fires warfighting function* is the related tasks and systems that provide collective and coordinated use of Army indirect fires, air and missile defense, and joint fires through the targeting processes (ADRP 3-0). Army fires systems deliver fires in support of offensive and defensive tasks to create specific lethal and nonlethal effects on a target. The fires warfighting function includes the following tasks:
- Deliver fires.
- Integrate all forms of Army, joint, and multinational fires.
- Conduct targeting.

(See ADRP 3-09 for doctrine on the fires warfighting function.)

2-218. Commanders ensure the coordinated use of indirect fires, AMD, and joint fires to create windows of opportunity for maneuver and put the enemy in a position of disadvantage. This is accomplished through the operations process, fire support planning, and targeting. These processes ensure the proper detection and delivery assets capable to produce the desired effects on the enemy are allocated against targets to enable friendly maneuver. The processes also allow for the rapid and responsive delivery of fires by building permissive and restrictive maneuver graphics, fire support coordination measures, and airspace coordinating measures. Commanders use long-range fires (rocket, naval, surface fire support, and rotary and fixed-wing air support) to engage the enemy throughout the depth of their AO. Units focus operations in their deep areas to set conditions that allow their subordinate units success in their operations within their respective AOs. A corps coordinates and synchronizes joint fires as its primary tool to shape an engagement in its deep area.

2-219. *Targeting* is the process of selecting and prioritizing targets and matching the appropriate response to them, considering operational requirements and capabilities (JP 3-0). Units may use the Army targeting process and the joint targeting cycle to integrate and synchronize fires into operations, creating the desired effects in time and space. Using targeting, fires cells recommend targeting guidance to the commander, develop targets, select targets for attack, and coordinate, integrate, and assign allocated joint, interagency, and multinational fires to specific targets and target systems. (See ATP 3-60 and JP 3-60 for more information on the Army targeting process and the joint targeting cycle.)

2-220. Commanders use a variety of methods and assets to achieve the desired effects on targeted enemy forces to enable friendly maneuver. Preparation fires, counter-fire, suppression fires, and EW assets provide commanders with options for gaining and maintaining fire superiority. Commanders use long-range fires (including rocket, naval surface fire support, and rotary and fixed-wing air support) to engage the enemy throughout the depth of their AO.

2-221. Coordinating and synchronizing joint fires is the primary tool corps use to shape enemy forces in the deep area to set conditions for their subordinate divisions. Division fires in deep areas set the conditions that allow subordinate BCTs to be successful in their operations.

2-222. Fires cells develop, recommend, and brief the echelon scheme of fires (including both lethal and nonlethal effects), recommend targeting guidance to the commander, develop targets, select targets for attack, and coordinate, integrate, and assign allocated joint and multinational fires to specific targets and target systems. The scheme of fires links organizations and systems capable of detecting and tracking enemy targets with fires organizations capable of producing the desired effects on those targets.

2-223. Corps fires planners develop radar coverage plans to enable effective counter-fire radar coverage over critical locations for the specific periods necessary for the successful conduct of operations.

Commanders combine ground maneuver with fires to mass effects, achieve surprise, destroy enemy forces, and obtain decisive results. The commander's guidance gives specified attack criteria for supporting fires assets, thus focusing the planning and execution efforts on those critical times and events. The specified attack criteria are a compilation of the commander's guidance, desired effects, high-payoff target (HPTs), and attack priorities. The amount of time available to plan an operation constrains the commander's ability to synchronize fire-support operations that employ well-matched effects of all available assets against HPTs.

FIELD ARTILLERY

2-224. Field artillery fires facilitate an attacking unit's maneuver by destroying, degrading, or neutralizing selected enemy forces and positions. Field artillery systems must take full advantage of available preparation time to create those effects. Fire plans may contain, but are not limited to—

- Targets that are confirmed or denied by reconnaissance and surveillance efforts.
- Designation of target sensor-to-shooter communication links.
- Possible use of preparation and deception fires to shape the enemy's defense.
- Air support to destroy HPTs on the objective and then shift to attacking enemy units, artillery assets, and C2 nodes.
- Proactive suppression or destruction of enemy air-defense capabilities.
- Preparation fires that shift just as the maneuver force arrives on the objective.
- Suppression and obscuration fire plans to support breaching operations.
- Pre-positioned ammunition backed by prepackaged munitions stocks capable of rapid delivery.
- Integration of nonlethal effects, such as electronic attack and MISO, into the attack guidance matrix.
- Integration of primary and alternate observers to engage high-priority targets.
- Fire support coordination and airspace coordinating measures, accounting for danger close and other technical constraints, to allow maneuver forces to get as close as possible to the objective before lifting fires.
- Signals for lifting and shifting fires on the objective, primarily by combat net radio and by visual signals as a backup means.

2-225. Fire support coordination measures should also facilitate the massing of fires, including close air support and air interdiction using kill box procedures, against HPTs throughout an AO. (See ATP 3-09.34 for more information on the employment of a kill box.)

2-226. The corps fires cell establishes conditions to support the conduct of successful operations by its subordinate divisions. Long-range fires are used to disrupt enemy C2 networks, sustainment nodes, long-range artillery and rockets, and IADSs. A tactical corps will have a USAF corps tactical air control party and required support from the JFACC's airspace control elements. Airspace control at the corps echelon to enable responsive effects in the deep area with Army systems provides unique challenges. A corps must coordinate with both the land component's BCD as its representative within the air component's air operations center and with the USAF theater air-ground system unit in direct support to the corps headquarters.

2-227. Commanders may need to position field artillery brigade and division field artillery assets in the close area to range enemy air-defense or fires assets due to the range superiority of some enemy artillery and air defense systems. This entails assuming risk. In this environment, friendly artillery assets must fire and displace quickly due to enemy counter-fire capabilities. Target acquisition radars cannot radiate constantly without risking destruction by enemy long-range systems. The effectiveness of single warhead precision-guided munitions and digital fires systems positioning, navigation, and timing may be disrupted by enemy cyberspace and electromagnetic activities. Priority fires in the close area are generally focused on enabling movement of ground forces by the destruction or degradation of enemy artillery, integrated air defense systems, command and control nodes, and sustainment facilities. Priority fires may also be massed against specific enemy formations to enable a penetration and rapid exploitation.

2-228. BCT, CAB, division, or corps fires cells should have developed procedures for identifying, prioritizing, and engaging time-sensitive and high-payoff targets, such as enemy C2 or long-range fire support systems, when they are identified. The priorities for fires may direct engagement of IADSs, headquarters, and sustainment hubs over the engagement of enemy maneuver forces. It will be very tempting for commanders to use joint fires or rocket artillery to target enemy maneuver forces instead of using them to conduct counter-fire. However, these assets are subject to enemy counter-fire during most close operations.

2-229. The field artillery brigade allocated to the corps augments the fires of divisions, provides counter battery fires, and performs corps-wide target acquisition functions. Fires planners at corps echelons develop and apply radar coverage plans to ensure counter-fire radar coverage. The types of missions assigned to field artillery brigades are general support, reinforcing, and general support reinforcing. These units may be attached to divisions or any BCT remaining under the corps commander's direct command. At the recommendation of the field artillery brigade commander, the corps commander assigns missions to the artillery battalions of any BCTs currently in reserve. In the offense, commanders give priority field artillery support to the decisive operation. If an operation is phased, priority for fires can vary from unit to unit. The same applies to the defense. A corps will usually provide additional field artillery brigade artillery battalions in a support relationship to the DIVARTY or field artillery battalions of the BCTs conducting the decisive operation.

2-230. Corps fires establish conditions to support the conduct of successful operations by subordinate divisions. Corps can choose to retain OPCON of some artillery assets. However, when a corps retains control of these assets, the corps commander assumes inherent risk, as using their long-range rocket munitions may restrict air operations inside division-controlled airspace during a decisive moment.

AIR AND MISSILE DEFENSE

2-231. The corps AMD element plans, provides early warning, and works with the subordinate division JAGICs and other headquarters to synchronize the use of airspace over the corps AO. The corps headquarters employs standard fire support coordination measures and airspace coordinating measures when synchronizing the use of that airspace using its direct support USAF tactical air control system element.

2-232. Air defense forces assigned to a corps may include air defense radars, weapons systems, and appropriate communications equipment. AMD assets may further be directed by the JFC to protect facilities located within a corps AO. The senior officer in the corps AMD section advises the corps commander on how the corps can best counter air and missile threats. (See FM 3-01 for more detail on air and missile defense planning and operations.)

SUSTAINMENT

Even the bravest cannot fight beyond his strength.

Homer

2-233. The *sustainment warfighting function* is the related tasks and systems that provide support and services to ensure freedom of action, extend operational reach, and prolong endurance (ADRP 3-0). The sustainment warfighting function includes the following tasks:
- Conduct logistics.
- Provide personnel services.
- Provide health service support.

(See ADRP 4-0 for doctrine on the sustainment warfighting function.)

2-234. The endurance of Army forces is primarily a function of their sustainment. Sustainment determines the depth and duration of Army operations. It is essential for retaining and exploiting the initiative. Sustainment provides the support necessary to maintain operations until mission accomplishment. The major categories of sustainment are logistics, personnel services, and health service support.

2-235. Sustainment assists the tactical commander in maintaining the tempo of operations. Commanders seek to take advantage of windows of opportunity and perform offensive and defensive tasks so that the enemy has little advance warning. Sustainment planners and operators must anticipate the tempo of operations and maintain the flexibility to support unit operations and extend the unit's operational reach.

2-236. A key to successfully extending operational reach is the ability to anticipate the requirement to push sustainment support forward, specifically in regards to ammunition, fuel, replacements, and water. Sustainment commanders must act, rather than react, to support requirements. The existence of habitual support relationships facilitates this ability to anticipate.

LOGISTICS

Mobility is the true test of a supply system.

Sir Basil Liddell Hart

2-237. Logisticians support operational tempo by delivering supplies and materiel as far forward as possible. They use throughput distribution and preplanned and preconfigured packages of essential items to do this. Logisticians maintain constant contact with operational units to determine requirements for supporting operations. Operational units also provide logisticians with support estimates for contingencies and requirements for cross-loading of supplies to prevent all of one type of supply from being destroyed by the loss of a single system.

2-238. Supplies and material should remain close to the maneuver force to ensure short response times for supplies and services. This includes uploading critical materiel—such as water, petroleum, oils, and lubricants and ammunition—in order to anticipate attempted occupation of a piece of terrain by more than one unit. Commanders must make risk decisions regarding logistics preparations and avoidance of enemy detection, since logistic preparations may give indications of friendly tactical plans.

2-239. The availability of supplies and materiel to sustain tactical unit operations becomes critical to extend operational reach as large-scale combat operations progress. Operational reach is reduced when supplies fail to keep up with the demand of tactical units. Slow or limited resupply may require commanders to use controlled supply rates for various classes of supply to reduce unit expenditures. When those controlled supply rates are not sufficient to continue operations, the force culminates.

2-240. During large-scale combat operations supply lines of communication are strained, and requirements for repair and replacement of weapon systems increase. Requirements for petroleum, oils, and lubricants increase during the offense. Conversely, requirements for munitions tend to be higher in the defense than in the offense. Sustainment units must be as mobile as the forces they support. One way to provide continuous support is to task organize elements of sustainment units or complete sustainment units with their supported maneuver formations as required by the mission.

2-241. The variety and complexity of possible situations arising during an attack requires sustainment operators to establish a flexible and tailorable distribution system in support of tactical commanders. There may be a wide dispersion of forces and lengthening of lines of communication. Required capabilities to support longer lines of communications include movement control, in-transit visibility, terminal operations, and mode operations.

2-242. Field maintenance assets move as forward as possible to repair inoperable and damaged equipment to return it to service as quickly as possible. Crews perform preventive maintenance checks and services as modified for the climate and terrain in which they find themselves. Battle damage assessment and repair restores the minimum essential combat capabilities necessary to support a specific combat mission or to enable the equipment to self-recover. Crews and maintenance and recovery teams conduct battle damage assessment and repair to rapidly return disabled equipment to battlefield service using field expedient components and means.

2-243. Establishing aerial resupply and forward logistics base camps may be necessary to sustain operations. This is especially true in the offense, if an attack transitions to exploitation and pursuit conducted at great distances from unit sustaining bases. Aerial resupply, either by rotary-wing or parachute, delivers critical supplies to the point of need during an entry operation, deep inland operation, or to a rapidly moving unit. The unit or support activity at the airlift's point of origin is responsible for obtaining the required packing, shipping, and sling-load equipment. It prepares the load for aerial transport, prepares the pickup zone, and conducts air-loading operations. The unit located at the airlift destination is responsible for preparing the landing zone to accommodate aerial resupply and for receiving the load. (See FM 4-95 for additional information on logistics.)

Personnel Services

2-244. During large-scale combat operations, the key subordinate functions of staffing the force that are important are personnel accountability and strength reporting. The subordinate functions of provide human resource services including postal operations, finance services, and casualty operations, continue during large-scale combat operations. Human resource planning and operations are the means by which human resources are addressed in the military decision-making process and in the attack operations plan. This includes casualty forecasts necessary to inform commanders and staffs. (See FM 1-0 for additional information on human resources support.) For personnel legal support, judge advocate legal services personnel will coordinate and provide personnel legal services.

Health Service Support

2-245. Large-scale combat operations place an incredible burden on medical resources due to the magnitude and lethality of the forces involved. Medical units can anticipate large numbers of casualties in a short period of time due to the capabilities of modern conventional weapons and the possible employment of weapons of mass destruction. These mass casualty situations can exceed the capabilities of organic and direct support medical assets without careful planning and coordination. Casualty evacuation must occur concurrently with operations. Units that cease aggressive maneuver to evacuate casualties while in enemy contact are likely to both suffer additional casualties while stationary and fail their mission.

2-246. Effective management of mass casualty situations depends on established and rehearsed unit-level mass casualty plans and detailed medical planning. There are a number of other variables which can ensure the success of a unit's mass casualty response plan. These include, but are not limited to—

- Coordination and synchronization of additional medical support and augmentation and their dispositions and allocations, such as medical evacuation support, forward resuscitative and surgical teams, combat support and field hospitals, casualty collection points, ambulance exchange points, and established Class VIII resupply.
- Predesignating casualty collection points.
- Quickly locating the injured and clearing them from the battlefield.
- Providing effective emergency medical treatment for the injured.
- Accurate triage and rapid medical evacuation of the injured to medical treatment facilities at the next higher role of care.
- Use of alternative assets when the number of casualties overwhelms the capacity of available medical evacuation systems.

(See FM 4-02 for additional information on health service support.)

PROTECTION

Petty geniuses attempt to hold everything; wise men hold fast to the most important resort. They parry the great blows and scorn the little accidents. There is an ancient apothegm: he who would preserve everything, preserves nothing. Therefore, always sacrifice the bagatelle and pursue the essential!

Frederick the Great

2-247. The *protection warfighting function* is the related tasks and systems that preserve the force so the commander can apply maximum combat power to accomplish the mission (ADRP 3-0). The protection warfighting function includes the following tasks:

- Conduct survivability operations.
- Provide force health protection.
- Conduct CBRN operations.
- Provide EOD support.
- Coordinate AMD.
- Conduct personnel recovery.

- Conduct detention operations.
- Conduct risk management.
- Implement physical security procedures.
- Apply antiterrorism measures.
- Conduct police operations.
- Conduct population and resource control.

(See ADRP 3-37 for doctrine on the protection warfighting function.)

2-248. Preserving the force includes protecting personnel (combatants and noncombatants) and physical assets of the United States and unified action partners, including the host nation. The protection warfighting function enables commanders to maintain their force's integrity and combat power. Protection determines the degree to which potential threats can disrupt operations and then counters or mitigates those threats. Protection is a continuing activity; it integrates all protection capabilities to safeguard bases and base camps, secure routes, and protect forces. Protection activities ensure maintenance of the protection prioritization list, which includes the critical asset list and defended asset list.

2-249. The fluidity and rapid tempo of large-scale combat operations pose challenges for the protection of friendly assets. Commanders deny the enemy a chance to deliberately plan, prepare, and execute an effective response to friendly actions by maintaining a high tempo of operations. Ensuring the enemy is reactive increases the survivability of the force. Techniques for maintaining a high tempo include using multiple routes, dispersion, maneuver by mobile forces isolating and bypassing enemy forces, rotating or relieving forces before they culminate, and simultaneously converging capabilities across multiple domains and across multiple enemy echelons.

2-250. Commanders, with input from their staffs, determine the degree to which potential threats and hazards can disrupt operations, and then they counter or mitigate those threats and hazards. Many of the assets grouped within the protection warfighting function, such as MP units, engineers, CBRN assets, and other forces with protection responsibilities, will probably be assigned or attached to BCTs in the close area to accomplish their mission, limiting the availability of their associated capabilities in the support or consolidation areas.

2-251. At the corps and division echelon, commanders account for many protection tasks. BCTs often lack the capacity of protection assets to perform protection tasks when assigned to the close area, and they require augmentation from corps or division forces to effectively perform the protection tasks listed in ADRP 3-0.

2-252. One important aspect of protection planning for corps and divisions involves the support and consolidation areas. If conditions in the support area degrade, it is detrimental to the success of operations. A degraded support area also inhibits the ability to shape the deep area for the BCTs involved in close operations. Therefore, the protection of support areas requires planning considerations equal to those in the close areas. When the support area is located inside a division consolidation area, the unit assigned responsibility for the consolidation area provides significant protection for support area units. (See FM 3-81 for more information on support areas.)

2-253. Commanders protect subordinate forces to deny the enemy the capability to interfere with their ongoing operations. That protection also meets the commander's legal and moral obligations to the organization's Soldiers. To help protect the force, commanders ensure that all protection tasks are addressed during the unit's planning, preparation, and execution, and they also constantly assess the effectiveness of those protection tasks. (See ADRP 3-37 for more information on protection tasks.)

SURVIVABILITY

2-254. *Survivability* is a quality or capability of military forces which permits them to avoid or withstand hostile actions or environmental conditions while retaining the ability to fulfill their primary mission (ATP 3-37.34). *Survivability operations* are those military activities that alter the physical environment to provide or improve cover, camouflage, and concealment (ATP 3-37.34). Survivability operations are critical to offensive, defensive, and stability operations, but they may be emphasized differently in each. Commanders may rely more on mobility and terrain to provide concealment during the offense, while stressing camouflage, survivability positions, and hardened facilities to avoid or withstand hostile actions during the defense.

2-255. All units have the inherent responsibility to conduct survivability operations and continually improve their positions. Although there are three general categories of threats and hazards (hostile actions, nonhostile actions, and environmental conditions), survivability is most concerned with avoiding or withstanding the threats posed by hostile actions and environmental conditions. The four tasks associated with survivability operations are constructing fighting positions, constructing protective positions, hardening facilities, and employing camouflage and concealment. All four tasks are often addressed in combination. Some additional factors which can enhance an organization's ability to avoid or withstand threats and hazards include dispersion, redundancy, leadership, discipline, mobility, situational understanding, terrain management, and CBRN planning. The loss or severe degradation of unit CPs and other key facilities by enemy attacks in the physical and cyberspace domains can prevent the successful execution of missions. A CP or key facility's size, immobility, and multispectral signature invites enemy attack and the resulting disruption of the friendly tempo of operations. Survivability of those critical assets that enable a high operational tempo are a top priority. (See ATP 3-37.34 for additional information on survivability operations.)

CHEMICAL, BIOLOGICAL, RADIOLOGICAL, AND NUCLEAR DEFENSE OPERATIONS

2-256. CBRN active defense consists of tasks taken to prevent a CBRN attack by destroying the weapon or its delivery system. CBRN passive defense prevents or minimizes friendly unit vulnerability to the effects of CBRN threats or hazards. Commanders integrate CBRN defensive considerations into all mission planning. CBRN passive defense principles cover hazard awareness and understanding, protection (including mission-oriented protective posture gear, detection equipment, warning, and reporting) and contamination mitigation (including avoiding contamination and performing decontamination). CBRN passive defense is focused on maintaining the force's ability to continue military operations in a CBRN environment while minimizing or eliminating the vulnerability of the force to the degrading effects of those CBRN threats and hazards. (See ATP 3-11.36 for additional information on CBRN passive defensive considerations.)

2-257. The purpose of CBRN reconnaissance and surveillance is to provide commanders with detailed, timely, and accurate CBRN intelligence and to gain situational understanding of CBRN treats and hazards. Implementing many CBRN passive defensive measures may slow the tempo, degrade combat power, increase logistics requirements, and may require the technical skills of low density resources. (See FM 3-11 for information on CBRN operations.)

EXPLOSIVE ORDNANCE DISPOSAL

2-258. EOD provides the supported commander the capability to render safe and dispose of all explosive ordnance, to include unexploded explosive ordnance, improvised explosive devices, and improvised explosives. EOD elements may dispose of all types of hazardous foreign or U.S. ordnance in the safest manner possible. Breaching and clearance of minefields is primarily an engineer responsibility. The EOD force serves as a combat multiplier by neutralizing unexploded ordnance and mines and booby traps that restrict unit freedom of movement and deny access to or threaten supplies, facilities, and other critical assets within the unit AO. (See ATP 4-32 for additional information on EOD operations.)

DETENTION OPERATIONS

2-259. Detention involves the detainment of a population or group that poses some level of threat to military operations. These operations inherently control the movement and activities of a specific population for reasons of security, safety, or intelligence gathering. The Army is the DOD executive agent for all detainee operations and for the long-term confinement of U.S. military prisoners. Detention operations include—
- Interning U.S. military prisoners. (See FM 3-39 for additional information on the battlefield confinement of U.S. military prisoners.)
- Conducting detainee operations (includes belligerents [privileged belligerents and unprivileged belligerents], retained personnel, and civilian internees). (See FM 3-63 for additional information on detainee operations.)

Chapter 2

- Supporting host-nation corrections reform. (See FM 3-39 and FM 3-63 for additional information on host-nation corrections reform.)

2-260. During major combat operations, enemy units, separated and disorganized by the shock of intensive combat, may be captured. The numbers involved may place a tremendous burden on friendly forces as they divert tactical units to handle detainees. The term detainee includes any person captured, detained, or otherwise under the control of DOD personnel.

2-261. Detainee operations are the range of actions taken by U.S. armed forces, beginning at the point of capture; through movement to a detainee collection point, detainee holding area, or theater detention facility; until detainee transfer, release, repatriation, death or escape. Detainee operations is a broad term that encompasses the capture, initial detention, screening, transportation, treatment and protection, housing, transfer and release of the wide range of persons who could be categorized as detainees. Soldiers participating in military operations must be prepared to process detainees. Actions at the point of capture—the point at which a Soldier has the custody of, and is responsible for safeguarding, a detainee—can directly affect mission success and could have a lasting impact on U.S. strategic military objectives.

2-262. In any conflict involving U.S. forces, the safe and humane treatment of detainees is required by international law. Respect for individual human rights and humanitarian concerns is the basis for the Geneva Conventions and the Law of Armed Conflict, which codify the ideal that Soldiers, even in the most trying of circumstances, are bound to treat others with dignity and respect. Failure to conduct detainee operations in a humane manner and according to international law can result in significant adverse strategic impacts for the U.S. military.

2-263. MP units performing detainee operations can preserve the combat effectiveness of the capturing unit by relieving it of the responsibility to secure and care for detainees. During the conduct of military operations, MPs must possess the capability to plan, execute, and support detainee operations across the range of military operations.

PERSONNEL RECOVERY

2-264. *Army personnel recovery* is the military efforts taken to prepare for and execute the recovery and reintegration of isolated personnel (FM 3-50). Army forces work together with the DOD and other unified action partners to recover individuals and groups who become isolated. Isolation refers to persons being separated from their unit or in a situation where they must survive, evade, resist, or escape.

2-265. Ground unit commanders adjust their personnel recovery standard operating procedures and plans to incorporate the current mission variables. A corps or division-level personnel recovery task force organizes, trains, and conducts rehearsals to search for, locate, identify, and recover isolated personnel. This task force includes not only search and rescue elements, but also security forces designed to protect search and rescue teams from enemy attack.

2-266. CAB unit commanders prepare to conduct rotary-wing personnel recovery operations to support their own unit operations and provide mutual support to other aviation units based on their unit's inherent capabilities. Such support is planned concurrently with ongoing operations and considers the capabilities of adjacent and supporting units. (See FM 3-50 for more information on personnel recovery operations.)

SECTION IV – TRAINING FOR LARGE-SCALE COMBAT OPERATIONS

Thus a victorious army wins its victories before seeking battle; an army destined to defeat fights in the hope of winning.

Sun Tzu

2-267. Training is the most important thing the Army does to prepare for operations, and it is the cornerstone of combat readiness. Effective training and unit leader development must be commander driven, rigorous, realistic, and to the standard and under the conditions that units are expected to have to fight in, which includes joint and multinational operations. Realistic training with limited time and resources demands that commanders focus their unit training efforts to maximize repetitions under varying conditions to build

Army Echelons, Capabilities, and Training

proficiency. Units execute effective individual and collective training based on the Army's principles of training as described in ADRP 7-0.

> **Readiness: Task Force Smith**
>
> As he boarded a C-54 transport aircraft at Itazuke Airfield on the morning of July 1, 1950, Lieutenant Colonel Charles B. "Brad" Smith, commander of 1st Battalion 21st Infantry Regiment, felt no premonitions of imminent disaster. Unit morale was high throughout the Eighth Army in Japan. The 24th Infantry Division's four infantry regiments were postured to meet General Walton H. Walker's demand they be fully ready for combat by 1 December 1950. The previous March, Smith's battalion had completed live-fire exercises and a series of force-on-force offensive and defensive training exercises. Although not yet trained as part of a regimental combat team, it was as well prepared to fight as any unit could be, given the resources available.
>
> Shortcomings in personnel, equipment, and training were downplayed in quarterly "combat effectiveness" reports. Major General William F. Dean, commanding the 24th Infantry Division, knew his division rated lowest in readiness. But he believed the mission in Korea would be quick and easy, and would not over-stress his brittle formations. He did not object when ordered to Korea at the end of June 1950. LTC Smith's boss, Colonel Richard Stevens, was similarly unconcerned, especially since he believed that "no division is any damn good until after its first fight."
>
> At the highest levels of command there was inadequate concern with readiness. General Douglas MacArthur knew his Army formations were weak, but like nearly every other American in Japan at the time, MacArthur believed that the sight of American Soldiers on the battlefield in Korea would cause the communist Korean People's Army to melt away.
>
> General Walker knew the Eighth Army was a hollow force. Having commanded the Desert Training Center in 1942 and a corps in combat in 1944 and 1945, Walker knew better than almost anyone how to build cohesive and effective teams. Moreover, he knew that most of the factors preventing him from making the Eighth Army a truly combat-ready formation were beyond his control. But he kept his concerns to himself, knowing that MacArthur would just replace him if he pressed hard about issues that affected the entire U.S. Army. MacArthur, for his part, accepted the necessity of putting under-manned, ill-equipped, and partially-trained formations into harm's way. Calling Task Force Smith an "arrogant display of strength," MacArthur justified the risk by calculating that a determined showing early in the campaign would convince the communists that the Americans had more and better resources available than was actually true.
>
> The North Koreans were not fooled. On July 5, 1950, the KPA 4th Division destroyed Task Force Smith after a seven-hour fight. Over 150 of LTC Smith's men paid the ultimate price in part because of the Army leadership's inability to explain the Army's role in national security, and to secure the resources necessary to accomplish that role. Another 38,000 men would die fighting in Korea, many in the first year of combat, before the Army was able to adequately train and equip its forces.

2-268. The Army prepares itself for large-scale combat operations continuously. There is no time to build readiness necessary to win once hostilities commence in the current operational environment. Army forces must demonstrate a credible level of readiness against regional peer threats to effectively deter adversaries and assure partners. Generating credible readiness is the most important shaping task for units as they train at home station and during combat training center (CTC) exercises. Readiness to successfully conduct decisive action tasks in the context of large-scale land warfare against regional peer threats is the primary focus of Army forces.

2-269. Current threats demand Army units capable of maneuvering with unified action partners and generating effects, both lethal and nonlethal, across multiple contested domains against enemies whose capabilities may be superior and who enjoy positions of relative advantage. This, in turn, requires that Army units train under realistic conditions that portray real threat capabilities in ways that stress units and leaders for the realities of large-scale combat operations. It is critical for leaders to learn hard lessons and adapt before combat, and that is most effectively accomplished during tough collective training to standard against realistic threats. This training is accomplished at home stations, at CTCs, and while deployed.

2-270. All components of the Total Army (including Active, Reserve, National Guard, and Department of the Army [DA] civilians) bear responsibility for achieving the required levels of collective readiness. Each component has resident capabilities that enable the others, and as such must be trained to the same standard against peer threats. Units not training for specific missions focus on training for proficiency in large-scale combat operations. There are numerous implications across the warfighting functions that impact training, education, and priorities of effort, and they apply across all components. The scarcity of resources, particularly available time in the Army Reserve and Army National Guard, requires a clear prioritization of effort that is informed by the missions particular units perform during large-scale combat operations.

MISSION-ESSENTIAL TASKS

2-271. Commanders rarely have enough time or resources to train all tasks. Each commander determines what essential supporting collective tasks must be trained to attain the required levels of objective training requirements for mission-essential task list (METL) proficiency. The concept of mission essential tasks provides commanders a process to provide their units a battle focus. Each mission-essential task aligns with collective tasks that support it. All company and higher units have a METL. Units with a table of organization and equipment have an approved and standardized METL based on the type of unit and focused on large-scale combat operations. Standard METLs can be found on the Army Training Network, Digital Training Management System, and Combined Arms Training Strategies websites.

TRAINING TECHNIQUES

To lead untrained people to war is to throw them away.

Confucius

2-272. Using an integrated approach of live, virtual, and constructive training at home station, CTC rotations, and during deployments builds confident and cohesive units able to adapt to their environment and defeat the enemy. Demanding and repetitive training builds Soldiers' confidence in their weapons and equipment, their ability to fight and overcome challenges, their leaders, and their teams. Teams train under conditions that emphasize change, uncertainty, degraded friendly capabilities, capable enemies, and austere conditions. Soldiers must conduct realistic training that prepares them for combat by including unexpected tasks and moral-ethical challenges that help develop agile, adaptive, and innovative leaders. Training scenarios should require Soldiers to make the right decisions consistent with the moral principles of the Army Ethic, including the Army Values. (See FM 7-0 for an expanded discussion of Army training concepts.)

2-273. The effectiveness of each CTC is directly related to its ability to realistically replicate the conditions friendly units will face in combat against well trained peer threats with advanced capabilities. The CTC's capabilities should be leveraged to prepare Army forces for activities and operations across the range of military operations. The opposing forces portray capable enemies with modern technology to generate combinations of conventional, irregular, and disruptive threats across each of the domains and the information environment. They challenge training units to the limits of their capabilities, and they provide opportunities for collective training at a scale necessary to exercise combined arms operations in both force-on-force and live-fire conditions. Home station training preparation for CTC rotations should be multi-echelon, use combined arms, leverage opportunities for conventional force and SOF interdependence, and be as demanding as resources and conditions allow. The paradigm of predictable deployments to known areas, against known enemies and supported by a mission specific predeployment mission rehearsal exercise, is not a viable way for preparing units to win in the current OE. Army forces train and prepare for winning battles and engagements under realistic conditions with little or no prior notification.

2-274. BCTs with enabling capabilities conduct CTC rotations with a focus on decisive action, force-on-force exercises, and live fire against a regional peer threat. Functional and multifunctional brigades, divisions, corps, and theater armies use mission command training program exercises to train the operations process and mission command. They train on large-scale combat operations with unified action partners under realistic conditions that reflect the challenges of unified action in a multi-domain environment. Given the dynamic and lethal nature of large-scale land warfare, division and corps headquarters require proficiency in the operations process and mission command to orchestrate the high-tempo operations required to create, recognize, and exploit windows of opportunity. Deployment, field craft, continuous reconnaissance in depth, targeting, synchronization of fires and movement, airspace control, combined arms breaching and gap crossing, CP displacement, security, and sustainment are all essential tasks that require continuous, repetitive training for proficiency.

2-275. *Multiechelon training* is a training technique that allows for the simultaneous training of more than one echelon on different or complementary tasks (ADRP 7-0). As each echelon conducts its mission analysis to determine the tasks to train, it provides a logic trail from individual Soldier tasks to brigade-level mission essential tasks. An effective logic trail clearly nests from one echelon to the next and effectively crosswalks the tasks up and down unit echelons. Cross-walking enables leaders to visualize how the top-down training guidance directly supports the bottom-up alignment of individual and collective tasks that support the higher echelon unit.

TRAINING CONSIDERATIONS BY WARFIGHTING FUNCTION

The instruments of battle are valuable only if one knows how to use them.

Charles Ardant du Picq

2-276. Commanders and staffs use the warfighting functions to focus unit training. Training considerations for each warfighting functions are discussed in paragraphs 2-277 through 2-320.

MISSION COMMAND

2-277. Adversaries and enemies will target mission command systems and processes. Units must adapt their mission command systems and processes to the realities of fighting peer threats. Conditions in large-scale combat operations require the smallest possible physical and electronic signature and the highest possible level of agility. Leaders need to be able to command their formations when communication networks are disrupted, while on the move, and without perfect situational awareness. Training to become proficient in the use of analog data tracking systems, voice communications, and unaided navigation techniques requires significant amounts of repetition, particularly when integrating all of the elements of combat power. Habitual relationships, practiced standard operating procedures, and the use of battle drills can mitigate some of the risk and friction inherent in lost situational awareness.

2-278. Training units and leaders to execute operations using the philosophy of mission command mitigates the effects of systems degradation while enabling subordinates to effectively execute offensive, defensive, and stability tasks in the rapid manner required to exploit opportunities that come with achieving positions of relative advantage. Effective leadership applied through a combination of education, application, and repetition develops the trust and understanding of the commander's intent necessary to act in the absence of orders. Subordinates are unlikely to act in the absence of orders in the face of capable enemies if they cannot do so during peacetime training. Developing the climate necessary to facilitate effective mission command requires daily emphasis.

2-279. Successful commanders build cohesive organizations with a strong chain of command, ésprit de corps, and good discipline. As a unit trains, leaders direct, coach, guide, and challenge subordinates to develop depth of knowledge, judgement, and understanding. These actions ultimately build trust among Soldiers and between Soldiers and their leaders. Commanders ensure that their subordinates know the fundamentals about their jobs and how to think; both are necessary for subordinates to "connect the dots" necessary for solving new problems. They develop their subordinates' confidence and empower them to make independent, situation-based decisions. Effective commanders develop subordinates with agile and adaptive approaches to problem solving as insurance against the ambiguity, adversity, and uncertainty found on every battlefield.

> ### Fight as You Train: Kasserine Pass
>
> After more than three years of war, German forces were experienced veterans operating with doctrine that was combat proven in a variety of operational environments. U.S. Army doctrine and training at the tactical level, while partially informed by observing other armies fight from a distance, was largely unproven. The First Armored Division was relatively well-trained, organized, and equipped for doctrinal methods untested in the realities of World War II. U.S. Army armor and mechanized doctrine, particularly its dependence upon tank destroyers instead of tanks to defeat enemy armor, quickly proved inadequate.
>
> On 14 February 1943, four German Kampfgruppen (battalion task forces) surrounded and nearly annihilated Combat Command A (brigade combat team) of the 1st Armored Division at Sidi Bou Zid. The U.S. II Corps commander, MG Lloyd Fredendall, desperate to restore the security of the corps' open right flank, ordered the 1st Armored Division to "concentrate tomorrow on clearing up the situation there and destroying the enemy."
>
> On the morning of 15 February, Combat Command C began a movement to contact toward Sidi Bou Zid without a clear intelligence picture of the enemy. An M4 Sherman tank battalion led the formation with M3 halftrack tank-destroyers screening each flank. A self-propelled artillery battalion, a mechanized infantry battalion, and an attached tank company (reserve) followed. The battalions were in almost perfect, doctrinally correct formations. With the Combat Command C commander located on a hilltop to observe the movement, the tank battalion commander, LTC Alger, had tactical control of the entire force as it was approaching its first major engagement.
>
> At 1230, Combat Command C found itself in a German ambush. Alger's tanks initially engaged German anti-tank guns to their front (east), but German Panzer IIIs and IVs, twice their number, suddenly appeared to their north and south. LTC Alger's battalion reoriented to face these converging threats, establishing hasty battle positions. The Shermans, although qualitatively superior to the majority of German tanks, quickly succumbed to flank and rear shots in the melee. The thinly armored tank destroyers, intended by doctrine and design to engage the enemy at long range, were at a severe disadvantage and were quickly destroyed in the confused, close-in fight. By 1740, Combat Command C withdrew, having lost 52 of its original 56 Sherman tanks and almost all of its tank destroyers. It had suffered heavy casualties and was combat ineffective seven hours into its first battle.
>
> In a matter of days, the U.S. Army discovered that it would need to quickly adapt its doctrine, equipment, and training to the realities of modern combined arms warfare. A better appreciation for the experiences of other armies and the realities of the operational environment would have likely informed a different approach to tactical operations and reduced the cost of learning by experience.

AIRSPACE CONTROL

2-280. Airspace will be contested during large-scale land warfare against peer threats. Effective air-ground operations will require that units train to integrate airspace use during planning and execution, in conditions where use of mission command information systems is uninterrupted, and during degraded conditions when units must rely on procedural control. Effective airspace control training requires significant coordination, as airspace control is inherently a joint task. Installations and units must work together to replicate the complexity of airspace control in training in order to ensure units can effectively execute simultaneous organic fires, joint fires, air-ground operations, information collection, and air defense.

2-281. Setting the conditions for effective home-station training requires collaboration between the operational force units, the installation range operations, and the aligned USAF air support operations squadron. Installations must be able to simulate division-assigned airspace, so that individual airspace users as well as BCTs can operate as if they were under the control of a division JAGIC. Divisions need to train with the joint command and control system linkages they would have in an actual conflict. For example, installations can develop common graphic control measures that are shared on a digital COP through the use of Global Area Reference System keypad-based UAS airspace coordinating measures to enable rapid airspace coordination for UASs. Installations and units enhance individual and staff skills by using the mission training center to train Soldiers on their individual mission command information systems and how to pass relevant data to other Army and USAF C2 systems.

2-282. Airspace control collective training is centered on the combined arms live fire exercise. The division working with the aligned air support operations squadron provides input to the air tasking order and airspace control plan and airspace control orders primarily through the submission of air support requests. These inputs support subordinate unit planning. Unit combined arms wargaming and rehearsals of the selected course of action identify airspace control issues and allow resolution by refining tactics, techniques, and procedures and establishing necessary procedural controls. Units can then provide their installation range control with their coordinated airspace requirements for their combined arms live fire exercises. Then the range control can analyze what the respective unit's needs are and build airspace to support the unit's training. Execution of a combined arms live fire exercise gives a unit experience with integrating Army fires, close air support, Army manned and unmanned aircraft, and air defense. Division JAGICs maintain skills by periodic CP exercises initially focusing on internal JAGIC exercises, but as the JAGIC gains proficiency, the division headquarters can add CP exercises that include the key members of the current operations integration cell and subordinate brigade air defense airspace management/brigade aviation element. Installations support training units by assisting with the integration of the air support operations squadron digital systems and training and simulation systems with the mission training center. When the division JAGIC is proficient, it should work with the installation range control to take control of portions of installation airspace to maintain proficiency in performing procedural control in division-assigned airspace.

CIVIL AFFAIRS INTEGRATION

2-283. Existing or potential civil vulnerabilities will be exploited by threat actors. CA teams should be incorporated into routine training and staff exercises. These teams expand the capabilities of supported units by providing expertise in coordinating, integrating, and synchronizing with the civil component and by planning and executing CA-related stability tasks. CA teams provide expertise in civil security, civil control, foreign humanitarian assistance, civil information management, and civil administration planning which are all essential to the continuous process of consolidating gains to ensure a lasting and sustainable outcome.

INFORMATION OPERATIONS

2-284. Replicating the information environment and its activity in training is both challenging and essential. Integrating information operations into all staff planning and exercises builds understanding of how to effectively shape the information environment and specific ways they can synchronize information-related capabilities (IRCs) to achieve desired effects. As a result, it also builds competency in leaders.

2-285. Within their authorities, staffs should continuously synchronize information-related capabilities, such as MISO and OPSEC, into daily, steady-state activities. Information operations conducted as part of routine operational readiness may accomplish strategic and operational level objectives. They may—

- Directly engage unified action partner key leaders to influence their tacit and explicit cooperation.
- Protect classified and other sensitive information from adversaries, while confidently and transparently sharing information with unified action partners that will lead to greater trust among stakeholders.

Chapter 2

- Develop a credible shared narrative (most importantly by aligning words, deeds, and images) that explains U.S. and unified action partner intentions and efforts while diminishing the impact of adversary narratives.
- Employ required available media platforms to reinforce and amplify messages to support an approved narrative and broadcast unified action partner achievements and accomplishments through words, pictures, and video and immediately respond to adversary propaganda and subversion attempts.

TACTICAL DECEPTION

2-286. Often, Army commanders will be faced with deciding when and where to employ MILDEC in support of tactical operations. TAC-D, a category of MILDEC, is deception activities planned and executed by, and in support of, tactical-level commanders to cause enemies or adversaries to take actions or inactions favorable to the commander's objectives. TAC-D is conducted to influence military operations in order to gain an advantage over an enemy or adversary, mask vulnerabilities in friendly forces, or to enhance the defensive capabilities of friendly forces. This is accomplished by either increasing or decreasing the ambiguity of the enemy decision maker through the manipulation, distortion, or falsification of evidence. Tactical MILDEC requires leaders to understand enemies' and adversaries' information collection capabilities and decision processes. TAC-D is unique to the tactical requirements of the local commander and not necessarily linked or subordinate to a greater joint MILDEC plan. (See FM 6-0 and JP 3-13.4 for additional information on MILDEC and TAC-D.)

2-287. Commanders train TAC-D in a variety of ways. The most important is by incorporating elements of deception into plans routinely to determine effective techniques that work for their particular units and in particular situations. There is no single applicable technique or procedure that results in effective TAC-D because each TAC-D situation is unique. Developing the habit of incorporating deception into operations requires repetitive experimentation. The human aspects of MILDEC require a strong appreciation of the cognitive aspects of the OE. Training planners to use elements of misdirection, like feints and demonstrations, which can exploit what an enemy expects to see or portray a strength as a weakness, is an example of incorporating TAC-D without requiring additional resources.

OPERATIONS SECURITY

2-288. Enemies and adversaries exploit unclassified information used to synchronize and coordinate activities among all unified action partners. The five step OPSEC process identifies friendly capabilities, activities, limitations, vulnerabilities, and intentions that require defined protective measures to eliminate or minimize indicators of friendly activities. OPSEC measures require training and enforcement of standards. They include—

- Enforcement of noise and light discipline.
- Camouflage of CPs and combat vehicles.
- Destruction of sensitive material no longer required.
- Encryption of all electronic transmissions.
- Reduction of footprint through proper disposal of trash.
- Maximum use of couriers when possible.
- Minimum use of personal electronic devices.
- Physical and electronic masking of CP locations.

2-289. When OPSEC is combined with traditional security measures, classified and unclassified information is protected from adversarial observation. TAC-D and OPSEC are complementary, and they are trained concurrently to protect indicators of friendly activities and deny information which could reveal a true operation. The measures identified to protect information during the OPSEC process do not expose TAC-D activities while promoting and exposing those indicators and information supportive of TAC-D. Tactical units may train to shape an enemy's perceptions by releasing false information about friendly forces' intentions, capabilities, or vulnerabilities. False information targets the enemy's intelligence, surveillance, and reconnaissance capabilities to distract the enemy's intelligence collection away from, or provide cover for, unit operations. A TAC-D releasing information normally identified sensitive during the OPSEC process is a relatively easy form of deception to use, and it is very appropriate for use at battalion and lower echelon units. To be successful, a balance must be achieved between OPSEC and TAC-D requirements. Units train OPSEC continuously to develop the discipline and good habits required in all OEs.

CYBERSPACE AND ELECTRONIC WARFARE

2-290. As U.S. forces increase their reliance on networks and networked capabilities, vulnerability to attacks in cyberspace and the EMS increases. Adversaries use cyberspace and electronic attack capabilities to support military operations. Threats to U.S. forces include state and non-state actors, criminals, and insider threats. Mitigating the threat includes properly establishing, maintaining, securing, and defending U.S. networks. Cyberspace-related training designed to increase threat awareness and educate the force on policies and acceptable uses highlights how the improper use of networks and networked capabilities can create risks to the mission. This training is part of a defense-in-depth approach to protect U.S. networks. Training enables the force to protect networks, and their associated data, along with enhancing freedom of maneuver in cyberspace and the EMS.

2-291. The Army depends on the DODIN to enable all warfighting functions. Understanding threats to networks, including likely adversary tactics, techniques, and procedures is critical. Units must train to identify key terrain in cyberspace in relation to their commander's priorities to enable a focused defense. Establishing a properly configured, monitored, and secured network increases the ability to detect malicious and unauthorized activity and enables mission command and the warfighting functions.

2-292. Incorporating integrated training with a realistic threat able to attack networks into unit training at home station and CTCs prepares units for actual missions. It also prepares units to understand the threat, the indicators of a contested environment, and penalizes failures to follow security protocols.

2-293. Training in a realistic environment allows staffs to practice planning, integrating, and synchronizing defensive and offensive cyberspace operations, DODIN operations, and EW operations. A realistic training environment provides the appropriate defensive and offensive aspects of cyberspace and EW operations, and it allows units to integrate actions and effects in support of maneuver commanders. Training of forces tasked to engage targets in cyberspace and the EMS is critical to mission success. This requires specialized technical training, instrumentation, and ranges at home station and combat training centers to allow realistic and combat-focused training.

SPACE INTEGRATION

2-294. Space-based capabilities enable military forces with positioning, navigation, and timing; satellite communications; space-based information collection; missile warning; space and terrestrial weather forecasting; and space control capabilities. Many mission command systems and fires and maneuver units rely on space-based capabilities, creating potential vulnerabilities.

2-295. Army units cannot rely on uninterrupted and uncontested access to space-based capabilities. Units should assume the enemy will deploy denied, degraded, and disrupted space operational environment (D3SOE) strategies. Therefore, all echelons must train on tactics to counter conditions of D3SOE.

2-296. An enemy who is able to contest the space domain and force D3SOE conditions will be able to provide some level of disruption to every warfighting function. Peer enemy forces are equipped to employ theater-wide D3SOE effects which are part of an antiaccess and area denial strategy. Even small factions and individual actors could cause enough D3SOE to have significant impacts on mission accomplishment.

2-297. Employing D3SOE may hamper mission command, maneuver, fires, intelligence, protection, security, and information operations. Unit training and staff exercises must regularly include a D3SOE. Repetitive training in D3SOE conditions is essential for all units to become competent while operating with reduced space-based capabilities.

2-298. While the Army space training strategy covers both operational and institutional aspects of training, the operational piece focuses on providing unit training to counter D3SOE during multi-echelon home station training and at CTCs. Training efforts are focused on ensuring operational units can initiate and maintain access to space capabilities and tactics to apply when operating in contested conditions. Using training devices to replicate D3SOE conditions is critical to providing realistic operational training, especially for positioning, navigation, and timing and satellite communications. Direct injection devices allow a large maneuver, fires, or other audience to experience D3SOE effects on their space-enabled systems during training and exercises. These training devices realistically replicate the effects of D3SOE, and they are critical for implementing this training.

MOVEMENT AND MANEUVER

2-299. Threat conventional capabilities dictate that Army formations be capable of maneuvering from positions of disadvantage in order to create opportunities for exploitation by other members of the joint force. Given the potential advantages some enemies have in the range of their direct and indirect fires weapons systems, it is important to train maneuver in ways that allow for rapid and synchronized concentration of friendly units into close combat with the enemy. Training in the use of terrain, limited visibility, obscuration, and rapid maneuver on multiple routes or axes, coupled with the ability to discriminately and effectively destroy targets while on the move at maximum direct fire ranges, can mitigate the effects of enemy stand-off advantages.

2-300. Units must also train under degraded conditions, including CBRN conditions, degraded networks, and degraded positioning, navigation, and timing. The ability to maintain a rapid operational tempo and close quickly with enemy long-range systems in order to force them to reposition, making them vulnerable to Army and joint fires, requires significant skill derived from repetitive, realistic, and demanding training. Units, particularly dismounted and reconnaissance units, must be able to conduct infiltrations in a dispersed manner and then rapidly concentrate to generate effects, if they are to avoid being targeted by fires.

Mobility and Countermobility

2-301. The ability to maneuver ground forces under the threat of long-range fires, either from the ground or the air, demands focused training in areas where Army units enjoy few advantages in capability. The ability to mitigate the effects of obstacles is fundamental to the ability of maneuver forces to close with and destroy the enemy and to the overall mobility (movement and maneuver) of the force. Units must train as a combined arms team, under degraded conditions, in the execution of the mobility tasks to include breaching, clearing and gap-crossing operations. (See ATP 3-90.4 for an in-depth discussion of combined arms mobility.)

2-302. The ability to defend gives commanders the time to create conditions favorable for offensive tasks to continue, to retain a position of relative advantage, to degrade enemy advantages, or to protect forces. Countermobility operations use or enhance the effects of obstacles to help shape terrain and isolate the battlefield during the execution of both the offense and defense. During the offense, countermobility operations must stress rapid emplacement and flexibility and typically focus on protecting friendly forces from an enemy counterattack. In the defense, countermobility operations use or enhance obstacles to halt or slow enemy movement and maneuver, canalize the enemy into engagement areas, and protect friendly positions. Commanders and staffs at all echelons must be able to effectively plan, synchronize and execute countermobility operations in support of the offense and defense. (See ATP 3-90.8 for an in-depth discussion of combined arms countermobility.)

Army Aviation Maneuver (Manned)

2-303. Peer threats are equipped with and employ significant air defense capabilities, which will force Army aviation to operate in novel ways, leveraging all domains to conduct effective attack, reconnaissance, aerial medical evacuation, and air assault operations in high threat environments. Aviation training must focus heavily on operations during periods of limited visibility, aircraft survivability equipment employment, terrain flight techniques, and gunnery operations that emphasize engaging targets at maximum stand-off ranges. Army aviation can conduct effective attack, reconnaissance, aerial medical evacuation, and air assault operations throughout the division and corps area of operations. Corps and division staffs must also develop expertise in airspace control in highly contested and congested airspace and be able to effectively plan, synchronize, and execute electronic and lethal suppression of enemy air defenses synchronized with aviation maneuver. (See FM 3-04 and ATP 3-04.1 for an in-depth discussion of Army aviation operations and employment.)

Army Aviation (Unmanned)

2-304. The use of UASs will be heavily contested by both enemy air defense and EW capabilities. Given the altitudes required to operate tactical and larger unmanned systems and the density of enemy air defense and EW capabilities in the close and deep battle areas, Army units will not be able to rely on freedom of action when employing UASs and must plan for significant attrition. Conducting detailed IPB and understanding enemy capabilities is essential to developing the best AOs for UASs to safely and effectively operate during the initial stages of a conflict. In contested environments, using maximum standoff ranges provided by Gray Eagles instead of the electro-optical sights increases stand off and survivability. This requires both operator and analyst training to maximize the effectiveness this capability. As an enemy is displaced from positions of advantage, Army forces must be trained to rapidly understand and exploit windows of opportunity where UASs can be effectively employed in the close, deep, and consolidation areas.

INTELLIGENCE

2-305. Organic Army information collection assets, both ground and air, will be contested during operations against capable conventional and hybrid peer threats. Furthermore, Army units cannot depend upon uninterrupted access to national intelligence assets due to likely disruption of the cyberspace and space domain-based enablers. Ground forces will need to fight for information by developing the situation in contact, and they must train to operate with less than perfect situational awareness at the tactical level. To avoid becoming surprised, units should assume during training that they are under continuous observation and develop tactics that account for the reality that the enemy may have equal or better situational awareness than they do.

2-306. The preparation of intelligence organizations and staffs to cope with lethal operational environments that support rapid decision making must occur as part of operations to shape. The ability to discern adversary warnings and enemy intentions and capabilities across all of the domains informs the commander's decisions and realistic assessment of operational and tactical risk. Balancing situational awareness with the requirement to maintain a high operational tempo must be practiced before the commencement of hostilities.

FIRES

2-307. The ability to deliver fires against peer threats requires units that can stealthily position themselves to create effects and then rapidly reposition or disperse to avoid effective counter-fire. The superior range and lethal effects of many enemy missile, rocket, and cannon systems requires significant training to overcome, and they require a combined arms approach to execution. Enemies able to contest the cyberspace domain can disrupt the mission command and targeting process for friendly artillery units. To prevent this, units require training in analog methods of employment. Army units cannot always depend upon Army aviation or joint fires to solve tactical problems, so unit organic fires elements (including tube, rocket, and mortar units) must become proficient in maneuvering into positions of advantage to shoot and then rapidly displace.

2-308. Degraded networks and reduced situational awareness demand training in the clearance of fires, effective airspace management, and the development of trust between firing and maneuvering units. Determining the balance between speed, accuracy, and fratricide avoidance requires repetitive training under realistic conditions. Doing so as part of a multinational coalition is even more challenging, and use of fires should be rehearsed during operations to shape as often as possible. Integration of fires systems with unified action partners can be time consuming, but it is vital to success during multinational operations.

Counterfire

2-309. Some peer threats can deliver fires over an extended distance and at a sustained volume. Threat forces also have the equipment and tactics to execute effective counter-fire. Training requirements for counter-fire and the displacement of weapons locating radars and sequencing are integrated into all combined arms training exercises. If acquisition radar systems are continuously radiating, they will be targeted by the enemy. Sensor management must be planned and rehearsed, and cueing schedules developed and actively pursued. Training for timely displacements of field artillery units is essential for successful operations. Field artillery units move well forward prior to an attack, with synchronized plans to displace by echelon to provide

Chapter 2

continuous fires to the supported force. Displacements should maximize continuous delivery of fires, and they should be completed as rapidly as possible.

Air Defense

2-310. Adversary antiaccess systems include IADSs designed to allow their air and missile forces to approach and engage friendly forward units. Fires units train to protect forces from rockets, air and missile threats (including UAS), and provide warnings, while defending other forces from air threats as they maneuver. Short range air defense units are integrated into maneuver forces during field training exercises whenever possible. Other units develop training plans to integrate direct fire weapons into existing air defense plans. (See FM 3-01 for more information on air defense.)

SUSTAINMENT

2-311. Sustainment units must train to support forces in an expeditionary, austere operational environment. Threats consider U.S. sustainment units HPTs. Commanders conduct training under realistic conditions that force sustainment leaders to prepare to overcome an adaptive, agile enemy. Sustainment units must train to overcome the challenges presented by interdicted road networks and degraded communications and information systems while providing operational reach, endurance, and freedom of action to maneuver forces. Training must also include the reconstitution and regeneration of combat forces. Sustainment leaders and staffs must be proficient tacticians and sustainers to anticipate requirements and ensure supported forces maintain combat and operational readiness.

2-312. Sustainment training is included as part of multi-echelon field training exercises. All units and their supporting sustainment elements train on resupply (including delivery of logistic packages), vehicle recovery, convoy operations, and establishing and operating both unit maintenance collection points and casualty collection points. Establishing forward air refueling points and training to refuel on the move are other examples of integrating sustainment into collective training events.

2-313. Large-scale combat operations will create high demands for logistics, personnel services, and Army health services. Sustaining units in the corps, division, and brigade AOs will incur significant risk due to the presence of conventional and hybrid threats. Sustainment units must be equipped, structured, trained, and prepared to execute reconnaissance and security tasks during large-scale combat operations to ensure they can complete sustainment missions when maneuver support is unavailable.

PROTECTION

2-314. Commanders must deliberately plan and train to engage enemy forces while preserving combat power. Readiness requires commanders to clearly articulate protection priorities as they relate to the adversary's and enemy's collection capabilities to observe friendly ground forces using space, cyberspace, air, and ground-based means. Units need to be proficient in the development of the protection prioritization list, which includes the defended asset list.

2-315. Army forces must enforce camouflage, light discipline, and minimal electromagnetic signatures in training. Detection risks destruction. The adversaries' use of UAS increases their ability to detect friendly forces. Therefore, Army units must be able to reposition rapidly to complicate enemy targeting efforts. Units must be proficient in the use of active and passive protection measures, and leaders must enforce these measures to ensure compliance. Discipline is foundational to protection.

2-316. All military activities have some inherent or organic protection capability (including survivability, antiterrorism measures, local security, and area security). Some protection is achieved through movement and maneuver by changing tempo, taking evasive action, or maneuvering to gain positional advantage in relation to a threat. Formations often acquire protection using terrain, weather, or darkness to mask movement.

2-317. Some terrain feature attributes offer more survivability than others, and they can complement force positioning. Commanders can gain protection from enemy indirect fire by dispersing CPs, while they can also gain protection from enemy reconnaissance by concentrating CPs in more permissive environments.

2-318. Protection of friendly networks requires developing and following policy and procedures. Adversary electronic warfare capabilities can disrupt both data and voice networks. Commanders enforce the proper execution of drills designed to mitigate attacks. Mitigating attacks enables mission command, fires, and information collection by friendly networks. Policies and procedures should also address establishing secure communications with appropriate multinational partners.

2-319. The likelihood of weapons of mass destruction being employed by an enemy is significant during large-scale combat operations, particularly against critical friendly mission command nodes, massed formations, and critical infrastructure. Commanders balance the need to mass effects against the requirement to concentrate forces and ensure as much dispersal as is tactically prudent to avoid presenting lucrative targets for enemy weapons of mass destruction. Army units must train effectively to operate under CBRN conditions. Operating during CBRN contamination should be a training condition, not simply a task.

2-320. Commanders and staffs must also synchronize, integrate, and organize additional protection capabilities and resources that preserve combat power. The integration and synchronization of area and local security activities, OPSEC, and cyberspace and EW operations with the protection warfighting function tasks enable commanders to protect forces, critical assets, information, and combat power. Camouflage and concealment activities are individual or unit responsibilities and governed by standard operating procedures. They also play a role in tactical deception to mislead the enemy's information collection capabilities. Merely hiding forces may not be adequate, as an enemy may need to "see" these forces elsewhere. In such cases, cover and concealment can hide the presence of friendly forces, but decoy placement should be coordinated as part of the deception in support of OPSEC.

This page intentionally left blank.

Chapter 3
Operations to Shape

The history of failure in war can be summed up in two words: Too Late. Too late in comprehending the deadly purpose of a potential enemy; too late in realizing the mortal danger; too late in preparedness; too late in uniting all possible forces for resistance; too late in standing with one's friends.

General Douglas MacArthur

This chapter provides an overview of operations to shape. It discusses operations assessments and describes threat activities prior to armed conflict. A discussion of shaping activities performed by Army forces follows. The chapter then describes Army organizations and their roles during operations to shape. The chapter concludes with a discussion of consolidate gains during operations to shape.

OVERVIEW OF OPERATIONS TO SHAPE

3-1. Operations to shape consist of various long-term military engagements, security cooperation, and deterrence missions, tasks, and actions intended to assure friends, build partner capacity and capability, and promote regional stability. Operations to shape typically occur in support of the geographic combatant commander's (GCC's) theater campaign plan (TCP) or the theater security cooperation plan. These operations help counter actions by adversaries that challenge the stability of a nation or region contrary to U.S. interests. Operations to shape occur across the joint phasing model. (See paragraphs 1-51 through 1-60.) Ultimately, operations to shape focus on four purposes:
- Promoting and protecting U.S. national interests and influence.
- Building partner capacity and partnerships.
- Recognizing and countering adversary attempts to gain positions of relative advantage.
- Setting conditions to win future conflicts.

3-2. Operations to shape roughly correlate with theater shaping activities described in JP 3-0. They also include unit home station activities, including maintaining operational readiness, training, and contingency planning. Combined exercises and training, military exchange programs, and foreign military member attendance at Army schools are examples of home station shaping activities.

3-3. Optimally, shaping activities ensure regions remain stable, a crisis does not occur, and there is no need for an escalation of force. Upon activation of a joint operation order (OPORD) for a crisis or a limited contingency operation Army operations to shape occur simultaneously within a joint operations area (JOA) or designated theater of operations and across the GCC's area of responsibility (AOR).

3-4. Figure 3-1 on page 3-2 depicts shaping and prevent activities in an environment of cooperation and competition. Army forces, as part of a joint team and a larger whole-of-government effort, conduct shaping activities to assure friends, build partners, and prevent, deter, or turn back escalatory activity by adversaries. Army operations to shape align with military engagement and security cooperation activities. Army prevent activities align with deterrence and crisis response and limited contingency operations. (See chapter 4 for a discussion of operations to prevent.)

Chapter 3

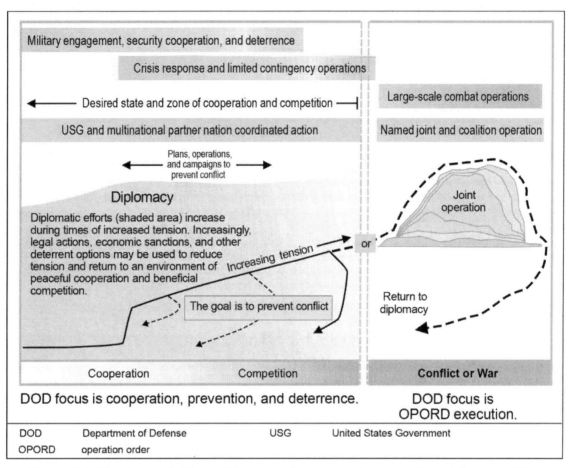

Figure 3-1. Shaping activities within an environment of cooperation and competition

3-5. Projecting U.S. military force requires extensive use of international waters, international airspace, space, and cyberspace. As such, shaping activities help assure operational access for crisis response and contingency operations despite changing U.S. overseas defense posture and the growth of antiaccess (A2) and area denial (AD) capabilities around the globe. The more the GCC can promote favorable access conditions in advance across the AOR and in potential operational areas, the better. Shaping activities involving Army forces in support of the GCC to promote favorable access include—

- Key leader engagements.
- Bilateral and multinational exercises to improve multinational interoperability and operations.
- Missions to train, advise, and equip foreign forces.
- Negotiations to secure basing and transit rights, establish relationships, and formalize support agreements.
- The use of grants and contracts to improve relationships with and strengthen partner nations.
- Designing interoperability into acquisition programs.
- Electromagnetic spectrum (EMS) mapping of adversary capabilities.

3-6. The theater army provides a critical role for the GCC in gaining operational access and positions of relative advantage throughout the AOR. Gaining positions of relative advantage during operations to shape involves analysis of current adversary intent and capabilities and potential future adversary intent and capabilities. Through this analysis, commanders can recommend actions that may impede adversary goals or convince adversaries to seek alternative courses of action more favorable to U.S. interests. Friendly actions could involve physical positioning of forces to block potential avenues of approach or transmission of messages that serve to counter an adversary's intent, decision making, and influence on populations.

3-7. Effective training and leader development form the cornerstone of operational readiness, and they are part of Army operations to shape. The priority focus for Army forces not committed to specific combatant commander requirements is building and sustaining readiness to conduct large-scale combat operations. Units, leaders, and Soldiers achieve the tactical and technical competence that builds mutual trust, esprit de corps, and adaptability by overcoming challenges through realistic training. (See chapter 2 for a discussion on training for large-scale ground combat).

> **Security Force Assistance and Long-Term Consolidation of Gains: Columbia**
>
> U.S. support to Colombia as it confronted the destabilizing challenges of the Revolutionary Armed Forces of Colombia (FARC) provides a useful example of how security assistance and security force assistance, as part of operations to shape, enabled a partner nation to defeat a decades-long insurgency, secure its population, and ultimately consolidate government control over the country. While there was no serious intention to initiate large-scale combat operations, the United States was willing to conduct a wide variety of other types of operations to achieve its aim of preventing the overthrow of the Colombian government. Through various long-term programs, the United States aided Colombian forces in their struggle against a protracted, bloody insurgency that at times controlled significant areas of the country. Those programs adapted to conditions on the ground and contributed to Colombian government successes against the FARC. They also created positions of relative advantage that the government, enabled by U.S. support, could exploit. The Army's direct efforts helped synchronize the combination of physical and cognitive means to achieve objectives such as retaining popular support of the government and expanding an alternative narrative to the FARC's ideals.
>
> Army forces played a significant role in the planning and execution of the U.S. supporting effort. Security force assistance helped develop competent law enforcement and military units. The United States provided technical and operational assistance to Colombian National Police and military forces, along with fully integrating the military approach into interagency efforts. Improved police and military operations weakened transnational drug cartels and the FARC, enhancing Colombia's security. This combination of high quantity and high quality engagement bolstered Colombian capacity. Decades of exposure to U.S. special operations forces (SOF) personnel, together with training and educational programs, gave Colombian security professionals newfound technical expertise, which enabled them to make use of advanced technologies and techniques provided by the United States. These capacity-building activities also implanted in rising generations of Colombian military and governmental personnel certain cultural attributes that made them more effective in current and future operations.

OPERATIONS ASSESSMENTS

All planning, particularly strategic planning, must pay attention to the character of contemporary warfare.

Carl von Clausewitz

3-8. Operations assessments are an integral part of planning and execution of any operation, fulfilling the requirement for identifying and analyzing changes in an operational environment (OE) and determining the progress of a campaign or operation. TCPs and country plans are continuously in some stage of implementation. Operations assessments helps combatant commanders and subordinate component commanders develop, adapt, and refine these plans.

Chapter 3

3-9. Collecting information to facilitate understanding of the operational variables provides the basis for operations assessments. A complete understanding of an OE may be hindered if the focus is solely on adversary information and actions. Additional information collection with a focus on the civilian population is often required for and is key to influencing and mobilizing the population and the consolidation of gains. People and populations within a region can present significant security challenges. Operations to shape are accomplished through a variety of missions, tasks, and actions, and they are often focused on understanding, engaging, influencing, changing, or countering human perceptions. This requires study and analysis to ensure the right decisions and actions are taken at the right time to get positive outcomes. The complexity of the human aspects of conflict are dynamic. Therefore, operations to shape must be persistent and consistent to maintain continuity, and they must be adjusted based on changing conditions.

3-10. Operations to shape provide opportunities for conventional forces and SOF interdependence, integration, and information sharing. Army special operations forces (ARSOF) largely focus on human aspects of an OE and have specific capabilities that facilitate this focus. By leveraging ARSOF use of four operational tenants (understanding human factors of the OE; incorporating human factors into planning; partnership activities and enablement; and operate with and through unified action partners), commanders can incorporate human aspects into their training, planning, and execution of operations to shape. (See ADRP 3-05 for detailed information on ARSOF operational tenets.)

THREATS

3-11. Threats desire to further their interests and achieve their goals without fighting. If threat actions are successful and are counter to U.S. interests, the United States may consider introducing armed forces to reverse or stabilize the situation. Threat goals prior to this introduction are centered on preventing this, and if that fails, to constrain that introduction in such a way as to prevent the success of a U.S. joint operation. Threats will focus on the following methods:

- Threat information warfare activities will manipulate the acquisition, transmission, and presentation of information in such a way that suits the threat's preferred decision outcomes.
- Threats will foster instability in key areas and among key groups in such a way that regional security does not match U.S. operational requirements.
- Threats will act upon partnerships to reduce the ability of the United States to operate in its preferred combined, joint, and interagency manner.
- The focus of pre-conflict preclusion activities is through nonlethal means. Threats will undermine relationships, raise political stakes, manipulate public opinion, and attack resolve in order to constrain or deny basing rights, overflight corridors, logistic support, and concerted allied action.

3-12. Adversaries may seek to establish conditions that limit or prevent U.S. access to a region. This includes forward positioning of robust and layered integrated air defenses, early warning surveillance radars, and electronic warfare capabilities (as demonstrated by Russian forces in Syria). Additionally, adversaries may seek to position intermediate range ballistic missiles, cruise missiles, fixed-wing aircraft, unmanned aerial systems, and naval surface and sub-surface forces to shape an OE in their favor. Positioning A2 systems provides adversaries a position of advantage to deny or disrupt U.S. access to a region in the event of hostilities, while providing leverage against friendly partner nations during steady-state operations. An adversary's ability to establish, maintain, and demonstrate robust A2 systems erodes partner-nation trust and confidence while bolstering adversary domestic narratives.

SHAPING ACTIVITIES

3-13. Shaping activities are continuous within an AOR. The combatant commander uses them to improve security within partner nations, enhance international legitimacy, gain multinational cooperation, and influence adversary decision making. This cooperation includes information exchange and intelligence sharing, obtaining access for U.S. forces in peacetime and crisis, and mitigating conditions that could lead to a crisis.

3-14. Shaping activities are directly tied to authorities provided in various titles of the United States Code (USC) and approved programs, and integrated and synchronized with the Department of State, other government agencies, country teams, and ambassadors' plans and objectives. The Department of State and the United States Agency for International Development develop the joint regional strategy to address regional goals, management, operational considerations, and resources. Each country team develops an individual country plan to address country context, joint mission goals, and coordinated strategies for development, cooperation, security, and diplomatic activities. Working with the Department of State and various country teams, the GCC and planners develop a theater strategy to influence regional and country conditions to achieve national objectives. The theater strategy is translated into a TCP. The TCP guides the shaping activities conducted throughout the AOR by joint forces.

3-15. The theater army significantly contributes to the planning, execution, and assessment of the GCC's TCP. Army forces conduct operations to shape with various unified action partners through careful coordination and synchronization facilitated by the theater army through the GCC, and when authorized, directly with the partner nation's military forces. Army forces provide security cooperation capabilities AOR-wide, including building defense and security relationships and partner military capacity through exercises and engagements, gaining or maintaining access to populations, supporting infrastructure through assistance visits, and fulfilling executive agent responsibilities. Military-to-military contacts and exchanges, joint and combined exercises, various long-term persistent military engagements, and other security cooperation activities provide the foundation of the GCC's TCP.

MILITARY ENGAGEMENT

3-16. *Military engagement* is the routine contact and interaction between individuals or elements of the Armed Forces of the United States and those of another nation's armed forces, or foreign and domestic civilian authorities or agencies to build trust and confidence, share information, coordinate mutual activities, and maintain influence (JP 3-0). Military engagement occurs as part of security cooperation, but it also extends to interaction with domestic civilian authorities. GCCs seek out partners and communicate with adversaries to discover areas of common interest and tension. This increases the knowledge base for subsequent decisions and resource allocation. Such military engagements can reduce tensions and may prevent conflict, or, if conflict is unavoidable, they may allow the U.S. to enter into it with greater access and stronger alliances or coalitions. Army forces support military engagement through key leader engagement and Soldier and leader engagement.

SECURITY COOPERATION

3-17. *Security cooperation* is all Department of Defense interactions with foreign security establishments to build security relationships that promote specific United States security interests, develop allied and partner nation military and security capabilities for self-defense and multinational operations, and provide United States forces with peacetime and contingency access to allied and partner nations (JP 3-20). These efforts may include Army forces participating in joint and multinational exercises and employing regionally aligned forces. Conducting security cooperation is one of the Army's primary stability tasks. Security cooperation is governed by various sections of Title 10, USC; Title 22, USC; and specific public laws addressing Department of Defense (DOD) interactions with other nations. (See JP 3-20 for more information on security cooperation.)

3-18. Commanders and staffs conduct security cooperation to develop allied and friendly military capabilities for self-defense and multinational operations, to improve information exchange and intelligence sharing, to provide U.S. forces with peacetime and contingency access, and to mitigate conditions that could lead to a crisis. Security cooperation activities include—

- Security assistance.
- Security force assistance (SFA).
- Foreign internal defense.
- Security sector reform.

(See FM 3-22 for a more detailed discussion of Army support to security cooperation.)

Chapter 3

Security Assistance

3-19. Security assistance is a group of programs the U.S. Government uses to provide defense articles, military training, and other defense-related services by grant, loan, credit, or cash sales. Security assistance programs are typically focused on the transfer of defense articles and services to eligible foreign governments, the provision of training and education to foreign military personnel, and the sale of construction services in support of partner nations' military establishments. These security assistance programs are discussed in JP 3-22.

Security Force Assistance

3-20. *Security force assistance* is the Department of Defense activities that contribute to unified action by the United States Government to support the development of the capacity and capability of foreign security forces and their supporting institutions (JP 3-22). Consistent with DOD policy for SFA, the Army develops, maintains, and institutionalizes the capabilities of its personnel to support DOD efforts to organize, train, equip, and advise foreign security forces (FSF) and relevant supporting institutions. *Security forces* are duly constituted military, paramilitary, police, and constabulary forces of a state (JP 3-22). Army forces conduct SFA activities in support of the combatant commanders' TCP when directed to do so in accordance with appropriate legal authorities. SFA brigades are Army organizations created in 2017 that are tailored to meet SFA requirements.

3-21. SFA and security assistance are different. SFA often works in conjunction with the various security assistance programs discussed in JP 3-22. However, the focus of SFA is on building the capacity and capability of FSF and their supporting institutions. SFA encompasses various activities related to the organizing, training, advising, equipping, and assessing of FSF and their supporting institutions, from tactical to ministerial levels. These activities contribute to unified action to generate, employ, and sustain FSF. SFA activities are conducted primarily to assist partner nations to build their capacity to defend against external and transnational threat actors. Army forces may also conduct SFA to assist partners to contribute to multinational operations; or organize, train, equip, and advise a nation's security forces and supporting institutions.

Foreign Internal Defense

3-22. *Foreign internal defense* is participation by civilian and military agencies of a government in any of the action programs taken by another government or other designated organization to free and protect its society from subversion, lawlessness, insurgency, terrorism, and other threats to its security (JP 3-22). Foreign internal defense (FID) includes the actions of nonmilitary organizations as well as military forces.

3-23. FID is a whole-of-government approach that supports partner development towards democratic governance and military deference to civilian rule. FID activities may employ the indirect use of military forces along with the diplomatic, informational, and economic means. FID principles are used to mitigate the need to deploy large numbers of U.S. military personnel and equipment. FID involves the support of a standing, host-nation government and its military or paramilitary forces. FID is a key supporting component of a host-nation's internal defense and development program. The focus of all U.S. FID efforts is to support the host-nation's internal defense and development program to build the capability and capacity of the host nation to achieve self-sufficiency. FID has been and remains an Army SOF core activity. (See JP 3-22 and ATP 3-05.2 for a detailed discussion of foreign internal defense. See ADRP 1-03 for the Army tasks associated with FID.)

3-24. SFA and FID have much in common because both enable friendly partners' capacity to provide for their own defense. FID activities directly support organizing, training, equipping, advising, and assisting FSF to combat internal threats. SFA prepares FSF to defend against external threats and to perform as part of an international force.

Security Sector Reform

3-25. *Security sector reform* is a comprehensive set of programs and activities undertaken by a host nation to improve the way it provides safety, security, and justice (JP 3-07). The overall objective is to provide these services in a way that promotes an effective and legitimate public service that is transparent, accountable to civilian authority, and responsive to the needs of the public.

3-26. Security sector reform (SSR) is an umbrella term that includes integrated activities in support of defense and armed forces reform; civilian management and oversight; justice, police, corrections, and intelligence reform; national security planning and strategy support; border management; disarmament; demobilizations and reintegration; or reduction of armed violence. The Army's primary role in SSR is supporting the reform, restructuring, or re-establishing the armed forces and the defense sector across the range of military operations.

3-27. With the support of the host nation, U.S. and partner military forces collaborate with interagency representatives and other civilian organizations to design and implement SSR strategies, plans, programs, and activities. DOD leads and provides oversight for these efforts through its bureaus, offices, and overseas missions. SSR facilitates security cooperation and SFA activities that build partner capacity. SSR involves reestablishing or reforming institutions and key ministerial positions that maintain and provide oversight for the safety and security of the host nation and its people. Through unified action, those individuals and institutions assume an effective, legitimate, and accountable role. SFA activities help provide internal and external security for host-nation citizens, under the civilian control of a legitimate state authority. Effective SSR enables a state to build its capacity to provide security. The desired outcome of SSR programs is an effective and legitimate security sector firmly rooted within the rule of law.

ADDITIONAL SHAPING ACTIVITIES

3-28. As part of operations to shape, Army forces participate in and conduct numerous other activities in support of the combatant commander's TCP. These include developing intelligence, countering weapons of mass destruction (CWMD), providing support to humanitarian efforts, conducting information operations, and organizing and participating in combined training and exercises.

Intelligence

3-29. The military intelligence brigade-theater (MIB-T) provides regionally focused collection and analysis in support of the GCC, theater army commander, and Army forces conducting operations to shape. The MIB-T is the anchor point for reachback intelligence support to tactical forces. Regionally aligned or assigned forces and other specified Army units require ready access to and seamless interaction with their associated combatant command's intelligence architecture. When Army units enter a GCC's AOR, the primary intelligence support is through the GCC joint intelligence center or joint intelligence operations center unless another support relationship is established in the GCC orders. The MIB-T provides this linkage. (For additional information on the MIB-T, see paragraphs 2-26 through 2-27.)

3-30. Identifying threat capabilities, strengths, weaknesses, and intent accurately is critical to providing commanders the timely indications and warnings necessary to ensure operational success. During operations to shape, the most important role of intelligence is to provide early and accurate warnings of changes to the OE and threats to enable commanders and senior government officials to make timely and informed decisions. Commanders focus information collection operations to provide early warning of adversary activity, recognize threats, and identify the need to make timely decisions.

3-31. In addition, commanders focus the intelligence enterprise on identifying strengths and vulnerabilities of threat capabilities, their area denial strategy or systems, their information warfare and subversion activities, and their ability to achieve parity or superiority across each of the five domains (land, air, maritime, space, and cyberspace) and the information environment. (See ADRP 2-0 for a discussion of the intelligence enterprise).

Countering Weapons of Mass Destruction

3-32. Shaping activities work toward strategic deterrence of weapons of mass destruction (WMD). Army forces shape an OE to dissuade or deter adversaries from developing, acquiring, proliferating, or using WMD. To prevent the use of WMD, Army forces must develop an understanding of the threat and materials that affect an area of operations (AO) as part of the countering WMD mission. The cooperation and partner activities mission is a collection of interrelated day-to-day activities to deny, dissuade, and prevent adversaries from obtaining or proliferating WMD. (See ATP 3-90.40 for more information on countering weapons of mass destruction.)

Humanitarian Efforts

3-33. The United States Agency for International Development is the lead U.S. government agency, responsible to the Secretary of State, for administering civilian foreign aid and providing humanitarian assistance and disaster relief. The United States Agency for International Development often works in concert with Army forces when Soldiers are tasked to provide assistance. It can supplement civil affairs (CA) forces conducting CA activities that the DOD conducts to build relationships and win the trust, confidence, and support of local populations.

> **Humanitarian Assistance: Republic of Cameroon**
>
> A humanitarian assistance operation conducted by a U.S. Army civil affairs (CA) unit in the Republic of Cameroon in Africa provided relief to a nation devastated by disease. In 1989, the American Embassy and the Ministry of Public Health in Cameroon proposed a campaign to inoculate citizens against meningitis, a disease that ravages that tropical country each year during the dry season. The embassy defense attaché office contacted United States European Command and plans were drawn to support a humanitarian assistance exercise in conjunction with CA support. In February 1991, a medical team from the 353d Civil Affairs Command, working in conjunction with the host nation, inoculated more than 58,000 people against meningitis and treated an additional 1,700 people for other ailments. This exercise not only accomplished its humanitarian goals, but it also provided an opportunity for the unit to train and use its language skills. At the same time, it strengthened the partnership between the United States and Cameroon while improving access and stability in a vital region.

Information Operations

3-34. The primary stability mechanisms Army units employ during operations to shape are influence and support. By influencing regional perceptions and improving the ability of partner nations to secure themselves through unilateral security partnerships and regional alliances, Army forces can isolate adversaries and thwart behavior that runs counter to U.S. interests. During operations to shape, Army forces gain access and establish the relationships, agreements, and contracts necessary for rapidly setting the theater should a larger force be required to deter hostility or conduct large-scale combat operations in the future.

3-35. Both as its own line of effort and in support of other lines of effort, information operations ensure information-related capabilities are synchronized to optimize the information element of combat power and shape the information environment. Commanders ensure development and propagation of actions and messages within the established theme guidelines of the higher echelon or supported commander and align these messages with actions and activities. These themes guide the development of activities and messages focused on aligned themes, messages, and actions to enhance trust and confidence in U.S. efforts and to portray partner nations as competent and capable. They also demonstrate U.S. and partner interoperability, and present an Army prepared to address challenges worldwide. (See JP 3-13 and FM 3-13 for doctrine on information operations.)

Combined Training and Exercises

> *When near he makes it appear that he is far away; when far away, that he is near.*
>
> Sun Tzu

3-36. Combined training and exercises with multinational partners play a key role in shaping an OE. Through training and exercises, Army forces build partner combat readiness and set conditions for future contingency operations. Multinational forces that maintain high levels of combat readiness demonstrate undeniable credibility to adversaries. It is this credibility that provides access, assurance, and deterrence. The developing country combined exercise program authorizes the Secretary of Defense, in coordination with Department of State, to fund developing country participation in combined exercises. Combined exercises familiarize both forces with the capabilities and shortfalls of the other force and develop procedures to leverage capabilities and mitigate shortfalls. Training exercises are conducted at all levels of command, from tactical units to large-scale combined task forces.

INTERAGENCY COORDINATION

3-37. Shaping activities encompass a wide range of actions where the military instrument of national power supports other instruments of national power, as represented by interagency partners. Shaping activities also involve cooperation with international organizations (for example, the United Nations) and other countries to protect and enhance national security interests, deter conflict, and set conditions for future contingency operations. Military engagement, security cooperation, and deterrence activities usually involve a combination of military forces and capabilities separate from, but integrated with, the efforts of interagency participants, and they are coordinated by ambassadors and country teams.

AMBASSADOR, UNITED STATES DIPLOMATIC MISSION

3-38. The U.S. diplomatic mission includes representatives of all U.S. departments and agencies physically present in the country. The American ambassador (chief of the U.S. diplomatic mission), often referred to as the chief of mission, is the principal officer in the embassy. This person oversees all U.S. government programs and interactions with and in a host nation.

3-39. The chief of mission is the personal representative of the President and the Secretary of State and reports to the President through the Secretary of State. The chief of mission ensures all in-country activities best serve U.S. interests and regional and international objectives. Depending on the size or economic importance of a country, the United States may maintain only an embassy and no consular offices. However, the United States may maintain one or more consular offices in some countries. Typically, Army elements conducting security cooperation activities coordinate with embassy officials, even in nations with only a consular office. Relationships with consular offices are determined on a case-by-case basis. The same entities and offices existing in an embassy are present or liaised at consular offices. (See FM 3-22 for a detailed explanation of this role in relation to Army operations.)

COUNTRY TEAM

3-40. The country team is the point of coordination within the host country for the diplomatic mission. The members of the country team vary depending on the levels of coordination needed and the conditions within that country. The country team is usually led by the chief of mission, and it is made up of the senior member of each represented U.S. department or agency, as directed by the chief of mission. The team may include the senior defense official or defense attaché, the political and economic officers, and any other embassy personnel desired by the ambassador.

3-41. The country team informs various organizations of operations, coordinates elements, and achieves unity of effort. Military engagement with a host country is conducted through the security cooperation organization. However, several other attachés and offices may be integral to security cooperation activities, programs, and missions as well. The country team provides the foundation of local knowledge and interaction with the host-country government and population. As permanently established interagency organizations, country teams represent the single point of coordination, integration, and synchronization of security cooperation activities supported by combatant commands and the theater army.

ARMY ORGANIZATIONS

3-42. The Army provides trained and ready forces to GCCs to assist them in shaping their AORs. A detailed discussion of the types of Army units by echelon is discussed in chapter 2. This section discusses key Army organizations and their primary roles during operations to shape.

THEATER ARMY

3-43. The theater army integrates landpower within theater engagement plans and security cooperation activities. Integrating landpower requires the theater army to train and prepare Army forces for operations and to coordinate training and readiness requirements with Service force providers. Integrating landpower also includes establishing and extending the network, sustainment infrastructure, intelligence enterprise, and protection capabilities that support operations throughout an AOR. Security cooperation occurs throughout an AOR, even when the primary effort directed by the GCC shifts to a named joint operation. The ability to manage landpower within a joint operation while continuing to support AOR-wide steady state activities, including security cooperation, underscores the requirement for a robust theater army.

Army Support to Theater Campaign Planning

3-44. The TCP is the GCC's vehicle for operationalizing the theater strategy. The TCP provides a framework within which the combatant commander conducts security cooperation activities and military engagement with regional partners through cooperative security and development. A TCP's main function is to provide guidance to coordinate steady-state components of contingency planning by conducting security cooperation activities across the AOR.

3-45. The theater army develops a support plan as an annex to the TCP. The support plan serves as the mechanism between planning, programming, budgeting, and execution processes by, with, or through the theater army. It is supported by Headquarters, Department of the Army; functional Army Service component commands; Army commands; direct reporting units; and the Reserve Components to resource security cooperation activities that shape an OE and achieve TCP objectives and *Guidance for Employment of the Force* end states. Army support planning usually addresses—

- Theater country-specific security cooperation section.
- Information collection activities.
- Ports of entry guarantees.

3-46. Synchronizing the narrative for Army forces with Army activities is a part of the Army's support plan. Actions and messages must be aligned for maximum effect. The theater army provides guidance on drafting broad mission statements from the TCP, aligned with local customs. The narrative supports the development of themes and specific messages Army forces deliver.

3-47. The theater army develops the force structure required to support the TCP. The theater army requests Army forces and the resources required to support them. These resources include sustainment, intelligence, protection, and any other capabilities required. The theater army provides support to forces participating in exercises, and it develops exercises to support the theater security cooperation plan. The theater army designs effective and efficient movement plans for land forces into and out of the theater of operations. The theater army also requests forces to support ongoing Army responsibilities for theater infrastructure development, primarily including intelligence, air and missile defense (AMD), sustainment, and communications. (See ATP 3-93 and FM 4-95 for detailed information on theater army support responsibilities.)

Host-Nation Support

3-48. The theater army provides a key role in host-nation support (HNS). *Host-nation support* is civil and/or military assistance rendered by a nation to foreign forces within its territory during peacetime, crises or emergencies, or war based on agreements mutually concluded between nations (JP 4-0). These formal agreements are between the governments of the United States and the host nation. However, real adjustments to existing U.S. Army deployment plans must be directly related to the actual details of HNS agreements and plans that define all specific tasks, priorities, and procedures for validation. Further, such adjustments will account for any political, economic, or diplomatic developments with respect to the host nation that increase the risk that the host nation will be unwilling or unable to fully comply with the HNS agreement. (See AR 570-9 and AR 11-31 for more information on HNS.)

3-49. Many HNS agreements have already been negotiated. HNS agreements may include pre-positioning of supplies and equipment, training programs outside the continental United States, and humanitarian and civil assistance programs. These agreements are designed to enhance the development and cooperative solidarity of the host nation and provide infrastructure compensation should deployment of forces to the host nation be required. The pre-arrangement of these agreements reduces planning times in relation to contingency plans and operations.

3-50. HNS agreements that the theater army may address are arranging labor support for port and terminal operations, using available transportation assets in country, using bulk petroleum distribution and storage facilities, supplying Class III and Class IV items, developing airspace procedures guides, and developing and using field services. The theater army initiates and continually evaluates agreements with multinational partners for improvement. Agreements should be specifically worded to enable planners to adjust for specified requirements. Additionally, planners should assess the risk associated with using HNS, and they should consider operational area security and operational requirements.

Operational Contract Support

3-51. The theater army also plans and coordinates operational contract support to land forces. Contracting support may include obtaining air and sea ports of entry. When directed, the theater army may contract for the establishment of intermediate staging bases, possible locations for pre-positioned stocks, and possible assembly areas in support of operations to prevent and win. Part of this process includes identification of Army and host-nation capability gaps, and how such gaps may be corrected or mitigated. The theater army conducts continual assessment of what resources may be required in support of both the host nation and deploying forces. When possible, operational contract support should be planned and coordinated in advance of an actual deployment. Additionally, HNS should be considered first before a decision is made to contract for required support. (See ATP 4-10 for more information on operational contract support.)

Foreign Disclosure

3-52. The theater army performs a central role in determining foreign disclosure rules. It is normally at the theater level where friction points to information sharing are identified and deconflicted. Commanders and unified action partners must receive combat information and intelligence products in time and in an appropriate format to facilitate shared understanding and support decision making. Timely dissemination of information is critical to the success of multinational operations. Dissemination is deliberate and ensures consumers receive combat information and intelligence products to support operations. Therefore, this critical role should enable information sharing to the maximum extent allowed by law, regulation, and government-wide policy.

ARMY CYBERSPACE ORGANIZATIONS

3-53. Cyberspace support is developed and employed during operations to shape. Cyberspace capabilities must be built, moved, and provided access to various nodes across cyberspace. These systems require constant maintenance and defense throughout operations to shape. Cyberspace capabilities should always be considered in contact with threats.

3-54. ARCYBER is the Army headquarters responsible for cyberspace operations to support joint requirements. It is the single point of contact for reporting and assessing Army cyberspace incidents, events, and operations in Army networks, and for synchronizing and integrating Army responses. When directed, ARCYBER conducts offensive and defensive cyberspace operations to ensure unified action partner freedom of action in cyberspace, and to deny it to adversaries. ARCYBER provides appropriate interactions, both as a supported and as a supporting command to Army Service component commands (ASCCs), including theater armies, Army commands, direct reporting units, and unified action partners. ARCYBER plans, coordinates, synchronizes, and directs an integrated defense within the Department of Defense information network-Army (DODIN-A) and, when directed, other portions of the Department of Defense information network (DODIN). ARCYBER is responsible for supporting joint information operations missions through operational control of the 1st Information Operations Command (Land).

3-55. The Army Cyberspace Operations and Integration Center is an operational element of the ARCYBER headquarters. It is the top-level control center for all DODIN operations and defensive cyberspace operations (DCO). The Army Cyberspace Operations and Integration Center provides DODIN operations reporting and situational understanding for the DODIN and the DODIN-A. The Army Cyberspace Operations and

Integration Center also provides worldwide DODIN operational and technical support across the strategic, operational, and tactical levels, in coordination with the theater armies.

3-56. Theater army cyberspace elements provide enablers for mission command and messaging to support the shaping of perceptions in concert with the GCC's TCP. The regional cyber center oversees DODIN operations and DCO internal defensive measures in theater. The theater cyberspace forces coordinate with national agencies to develop cyberspace awareness, potential tools for future use, and better protection for friendly networks. They act continuously to defend friendly networks against cyberspace attacks. (See JP 3-12[R] and FM 3-12 for more information on cyberspace elements.)

ARMY SPECIAL OPERATIONS FORCES

3-57. In addition to contributions through standard security cooperation programs and operations to support shaping activities, ARSOF have unique missions that are conducted in support of the shaping objectives. *Preparation of the environment* is an umbrella term for operations and activities conducted by selectively trained special operations forces to develop an environment for potential future special operations. (JP 3-05). Preparation of the environment includes operational preparation of the environment and special operations advance force operations, and preparation of the environment is supported by intelligence operations. Intelligence typically builds on the information provided by operational preparation of the environment and special operations advance force operations, compiling it with other sources of information to provide the intelligence picture. The information provided by preparation of the environment can enhance intelligence preparation of the battlefield (IPB) to support subsequent military operations.

CORPS AND BELOW

3-58. Army units at the corps and lower echelons execute shaping tasks and provide the forces for security cooperation. Army forces may support SFA, FID, or security assistance by participating in multinational exercises, medical and other civil-military operations, development assistance, and training exchanges. Army forces at corps echelons and below directly engage with partner forces, governmental and nongovernmental organizations, and civilian populations to accomplish their mission, build rapport, and improve conditions to promote stability.

3-59. Corps and lower echelon SFA tasks include general developmental tasks to organize, train, equip, rebuild, build, advise and assist, and assess. These tasks represent SFA capability areas. Each element of these tasks can be used to develop, change, or improve the capability and capacity of FSF and strengthen the resolve of partner nations. By conducting an assessment of FSF through the lens of U.S. interests and objectives, coupled with shared interests of multinational partners, U.S. forces can determine which area or areas within the organize, train, equip, rebuild and build, advise and assist, and assess construct to use to improve the FSF to the desired capability and capacity. Planners should determine what is in the best interests of the multinational partners involved to achieve shared objectives. (See FM 3-22 for a more detailed description of activities for building FSF.)

Regionally Assigned and Aligned Forces

3-60. Regionally assigned or aligned forces are those forces that provide a combatant commander with scalable, tailorable capabilities to shape OEs. They are those Army units assigned or allocated to combatant commands, or those Army capabilities distributed and prepared by the Army for combatant command regional missions. A corps headquarters prepares, exercises, and ensures readiness of its assigned forces to support security cooperation using the *Guidance for Employment of the Force* as a planning tool for prioritization of resources and support.

3-61. It is the combination of combat readiness, effective information operations, and security cooperation competence that lends credibility to Army regionally assigned and aligned forces. Regional assignment and alignment provides an effective approach to nontraditional threats. Forces organized under this concept can provide persistent presence for combatant commanders and the capability to assure partners and deter adversaries. Regionally aligned forces comply with combatant command requirements to understand the cultures, geography, languages, and militaries of the countries where they are most likely to be employed. They also develop expertise about how to impart military knowledge and skills to others. While all forces must be prepared to execute large-scale combat operations, during shaping they primarily conduct theater security cooperation activities, and they often participate in bilateral or multilateral military exercises to improve interoperability, build shared trust, gain greater access, and build capacity. (See ATP 3-92 for additional details about the role of the corps headquarters during operations to shape.)

Interdependence, Interoperability, and Integration of Conventional and Special Operations Forces

3-62. Many criteria determine when conventional forces, SOF, or a combination of the two are appropriate to conduct security cooperation activities. Both force levels and force characteristics suggest optimal, acceptable, and desirable force package options in planning and resourcing. Additional considerations include the unique capabilities of each force. For example, ARSOF provide information and intelligence to the Army to help build shared understanding of an OE. Options for the deployment of a modular brigade augmentation, a select number of conventional military transition teams, or SOF depend on the overall U.S. national policy, the conditions of an OE, the priorities of the internal defense and development strategy supported by the TCP and detailed in the theater security cooperation plan, and forces available. A clear understanding of command and support relationships between conventional forces and SOF should be in place prior to deployment of such forces. Additionally, means of communications and requirements for communication should exist where SOF may be operating in a conventional force's AO.

3-63. Rarely will Army (or joint) forces be homogenous in terms of conventional forces or SOF, as they will normally be comprised of both. Recent operational experience has shown that the supported commander may be the conventional force or SOF commander.

3-64. Regardless of the level of integration, there is also an interdependence tied to the objectives each force is supporting. Both may be conducting shaping activities in the same country simultaneously or sequentially. Each must account for the other's actions and their impact on the environment. Accounting occurs during the review of trip reports, after action reports, and other post-mission reports conducted at theater echelon or below. At the theater echelon this accounting occurs between theater special operations command and theater Army through lateral staff activities and at the GCC staff, as they track progress towards achieving objectives laid out in the theater security cooperation plan.

3-65. An SFA mission may exceed the capacity of available SOF. When required to train a large number of FSF in a short time, planners at the theater army level determine the number of conventional forces and SOF as part of their mission analysis aimed at training enough forces in a given time frame. A corps is usually responsible for the conventional forces assigned to these missions, and it is thus responsible to ensure the forces tasked with these missions are trained and capable of training FSF.

3-66. Corps headquarters maximize use of available talent to build interdependence in their assigned forces. Corps headquarters have the ability to design exercises that integrate conventional forces and SOF working together. The type of training depends on the nature of the FSF being trained, and commanders should consider transitioning the initial mission from SOF to a conventional force that can teach FSF using the crawl-walk-run approach.

3-67. An additional consideration is interoperability of forces. Interoperability is often measured by the ability of multinational formations to execute secure communications, process digital fire missions, and share a common operational picture. Army forces train with unified action partners to ensure interoperability. Working with unified action partners is critical to the Army's ability to build credible deterrence in any theater. Planners must consider procedural systems that facilitate interoperability when technical capabilities are not compatible.

Divisions

3-68. During operations to shape, the role of a division headquarters is not significantly different than that described for a corps headquarters. Division headquarters are often tasked to be the primary interface for the Army with various unified action partners during operations to shape. When regionally aligned, a division with a tailored package of its subordinate brigades and other enablers—both Regular Army and Reserve Component—is allocated to a combatant commander to help execute that combatant commander's TCP. Examples of additional enablers include CA units, military intelligence units, digital liaison detachments, maneuver enhancement brigades, military police (MP) units, chemical, biological, radiological, and nuclear (CBRN) units, and engineer units.

3-69. A regionally aligned division will normally work with multinational partners in a security cooperation context over extended periods. This division will help partners through conducting various activities such as exercises, training, equipping, education, conferences, and military staff talks to shape the environment by building partner capacity to prevent and deter conflict. Division support to security cooperation helps to shape regional stability by—

- Building defense relationships that promote U.S. security interests.
- Developing friendly military capabilities for self-defense and multinational operations.
- Providing division and other U.S. forces with peacetime and contingency access to host nations to prevent and deter conflict.

3-70. Division headquarters provide an interface between the brigades tasked to conduct specific operations to shape and the corps staff managing the missions, movement, operational readiness-related equipment issues, and training requirements across echelons. Division headquarters provide training resources and oversight to brigades at home station, and they play a critical role in the development of leaders down to the battalion level. Division headquarters should play a continuous positive role in the readiness of units over which they have influence, protecting their time and ensuring that demanding, realistic training remains the first priority.

3-71. Division headquarters further provide subject matter expertise to assist brigades as they prepare for missions or capstone training exercises. Low density military occupation specialties or specialized units may benefit from the division consolidation of training events. Divisions usually have subject matter experts in the low density specialties who develop, implement, execute, and evaluate training programs across several echelons. (See ATP 3-91 for a further discussion of the division roles and responsibilities during operations to shape.)

Brigades

3-72. Brigades deploying in support of operations to shape work closely with theater army and country team staffs through their higher echelon headquarters. Commanders ensure that readiness priorities align with their missions and that they are both realistic and relevant. (See TC 3-05.3 for useful references for the execution of tactics, techniques and procedures for security cooperation activities.)

3-73. The mission-essential task list (METL) provides focus for training. The credibility of Army forces stems first and foremost from their capability to wage large-scale combat operations. Brigade combat teams (BCTs) and multifunctional brigades generally provide most of the Soldiers performing operations to shape. Those operations may require proficiency in tasks not directly linked to the METL, so commanders determine where to assume risk during preparation in order to ensure flexibility across the force. Realistic training with limited time and resources demands that commanders focus their unit training efforts to maximize training proficiency. (See chapter 2 for additional information on training.)

3-74. Maintaining and training Army systems is part of readiness. Operational readiness reflects the professionalism of a force able to maintain its tools. Properly maintained equipment and weapons, coupled with tough, realistic training, is what creates cohesive, adaptive, and effective teams. Leaders ensure equipment readiness and accountability through technical competence, discipline, a commitment to stewardship of resources, and the integration of maintenance into training and operations. Leaders and units must be prepared for employment before a crisis; during a crisis they will be committed to missions as they are, not as how they might wish to be.

CONSOLIDATE GAINS

3-75. The primary consolidation of gains activities during operations to shape are those taken to ensure that consolidation of gains tasks are adequately addressed in operation plans (OPLANs) developed in support of GCC plans. The understanding of an OE and situational awareness developed during the execution of shaping activities should be used to develop these OPLANs, which in turn should be continuously updated as conditions change in various AOs. When Army forces are aligned to a region for a specified period of time, those forces have a responsibility to review standing OPLANs and provide updated information for them. This is often accomplished by sending planners to the theater army as part of a short-term OPLAN development effort.

3-76. Successful execution of security cooperation activities help achieve strategic and operational objectives during operations short of armed conflict. Professionalizing police forces early, inculcating the rule of law in military forces, enforcing discipline, and assisting with building strong governance are parts of the continuum of consolidating gains that ultimately result in regional stability.

3-77. During training, units plan, execute, and are evaluated on their ability to provide area security and perform various stability tasks. Integrating stability tasks reinforces the importance of planning, allocating resources, and evaluating consolidation of gains efforts.

This page intentionally left blank.

Chapter 4
Operations to Prevent

The ultimate determinant in war is a man on the scene with a gun. This man is the final power in war. He is in control. He determines who wins. There are those who would dispute this as an absolute, but it is my belief that while other means may critically influence war today, after whatever devastation and destruction may be inflicted on an enemy, if the strategist is forced to strive for final and ultimate control, he must establish, or must present as an inevitable prospect, a man on the scene with a gun. This is the Soldier.

Rear Admiral J. C. Wylie

This chapter provides an overview of operations to prevent. It describes the threat and discusses major activities Army forces conduct during operations to prevent. The chapter continues with a description of the roles and responsibilities of the theater army, corps, divisions, and brigades during operations to prevent. The chapter ends with a discussion of consolidation of gains.

OVERVIEW OF OPERATIONS TO PREVENT

The price of greatness is responsibility.

Winston S. Churchill

4-1. The purpose of operations to prevent is to deter adversary actions contrary to U.S. interests. They are typically conducted in response to activities that threaten unified action partners and require the deployment or repositioning of credible forces in a theater to demonstrate the willingness to fight if deterrence fails. As part of crisis response or limited contingency operations, operations to prevent are tailored in scope and scale to achieve a strategic or operational level objective. They may be conducted as a stand-alone response to a crisis, as in a non-combatant evacuation operation, or as part of a larger joint operation.

4-2. When crises develop and the President directs, geographic combatant commanders (GCCs) employ forces to deter aggression and signal U.S. commitment. For example, the United States may deploy forces to a region and conduct a combined joint training exercise with the host nation to send a strong message to an adversary. In other instances, Army forces may deploy and employ air and missile defense (AMD) capability to an area in support of a friendly nation or to protect forward stationed forces. If a crisis is caused by an internal conflict that threatens regional stability, U.S. forces may intervene to restore stability.

4-3. Prompt deployment of credible land forces in the initial phase of a crisis can preclude the need to deploy larger forces later. Effective early intervention can also deny adversaries time to set conditions in their favor. However, deployment alone does not guarantee success. Achieving successful deterrence involves convincing adversaries that the deployed force is able to conduct decisive action.

4-4. Given the tempo of modern conflict, Army forces will probably conduct operations to prevent under severe time constraints. Expeditionary Army forces reduce the risk of aggression by maintaining the ability to deploy rapidly and arrive ready to conduct operations. These actions alone may disrupt an adversary's assumptions, plans, or timelines. Operations to prevent create the conditions required to quickly transition, if necessary, into large-scale ground combat (see chapters 5 through 7). The ability of an Army force to prevent stems from an adversary's realization that further escalation would result in military defeat.

Chapter 4

> **Prevent: OPERATION DESERT SHIELD**
>
> Iraqi ground forces invaded Kuwait on August 2, 1990 with the objective of annexing the country and securing its vast oil fields. The United Nations had been monitoring the Iraqi troop build-up for weeks and had warned Saddam Hussein not to follow through on his plans. Instead, Hussein made a grave mistake and assumed the United States and the United Nations would stand aside. President H. W. Bush immediately denounced the Iraqi attack and promised to liberate Kuwait. On 7 August the President ordered the organization of OPERATION DESERT SHIELD in response to the Iraqi occupation of Kuwait.
>
> The United States rapidly deployed the 82d Airborne Division to northern Saudi Arabia to deter any further advance by Iraqi forces. The presence of this American division ended any Iraqi ambitions to drive further south. In the weeks and months that followed, the XVIII Airborne Corps and the VII Corps deployed into the theater. By January 1991, the coalition was composed of units from 32 countries, including the United States, Great Britain, France, and most of the surrounding Arab nations. The combined strength of these nations forced the Iraqi units into a static defensive posture. The coalition now held the initiative.
>
> The conflict moved closer to war in late November 1990 when the United Nations approved the use of military action to liberate Kuwait. When attempts to broker an agreement with the Iraqi government proved unsuccessful, the U.S. Congress in early January 1991 authorized the employment of U.S. forces in offensive military operations against Iraq. When the deadline of midnight of January 16, 1991 passed, the defensive OPERATION DESERT SHIELD transitioned to the offensive operation known as OPERATION DESERT STORM.

THREATS

4-5. During operations to prevent, peer threats will shift resources to information warfare and preclusion. Supporting efforts will include systems warfare and sanctuary. A threat force's capability, capacity, operational reach, and overall desired end state shape its strategy in generating threats across multiple domains. Determining what threat forces perceive as important will inform U.S. understanding of a threat's desired end state, associated courses of action, and employment of forces. (See paragraphs 1-37 through 1-50 for a discussion of threat methods.)

4-6. Understanding how threat forces will conduct operations will influence how the GCC employs forces and capabilities. A threat may accelerate operational timelines, use information warfare to attack U.S. capabilities, or introduce additional forces. Threat forces will consider four key areas when designing operations to mitigate U.S. deterrence efforts to ensure they do not interfere substantively with their interests. Potential threat activities include—

- Attempts to reduce the perceived risk to threat forces while increasing the perceived risk to any action taken by the United States.
- Limited attacks to expose friendly force vulnerabilities. These attacks may also degrade the deterrence value of deployed forces and destroy credibility among current and potential partners.
- Deception operations to conceal their real intent. This deception may cause the United States to build deterrence capacity aimed at deterring an unimportant action, or one that never takes place.
- Attempts to slow and disrupt deployment to limit the ability of U.S. forces to build combat power.

PREVENT ACTIVITIES

[Y]ou don't have to make them see the light, just make them feel the heat.

President Ronald Reagan

4-7. The intent of operations to prevent is to deter adversary actions to de-escalate a situation. Prevent activities enable the joint force to gain positions of relative advantage prior to future combat operations. Operations to prevent are characterized by actions to protect friendly forces and indicate the intent to execute subsequent phases of a planned operation. With the shift from shaping to deterrence, the theater army shifts to refining contingency plans and preparing estimates for landpower based on GCC's guidance. The theater army and subordinate Army forces perform the following major activities during operations to prevent:

- Execute flexible deterrent options (FDOs) and flexible response options (FROs).
- Set the theater.
- Tailor Army forces.
- Project the force.

Flexibility: 8th U.S. Army in Korea

The aftermath of the Korean War had an enormous impact on the geopolitical climate and tensions between the superpowers during the Cold War. Despite the official cessation of hostilities with the 1953 Korean Armistice Agreement, the Republic of Korea remains in a state of war with the Democratic People's Republic of Korea to the north. Since 1953, North Korea has periodically conducted violent provocations that indicate it has not abandoned its long-term goal of a unified peninsula under the Democratic People's Republic of Korea control. In an effort to support an ally, and as an extension of the Cold War containment policies, the United States created a permanent Army garrison in the Republic of Korea under the command of the Eighth Army.

In an effort to remain flexible and able to respond to enemy movements effectively in the context of other geopolitical demands on Army forces, the structure of the Eighth Army has changed numerous times. The complex, tenuous nature of the Korean Armistice Agreement requires Eighth Army to be ready for an escalation of Army forces in response to crises on the peninsula at any time. As a subordinate of U.S. Forces Korea under armistice conditions, it forms the basis of U.S. Army forces that would serve under Combined Forces Command and the United Nations Command, should hostilities commence on the peninsula. As such, it consistently assesses the security environment while ensuring the readiness of Army forces in the Republic of Korea.

Eighth Army remains forward-deployed as a bulwark against North Korean aggression. It retains mission command of forward deployed Army forces, rotational Army forces, and Army forces participating in exercises. Its role in ensuring the readiness of Army forces is a preventive hedge against the renewal of hostilities on the ground.

EXECUTE FLEXIBLE DETERRENT OPTIONS AND FLEXIBLE RESPONSE OPTIONS

4-8. A *flexible deterrent option* is a planning construct intended to facilitate early decision making by developing a wide range of interrelated responses that begin with deterrent-oriented actions carefully tailored to create a desired effect (JP 5-0). A *flexible response* is the capability of military forces for effective reaction to any enemy threat or attack with actions appropriate and adaptable to the circumstances existing (JP 5-0). While FDOs are primarily intended to prevent a crisis from worsening and allow for de-escalation, FROs are generally punitive in nature. FDOs are preplanned, deterrence-oriented actions carefully tailored to bring an issue to early resolution without armed conflict, and they can be initiated before or after unambiguous warning of threat action. FROs can be employed in response to aggression by an adversary, and they are intended to facilitate early decision making by developing a wide range of actions carefully tailored to produce desired effects. FDOs and FROs must be carefully tailored regarding timing, efficiency, and effectiveness. Care should be taken to avoid undesired effects such as eliciting an armed response should adversary leaders perceive that friendly FDOs or FROs are being used as preparation for a preemptive attack.

4-9. FDOs and FROs serve three basic purposes. First, they provide a visible and credible message to adversaries about U.S. will and capability to resist aggression. Second, they position U.S. forces in a manner that facilitates implementation of the operations plan or contingency plan if hostilities become unavoidable. Third, they provide options for decision makers during crises. They allow for measured increases in pressure to avoid unintentionally provoking full-scale combat and to enable decision makers to develop the situation and gain a better understanding of an adversary's capabilities and intentions.

4-10. FDOs and FROs are elements of contingency plans executed to increase deterrence in addition to, but outside the scope of, the ongoing joint operations. Key goals of FDOs and FROs are—

- Communicate the strength of U.S. commitments to treaty obligations and regional peace and stability.
- Confront adversaries with unacceptable costs for their possible aggression.
- Isolate adversaries from regional neighbors and attempt to split the adversary coalition.
- Rapidly improve the military balance of power in the area of responsibility (AOR) without precipitating armed response from the adversary.
- Develop the situation without provoking adversaries to better understand their capabilities and intentions.

4-11. FDOs and FROs are developed for each instrument of national power—diplomatic, informational, military, and economic—but they are most effective when combined as shown in figure 4-1. Combatant commanders execute them when approved by the President and the Secretary of Defense. Army forces conduct tasks as assigned within the directed FDO or FRO. (See JP 5-0 for a further information on FDOs and FROs.)

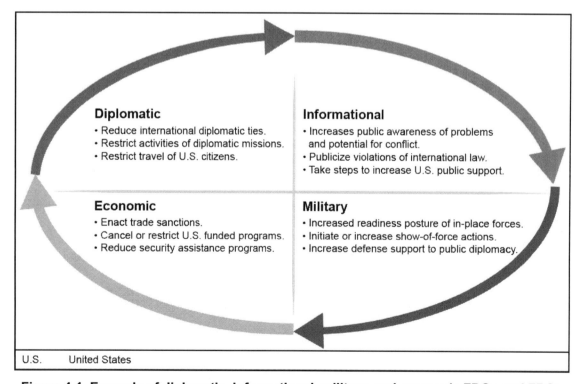

Figure 4-1. Example of diplomatic, informational, military, and economic FDOs and FROs

SET THE THEATER

4-12. Set the theater is a geographic combatant command responsibility that is continuous and encompasses whole-of government initiatives, including bilateral or multilateral diplomatic agreements. Set the theater describes the broad range of actions conducted to establish the conditions in an operational area for the execution of strategic plans. The theater army generally has responsibility for the planning and coordination of Army capabilities to meet the combatant commander's intent to set the theater.

4-13. The purpose of setting a theater is to establish favorable conditions for the rapid execution of military operations and the support requirements for a specific operation plan (OPLAN) during crisis or conflict. Setting the theater involves significant sustainment, AMD, engineering, information collection, and communications activities focused on countering a specific threat. The theater army executes many of its responsibilities through the theater sustainment command (TSC), some during operations to shape and other actions during operations to prevent. Setting the theater may also involve—

- Providing Army headquarters to meet various joint and multinational command and control requirements, (for example, providing a corps headquarters to serve as the base for a joint task force [JTF] headquarters and extending the communications infrastructure).
- Providing force protection, including improving AMD.
- Forward-stationing and rotational deployment of Army forces.
- Modernizing forward-stationed Army units.

4-14. The theater army and its supporting commands assess the adequacy of infrastructure in an operational area to support anticipated military operations, determine requirements for additional infrastructure, and manage infrastructure development programs assigned to Army forces for execution. The theater army and the TSC develops these plans in close collaboration with the combatant command's logistics directorate of a joint staff (J-4) and the Army Corps of Engineers. Infrastructure development activities may include identifying requirements for forward basing and air, land, and sea transit rights through the sovereign territories of partner or neutral nations. However, the Department of State and the appropriate U.S. diplomatic mission must negotiate any bilateral or multilateral agreements.

4-15. As the combatant commander shifts priorities to a specific nation or region, the theater army focuses on setting the theater. In conjunction with the GCC staff and interagency partners, the theater army identifies locations to develop or improve bases and base camps in the joint operations area (JOA) for sustainment, protection, and infrastructure development. The functions inherent in establishing the JOA include identifying responsibility for Army support to other Services (ASOS) and agencies, land transportation, inland petroleum pipeline operations, and common-user logistics. The associated functions of theater opening, port and terminal operations, and reception, staging, onward movement, and integration (RSOI) are critical to the initiation of military operations.

4-16. The theater army also prepares to support joint command and control in a JOA. Army contributions to joint command and control include establishing, maintaining, and defending the communications and network architecture to support forces operating within an operational area and maintaining connectivity between land-based forces and other forces. The Army is designated as the Department of Defense (DOD) combatant command support agency for theater communications and network architecture. Army forces execute these responsibilities primarily through the assigned signal command.

TAILOR ARMY FORCES

4-17. Force tailoring combines two complementary requirements—selecting the right forces and deploying the forces in the optimal sequence. The first—selecting the right force—involves identifying, selecting, and sourcing required Army capabilities and establishing their initial task organization to accomplish the mission. The result is an Army force package matched to the needs of the combatant commander. The second requirement of force tailoring establishes order of deployment for the force package, given the available lift and the combatant commander's priorities. Tailoring the force is a complicated and intensively managed Army-wide process, and the theater army plays a critical role in it.

> **Force tailoring**
>
> The process of determining the right mix of forces and the sequence of their deployment in support of a joint force commander (ADRP 3-0).

4-18. Most Army conventional operating forces are designated as "Service Retained" forces in the Global Force Management Implementation Guidance assignment tables and are primarily based in the continental United States (CONUS). United States Army Forces Command (FORSCOM), the largest of the Army commands, commands Active Component conventional forces (Regular Army, mobilized Army National Guard, and mobilized Army Reserve), executes training and readiness oversight of Army National Guard forces under state command, and does the same for non-mobilized Army Reserve units.

4-19. Based upon the landpower requirements developed by the theater armies and validated by the Joint Staff, the Department of the Army and FORSCOM develop force packages based on cyclical readiness. This includes forces for contingencies and forces needed to support security cooperation activities. Wherever possible, the Department of the Army identifies regionally aligned forces that concentrate on missions and capabilities required for a particular AOR. Regionally aligned forces begin planning for their mission in conjunction with the theater army staff. The tailored force package is task-organized by FORSCOM to facilitate strategic deployment and support the gaining joint force commander's (JFC's) operational requirements. FORSCOM is not the sole provider of Army forces; other supporting Army Service component commands (ASCCs) may contribute forces. The result is a set of trained and ready Army forces intended either for contingencies or for planned deployments, such as a rotation of forward-based forces.

4-20. The theater army works closely with FORSCOM to match the composition of the force with the forces identified in theater security cooperation plans or contingency plans for a crisis. FORSCOM modifies force packages as needed. The theater army commander identifies the major task organization and predeployment training required for a mission. FORSCOM then modifies force packages and training as needed. Whenever possible, FORSCOM (or the supporting ASCC) attaches forces to its gaining higher headquarters during deployment (for example, attaching brigade combat teams [BCTs] to a different gaining division headquarters). If geography or the sequence of deployment makes this impractical, the theater army executes task organization changes when forces arrive in the AOR. The gaining theater army commander modifies administrative control (ADCON) as required based upon the organization of the JTF and the support structure available in the theater. Figure 4-2 illustrates force tailoring.

4-21. The theater army also recommends the optimum deployment sequence for Army forces to the GCC's staff. The GCC's staff may modify this recommendation in coordination with FORSCOM and USTRANSCOM based upon factors such as available lift, location and readiness of deploying forces, and surface transportation requirements. Since the initial deployment may not match the situation developing in the JOA, the theater army refines the task organization based upon the ARFOR's requirements. The theater army adjusts support provided by theater assets to match the requirements of the forces on the ground.

4-22. Forces are allocated to the JFC from FORSCOM and supporting ASCCs. *Allocation* is the distribution of limited forces and resources for employment among competing requirements (JP 5-0). In addition to allocated forces, theater armies provide Army forces from theater-assigned forces. Army sustainment units (logistics and medical) normally have a support relationship with the deployed Army forces in the JOA. Other Army theater forces such as military police, aviation, engineers, or civil affairs (CA) units may be attached or under operational control (OPCON) to divisions or corps headquarters. Other units remain OPCON to the theater army and provide direct or general support to the ARFOR.

4-23. The organization established in force tailoring is not necessarily the same as the task organization for combat. It is a macro-level organization established to control the forces through deployment and RSOI. The gaining operational commander, typically the joint force land component commander, modifies this organization depending upon the situation. Once deploying Army forces have completed RSOI, the OPCON passes to the JFC and gaining functional component commander in the JOA or theater of operations. That commander further task organizes the force for land operations as needed.

4-24. Army headquarters provide the command structure to plan, prepare, execute, and assess operations with the joint force and partner nations. Army forces support partners with situational understanding, area security, and sustainment. Army forces ensure combatant commanders possess the ability to scale-up and sustain land forces rapidly through force projection, forward positioning, placement and maturity of theater infrastructure capabilities, and the use of Army pre-positioned equipment and supplies. Army headquarters help impose order in chaotic situations and synchronize plans, programs, and efforts necessary to accomplish the mission.

Operations to Prevent

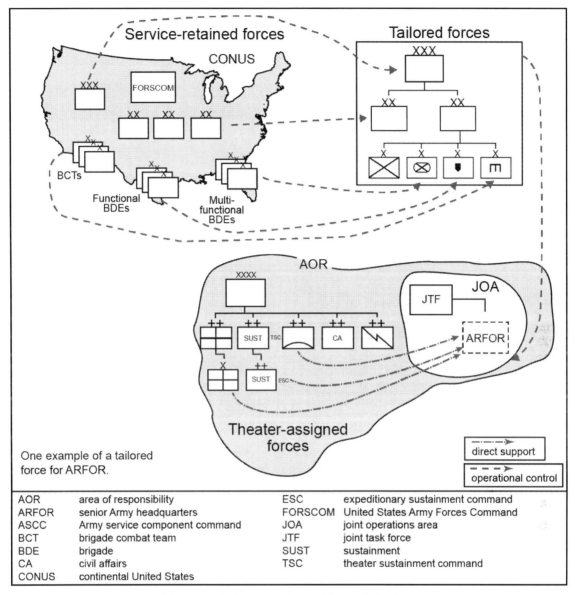

Figure 4-2. An overview of force tailoring

PROJECT THE FORCE

4-25. Projecting the force into an operational area is essential to effective operations to prevent conflict. *Force projection* is the ability to project the military instrument of national power from the United States or another theater, in response to requirements for military operations (JP 3-0). Speed is important—force projection is a race between friendly forces and the enemy or adversary. The side that achieves an operational capability first (a position of relative advantage) can seize the initiative. Speed in force projection is decisive when a combat-ready force deploys to an operational area before an enemy or adversary is ready. Force projection encompasses mobilization, deployment, employment, sustainment, and redeployment. (See table 4-1 on page 4-8 for explanations of these terms.)

Table 4-1. Force projection terms

Term	Descriptions
Mobilization	Mobilization is the process of bringing the armed forces to a state of readiness in response to a contingency. Upon alert for deployment, commanders ensure Army forces are manned, equipped, and meet all Soldier readiness criteria.
Deployment	Deployment is the movement of forces to an operational area in response to an order. Sustainment is crucial to the deployment of forces. Joint transportation assets, including air and sealift, provide the movement capabilities for the Army.
Employment	Employment encompasses a wide array of operations. The operations include, but are not limited to, entry operations, offensive operations, defensive operations, security operations and stability operations.
Sustainment	Sustainment provides logistics, personnel services, and health service support to maintain forces until mission completion. It gives Army forces their operational reach, freedom of action, and endurance.
Redeployment	Redeployment is the return of forces and materiel to the home or mobilization station or to another theater. It requires retrograde of logistics, personnel services, and health service support and reuniting unit personnel and equipment at their home station.

SUSTAINMENT PREPARATION

4-26. Sustainment is a critical aspect of force projection with sustainment preparation of an operational environment (OE) serving as the base for sustainment planning. Corps, division, and brigade planners focus on identifying the resources available in the operational area for use by friendly forces and ensuring access to them. The theater army is a key partner in providing this information to deploying units. A detailed estimate of requirements allows planners to advise the commander of the most effective method of providing adequate and responsive support, while minimizing the sustainment footprint. Sustainment preparation of the operational environment includes—

- Geography.
- Supplies and services.
- Facilities.
- Maintenance.
- Transportation.
- General skills.
- Operational contract support.
- Theater support contracts.
- External support contractors.
- System support contracts.
- Army pre-positioned stocks.
- Host-nation support.

GEOGRAPHY

4-27. Planners collect information on climate, terrain, and endemic diseases in an area of operations (AO) to determine equipment requirements, medical support requirements and personnel support requirements. Certain climates may require specialized equipment and fuel or place an increased demand on repair parts. As an example, in some areas of the world water re-supply presents challenges. Planners identify the need for the early deployment of well-digging, water production, purification, and distribution assets and units. Certain areas also require specialized medical care or capabilities, such as forward surgical teams, preventative medicine, veterinary services, and environmental health services.

SUPPLIES AND SERVICES

4-28. Planners collect information on supplies and services available in their operational area. Commonly available supplies include subsistence items, bulk petroleum, and barrier materials. Commonly available services include bath and laundry, sanitation, and water purification.

FACILITIES

4-29. Planners collect information on the availability of warehousing, cold-storage facilities, production and manufacturing plants, reservoirs, administrative facilities, hospitals, sanitation capabilities, and hotels. Availability of these facilities reduces the requirement for the deployment of U.S. assets.

MAINTENANCE

4-30. Planners should analyze current Army maintenance capability during the prevent phase of operations. Army maintenance units should always be the first considered to provide maintenance support. If analysis indicates that sufficient Army maintenance capability does not exist to meet mission requirements, planners examine the availability of other Service, multinational, and host-nation maintenance capabilities. At this point, planners also identify a need for and request contracted maintenance support when feasible and available. Units determine what items are included as part of the shop and bench stock of their combat repair and field maintenance teams and the types and numbers of on-board spares. These determinations are based on—

- The types and tempo of projected operations.
- The types and frequency of combat damage and maintenance failures.
- The quantities needed to support authorized table of organization and equipment and commercial off-the-shelf equipment.

4-31. For example, if insurgents in a projected armored brigade combat team (ABCT) AO make extensive use of land mines, the maintenance teams within the brigade support battalion request approval from the TSC's material management center to stock complete assemblies of road wheel arms for all of the different models of tracked vehicles within the ABCT.

TRANSPORTATION

4-32. Transportation planners collect information on road and rail nets, truck availability, bridges, airfields, ports, cargo handlers, petroleum pipelines, and material handling equipment. They examine traffic flow to identify potential choke points and control problems as well as any host-nation administrative requirements for the handling and transport of hazardous materials.

GENERAL SKILLS

4-33. Corps, division, and brigade planners collect information on the civilian population in their anticipated AO. Generally their units will need skilled translators and laborers.

4-34. Negotiating host-nation support and theater-support contracting agreements may include pre-positioning of supplies and equipment, civil augmentation program support contracts, overseas training programs, and humanitarian and civil assistance programs. Where possible, these agreements enhance the internal development and cooperative solidarity of the host nation and provide necessary infrastructure should deployment of forces to the host nation be required. Prearranging these agreements reduces planning times in relation to contingency plans and operations.

4-35. Negotiations and agreements enable access to host-nation support resources identified in the requirements determination phase. Negotiation facilitates force tailoring by identifying available resources within the projected AO, such as commercial pipeline construction, trucking companies, and common supplies. This precludes the need to deploy U.S. military capabilities, ship tonnage, and barrier materiel from the United States to support operations.

Chapter 4

OPERATIONAL CONTRACT SUPPORT

4-36. Operational contract support (OCS) is important in the sustainment of operations and helps obtain support for division or corps stability-focused operations. OCS augments other support capabilities by providing an additional source for required supplies and services. Because of the importance and specific challenges of OCS, commanders and staffs need to understand their role in planning for and managing contract support in an AO. (See ATP 4-10 for information regarding OCS roles and responsibilities.)

THEATER SUPPORT CONTRACTS

4-37. Army expeditionary contracting units deploy to an operational area and provide the ability to rapidly contract for logistics and other support functions, mostly from local commercial sources, within a theater of operations. Expeditionary contracting officers under the acquisition authority of the Army Contracting Command perform this function. The Army Contracting Command has a contracting support brigade allocated to each theater and provides contracting battalions and tactical contracting teams to deployed units. Contracting may become a key enabler during the early phases of a campaign. Theater-support contractors acquire and provide goods, services, and minor construction support, usually from the local commercial sources, to meet the immediate needs of operational commanders.

4-38. Theater support contracts are associated with contingency contracting. Sustainment brigades request theater support contracts through a supporting contracting battalion or team office. When this support involves commodities or services support, the supported unit is required to provide contracting officer representatives for contracted services and receiving officials for contracted logistic commodities and services provided in the affected AO. (See ATP 4-92 for more information on theater support contracts and expeditionary contracting in general.)

EXTERNAL SUPPORT CONTRACTORS

4-39. External support expeditionary contracts are prearranged contracts or contracts awarded during the conduct of operations. External support service contracts provide a variety of logistics and other non-combat related services and support. The largest and most commonly known external support contract is the Army's logistics civil augmentation program. The logistics civil augmentation program provides supply services, such as storage, warehousing, and distribution, for the nine classes of supplies. The logistics civil augmentation program also commonly provides base support services such as postal, minor construction, and facility maintenance. Deploying units must provide trained contracting officer representatives and receiving officials for any services provided by external support contractors. (See ATP 4-10.1 for additional information on the logistics civil augmentation program.)

SYSTEM SUPPORT CONTRACTS

4-40. System support contracts are prearranged contracts for technical support of specific systems. System contractors provide support in garrison, and they may deploy with the force to both training and deployed operations. System support contracts provide either temporary support during the initial fielding of a system, called interim contracted support, or long-term support for selected materiel systems, often referred to as contractor logistics support. Deploying units work with a supporting program manager representative or the deployed Assistant Secretary for Acquisition, Logistics, and Technology forward staff to coordinate this support. (See ATP 4-70 for more information on the Assistant Secretary for Acquisition, Logistics, and Technology forward organization and operations.)

ARMY PRE-POSITIONED STOCKS

4-41. Planning for and preparation of Army pre-positioned sets of equipment is essential in facilitating strategic and operational reach. Pre-positioning stocks in potential theaters of operations provides the capability to resupply forces until the establishment of systems. The Army locates its pre-positioned stocks at or near points of planned use or at other designated locations to reduce reaction time. Alternatives include pre-positioning stocks—

- Afloat, including port construction equipment and materiel.
- At an intermediate staging base.
- Assembling stocks in tailored packages for deployment with projected forces.

4-42. The categories of Army pre-positioned stocks are pre-positioned unit sets, Army operational project stocks, Army war reserve sustainment stocks, war reserve stocks for allies, and activity sets. Deploying units work with a supporting Army field support brigade to plan and coordinate Army pre-positioned stock reception support for their operations. (See ATP 4-91 and ATP 3-35.1 for more information regarding Army pre-positioned stocks.)

HOST-NATION SUPPORT

4-43. Host-nation support (HNS) will often be critical to the success of a multinational operation. Centralized coordination of HNS planning and execution will ensure that limited HNS resources are allocated most effectively to support the multinational force commander's priorities. The more limited HNS resources are in the operational area, the greater the requirement for centralized management.

4-44. To assist the multinational force commander in HNS coordination activities, a HNS coordination center may be established. One of the most important functions of the HNS coordination center is to assist the multinational force commander and legal counsel in developing technical agreements that involve logistic matters such as infrastructure, finance, purchasing and contracting, engineering, environment, hazardous material storage, landing and port fees, medical, border customs, tariffs, and real estate.

4-45. In order to effectively plan and coordinate HNS allocation, the HNS coordination center needs up-to-date information on HNS logistic capabilities and ongoing HNS allocation to multinational force contingents throughout the operation. In order to facilitate HNS planning and operational coordination, GCCs and alliance organizations should develop an HNS catalog or database of HNS capabilities in advance of operations. The theater army and theater sustainment commander work closely with the HNS coordination center for the planning and coordination for HNS.

DEPLOYMENT

4-46. Proper planning establishes what, where, and when forces are needed for successful deployment outcomes. How the GCC intends to employ forces is the foundation of the deployment structure and timing. Corps and division staffs examine all deployment possibilities and conduct parallel planning. The timing and amount of combat power delivered directly influences the GCC's or other JFC's courses of action. There are four principles that apply to the range of activities encompassing deployment. These principles are precision, synchronization, knowledge, and speed. (See JP 3-35 for a discussion of these principles.)

4-47. The objective of deployment planning is to synchronize deployment activities to facilitate effective execution in the operational area. The steps used in deployment planning are—
- Analyze the mission.
- Structure forces.
- Refine deployment data.
- Prepare the force.
- Schedule movement.

4-48. Successful deployment planning requires knowledge of the unit's deployment responsibilities, an understanding of deployment, and an appreciation of the link between deployment and employment. (See ATP 3-35 for a discussion of these deployment phases and planning in detail.)

MOVEMENT

4-49. The receipt of the Military Surface Deployment and Distribution Command port call message and the United States Air Force (USAF) Air Mobility Command air tasking order initiates port of embarkation operations and specifies the dates when units arrive at the port. An *air tasking order* is a method used to task and disseminate to components, subordinate units, and command and control agencies projected sorties, capabilities and/or forces to targets and specific missions (JP 3-30). At the installation staging areas, units verify movement data, and they inspect equipment to ensure that it is correctly configured for movement by the designated mode of transportation. Unit vehicles and cargo move to the ports of embarkation by convoy or commercial surface transport.

4-50. The installation coordinates or provides support to help the deploying force by using non-deploying units, installation resources, or contracted support. Support includes load teams, materials handling equipment, maintenance teams, arrival and departure airfield control groups, and deployment support teams. Deploying units and installations identify other support requirements during their deployment readiness exercises. (See AR 525-93, chapter 3) They write those additional requirements into unit and installation deployment support plans.

4-51. Deploying units configure for deployment by preparing their vehicles and aircraft for road, rail, air, inland water, and sea movement as appropriate. ATP 4-16 describes the process associated with surface movement control. ATP 4-14 addresses Army rail operations. ATP 4-15 addresses Army watercraft operations.

4-52. USTRANSCOM operates the Defense Transportation System and provides common user strategic transportation to support the GCC deployment requirements. The port-to-port phase begins with strategic lift departures from ports of embarkation and ends with lift asset arrival in the designated theater ports of debarkation. The GCC synchronizes the arrival of airlift and sealift force packages so that vessels can berth at a wharf or offloaded in-stream with minimal delay. This is fundamental to successful port-to-port movement and ensures the reception and clearance of cargo from the port in a timely manner.

4-53. Deploying units arriving at the port of embarkation complete and finalize all unit movement responsibilities. Critical information, such as movement schedules, manifests, and load lists are provided to control nodes and forward support elements to facilitate efficient onward movement of the corps' or division's deploying headquarters, brigades, and battalions to their respective ports of debarkation.

PROTECTION DURING TRANSIT

4-54. *In-transit visibility* is the ability to track the identity, status, and location of DOD units, and non-unit cargo (excluding bulk petroleum, oils, and lubricants) and passengers; patients; and personal property from origin to consignee or destination across the range of military operations (AR 700-80). It is critical for tracking the identity, status and location of all DOD units and cargo, and it provides critical input to mission command information systems. Successful in-transit visibility begins during the predeployment phase and continues throughout operations. Accurate weighing, measurement, marking, and tagging of all cargo, containers, and equipment directly contributes to a more complete and timely operational picture for the commander. (See JP 4-01 and ATP 3-35 for more information on in-transit visibility.)

4-55. The U.S. Army Installation Management Command is a critical protection enabler during unit pre-deployment activities through the fort-to-port phase. This command manages most Army installations, and garrison commanders ensure the protection of corps or division headquarters and headquarters battalion and subordinate brigades as they deploy. Installation provost marshals, with military police and civilian Department of the Army security forces, protect deploying unit assets as they prepare to deploy. In addition, installation safety personnel, medical personnel, and information management personnel protect deploying unit Soldiers and information. Corps and division protection cells coordinate closely with installation staff personnel to identify information and assets that need protection and apply appropriate protection and security measures consistent with their collective threat analysis.

4-56. Threats may try to impede or prevent unit deployments. Coordination for the physical security of deploying unit assets is required when they move to ports of embarkation. Physical security is required for those assets while they await transport at those ports of embarkation, during their strategic movement, and once they arrive at their ports of debarkation. Physical security remains a concern while those assets are matched with deploying unit Soldiers in unit staging areas, and while units and supplies move along routes designated for use by displacing units for onward movement into their areas of operations. This coordination occurs between the corps and division, local law enforcement agencies, port security personnel, strategic lift assets, and with the GCC and other joint force commanders. This coordination may also involve host-nation security forces.

4-57. Corps and division protection cell staffs coordinate with the port readiness committees at each port used by their subordinate deploying units. These committees provide deploying commanders a common coordination structure for the Department of Defense (DOD), the U.S. Coast Guard, and other federal, state, and local agencies at the port level. These committees are also the primary interface between the DOD and other officials at the ports during the movement of military equipment.

4-58. USTRANSCOM and the Military Surface Deployment and Distribution Command administer a defense transportation security program to provide standardized transportation security measures and procedures, constant oversight, and central direction in coordination with other DOD activities and port authorities. Commanders plan for protection measures for their units and equipment to the port in CONUS, while Military Surface Deployment and Distribution Command coordinates for security within the port.

4-59. Units ensure the security of their equipment and accompanying supplies throughout their movement from the installation, at railheads, while en route to ports of embarkation, and while in-transit to ports of debarkation. Corps and division protection cells ensure that all contract processes for transportation movements meet DOD security requirements. These protection cells coordinate with the installation transportation officer in CONUS or the movement control team outside of CONUS and authorized railroad or commercial truck carriers on guard and escort matters. Some government and commercial carriers provide limited security measures to protect in-transit equipment and supplies. These measures include the use of contract security personnel and secure transfer facilities to safeguard unit assets.

4-60. If the echelon protection cell determines that carrier security measures are insufficient given the threat, corps and division headquarters have a number of options for increasing the protection and security of deploying units and their associated equipment and supplies. These options include directing subordinate units to provide Soldiers to guard equipment and supplies throughout the deployment process. Commanders can also direct units to have Soldiers carry certain sensitive items as part of their individual loads throughout the deployment process.

4-61. Corps and division commanders submit protection plans through the respective ARFOR or the theater army headquarters to the GCC responsible to protect all military forces in the AOR within which deploying forces will operate before their arrival overseas. These plans match the guidance developed by the GCC, who coordinates and approves various deploying unit protection plans.

RECEPTION, STAGING, ONWARD MOVEMENT, AND INTEGRATION

4-62. RSOI delivers combat power to the JFC in a theater of operations or a JOA. RSOI support, whether provided by theater support contracts, external support contracts (primarily the logistics civil augmentation program), regionally available commercial host-nation support, or military assets, needs to have sufficient capabilities and capacity to provide the support required by arriving units at the various ports of debarkation. Effective RSOI matches personnel with their equipment, minimizes staging and sustainment requirements while transiting these ports of debarkation, and begins onward movement as quickly as possible. Deploying units need to understand and implement previously developed plans to accomplish integration and maintain combat readiness upon their arrival. (See JP 3-35 and ATP 3-35 for a discussion of factors and considerations associated with the conduct of RSOI.)

Reception

4-63. As the initial step of introducing combat power on the ground, reception determines success or failure of the RSOI operation. The theater army or a designated sustainment unit (typically the TSC) implements reception from strategic lift activities at or near designated air and seaports of debarkation. While the reception plan for each theater varies, reception capacity should at least equal planned strategic lift delivery capability. At a minimum, the theater army considers the impact of joint command and control (C2), movement control, and port operations as it develops the RSOI plan.

Staging

4-64. Staging is that part of the RSOI operation that reassembles and reunites unit personnel with their equipment and schedules unit movement to the tactical assembly area, secures or uploads unit basic loads,

Chapter 4

and provides life support to personnel. Units require extensive support when staging, including maintenance, supply, and life support.

Onward Movement

4-65. Commanders are influenced by many external factors during their deployment planning. These include the availability of strategic lift assets and the requirements of their future higher headquarters. These factors determine the sequence in which units move from their staging areas, where they completed reception and staging operations, onward to their respective AOs. Units integrate into a joint force once they complete this onward movement. Plans for the movement of the corps or division headquarters and its attached and supporting divisions or brigades into the AO maintain a balance between security and flexibility.

4-66. During this preparatory period, corps and division commanders rely heavily on their transportation staff and their integrating cells to finish coordinating their movement plans to their projected AO with all necessary military and civilian agencies. The movement control system may be operated by the U.S. military, the host nation, or by a combination of military and civilian agencies. The appropriate staff cells and elements consider the number of suitable routes and lift assets available to meet the movement requirements of subordinate units. Other considerations include—
- Road and route improvement and maintenance.
- Construction of routes.
- Clearance of obstacles, including explosive hazards.
- Repair of bridges and culverts.
- Bridging rivers or dry gaps.
- Establishment of security along routes.
- Traffic control to permit freedom of or restriction of civilian movements along routes.
- Communications architecture.

Integration

4-67. When a deploying unit replaces another unit, a relief in place must occur during integration. Command of combat-ready units is transferred to the operational commander. Integration is complete when the GCC, other JFC, or land component commander (LCC) establishes positive control over arriving units. This usually occurs in forward assembly areas when those units are capable of performing assigned missions.

4-68. If the deploying division or brigade is the first U.S. force into an AO, there may be a need to deploy an advance party weighted with security, protection, logistic, and engineering support capabilities. As part of the unit movement preparations, the advance party is resourced from the division, brigade, or attached assets, or it is provided by outside units. This is particularly true if the predeployment survey determines that the AO does not have the infrastructure to support division or brigade operations. In other circumstances, it is necessary for the assistant division commander or deputy brigade commander and a small group of specialized key personnel to lead this advance party. These personnel will set the groundwork for the rest of the division or brigade by performing face-to-face coordination with local civilian or military leaders.

4-69. Unit deployment operations end when all deploying elements complete their RSOI within the assigned JOA. The specifics of RSOI reflect the specific circumstances of the operational and mission variables prevailing in that JOA. (See JP 3-35 for an outline of joint RSOI doctrinal requirements.)

ARMY FORCES DURING OPERATIONS TO PREVENT

4-70. The Army provides the GCC with trained and ready forces to assist with preventing conflict, and to seize the initiative in combat, if required. This section addresses the activities and functions of the theater army, corps, divisions, and BCTs during operations to prevent.

THEATER ARMY

> *[I]t is an unfortunate fact that we can secure peace only by preparing for war.*
>
> President John Fitzgerald Kennedy

4-71. The theater army enables the GCC to employ land forces within the AOR and into specific operational areas. The theater army commands all Army forces in the AOR until the GCC attaches selected Army forces to a joint headquarters. When that happens, the theater army divides its responsibilities between the Army component (the ARFOR) in the JOA or theater of operations and Army forces operating in other parts of the AOR. The theater army is organized, staffed, and equipped to be the ASCC for the GCC. It has limited capabilities to perform two other roles as a JTF headquarters or a joint force land component headquarters for limited contingency operations. Table 4-2 on page 4-16 summarizes some of the theater army responsibilities.

4-72. As the ASCC to the GCC, the theater army commander remains responsible to Headquarters, Department of the Army for service specific requirements within the AOR. This falls under the ADCON chain of authority. This authority establishes a hierarchy for Army support to deployed forces without modifying the operational chain of command that runs from the combatant commander through subordinate JFCs. For example, theater army commanders may—

- Establish centers in the AOR to train individual replacements.
- Complete collective training, theater orientation, and theater acclimation.
- Manage force modernization of Army forces before their employment by the JFC in the JOA.

4-73. In all joint operations, sustainment is a Service responsibility except as specified by DOD combatant command support agent directives, combatant commanders' lead Service designations, or inter-Service support agreements. Combatant commanders direct theater army commanders to provide common-user logistics and ASOS, agencies, or multinational forces, as required. However, shared sustainment responsibility or common-user logistics is more effective, especially for joint operations. Directive authority for logistics is the additional authority used by combatant commanders to eliminate duplicated or overlapped sustainment responsibilities. The theater army coordinates with the combatant command staff to determine joint sustainment requirements, identify responsibilities, and enable commanders to exercise mission command for sustainment. (See ADRP 4-0 for more information on sustainment.)

Theater Army Contributions to Joint Command and Control

4-74. During operations to prevent, the theater army contributions to joint C2 include establishing, maintaining, and defending the communications network architecture to support Army and joint forces operating within the JOA or theater of operations. Army forces execute these responsibilities primarily through the signal command (theater) assigned to support the AOR.

4-75. Theater communications include DODIN operations and satellite communications. DODIN operations provide network and information system availability, information protection, and information delivery across strategic, operational, and tactical boundaries. Army space operations focus on defensive space control activities to protect space-based mission command capabilities.

4-76. At the joint level, cyberspace operations are operational missions accomplished by the commander, USCYBERCOM. The USCYBERCOM commander provides the command and control and situational awareness required to operate and defend the DODIN. ARCYBER, as the ASCC for USCYBERCOM, extends the command and control of the network through its operational control relationship with each regional cyber center and through the Army orders process to each of the signal commands (theater).

Table 4-2. Theater army responsibilities

Theater army responsibilities include—			
Executing combatant commander's daily operational requirements by—	Setting the theater by—	Setting the joint operations area by—	Serving as the core of a JTF or land component command for immediate crisis response and limited contingency operations by—
<ul><li>Maintaining ADCON of Army forces.</li><li>Executing Army executive agent functions.</li><li>Performing common-user logistics functions.</li><li>Providing theater security cooperation.</li><li>Assessing and developing infrastructure.</li><li>Developing a concept plan and contingency or operation plan.</li><li>Providing regional intelligence collection and analysis.</li><li>Providing communications architecture.</li><li>Providing land-based theater air and missile defense.</li><li>Conducting detention operations.</li><li>Conducting dislocated civilian operations.</li><li>Producing target materials.</li><li>Developing targets.</li></ul>	<ul><li>Establishing favorable conditions through exercises and support.</li><li>Supporting the formation of bilateral or multilateral diplomatic agreements.</li><li>Establishing area of responsibility transit rights.</li></ul>	<ul><li>Providing chemical, biological, radiological, and nuclear support.</li><li>Providing theater-specific training.</li><li>Conducting force modernization.</li><li>Conducting sustainment to include providing—<ul><li>Theater opening.</li><li>Port or terminal operations.</li><li>Reception, staging, onward movement, and integration.</li><li>Theater integration.</li><li>Medical operations.</li><li>Army support to other Services.</li><li>Common-user logistics.</li></ul></li></ul>	<ul><li>Conducting foreign humanitarian assistance.</li><li>Conducting disaster relief.</li><li>Providing immediate crisis response.</li><li>Replicating corps or division headquarters for major exercises.</li><li>Transitioning to follow on mission command headquarters.</li></ul>
ADCON administrative control ASCC Army Service component command		JTF joint task force	

Theater Army Contributions to Medical Support

4-77. The medical command (deployment support) (MEDCOM [DS]) provides health service support for the deployed joint force on an area basis. Army medical units form the backbone of the joint medical support and evacuation system. This system provides advanced medical care to all deployed joint forces, interagency personnel, and multinational forces specified by the GCC. The MEDCOM (DS) oversees theater-wide health

Operations to Prevent

service support and exercises operational control of certain medical units through its operational command post. The operational command post controls one or more medical brigades, which in turn control multifunctional medical battalions. The distribution and capability of medical units depends upon the density of U.S. forces, available infrastructure, and evacuation capabilities. The MEDCOM (DS) may place a task-organized medical unit in direct support of an Army division in anticipation of combat operations.

4-78. The MEDCOM (DS) synchronizes all Army Health System operations and provides mission command of medical brigades (support), multifunctional medical battalions, and other Army Health System units providing force health protection and health service support to tactical commanders. The MEDCOM (DS) is composed of an operational command post and a main command post that can deploy autonomously into an AO. MEDCOM (DS)—

- Task-organizes medical elements based on specific medical requirements.
- Monitors threats within each AO and ensures required medical capabilities to mitigate these health threats are available.
- Maintains situational awareness of medical infrastructure, treatment, and evacuation capabilities.
- Accomplishes Title 10, United States Code (USC) responsibilities and Army support to other Services for the AO.
- Partners and trains with host-nation and multinational health system units.
- Establishes a command relationship with the theater army and the GCC, linking the theater sustainment command to the medical logistics management center for coordination and planning.

Theater Army Contributions to Protection

4-79. Theater army staffs have the responsibility to plan and oversee protection requirements. The Unified Command Plan directs force protection responsibility for all Title 10, USC DOD forces stationed in, operating in, residing in, or transiting an AOR to the GCC. In support of the GCC, this responsibility falls on the theater army commander for all Title 10, USC Army forces in the AOR. The tasks of the theater army include the exercise of tactical control over Title 10, USC Army forces stationed in, operating in, residing in, or transiting the AOR.

4-80. The theater requirements for AMD, for example, often exceed the available capabilities. Mission requirements, informed by sustainment considerations, help determine the allocation of AMD resources. The theater army staff coordinates with the GCC and is supported by the GCC's aligned AMD command. The staff estimates the protection assets necessary for both the campaign and an increased threat across the AOR in accordance with the commander's priorities, forces available, and the adversary's perceived intent.

4-81. The theater army staff also requests forces to meet the GCC's priorities. This planning necessarily involves coordination with multiple host-nation militaries and the Department of State. If the ADCON of these distributed task forces exceeds it organic capacity, the theater army may request additional staff to support its ADCON requirements.

4-82. If the JFC establishes a joint security area within the JOA, the ARFOR normally determines its structure and its controlling headquarters. The options for the joint security area depend on the threat and the mission variables, particularly the forces available. If the threat to the joint security area is low to moderate, the theater army commander may tailor the ARFOR with a unit specifically task-organized to control the joint security area. If the threat to theater bases, base camps, and lines of communications is unknown, the theater army may tailor the ARFOR with additional maneuver units to control that joint security area. If hostile conditions exist, the theater army may assign the mission to an Army division with BCTs and one or more maneuver enhancement brigades. (See JP 3-10 for more information.)

4-83. Military deception (MILDEC) is a key aspect of protection. By their very nature, MILDEC activities seek to hold an adversary's attention. Commanders should seek opportunities for friendly forces to deceive. Most adversaries have access to space-based capabilities that provide at least limited observation of friendly forces and the capability to create or access human intelligence networks. Friendly deception activities should account for threat collection capabilities.

Chapter 4

Risk Management

> *It is discipline that makes one feel safe, while lack of discipline has destroyed many people before now.*
>
> Xenophon

4-84. Both in planning and during execution of operations to prevent, the theater army and subordinate headquarters consider risks to the force, risks to mission, and risks of escalation. Leaders must be comfortable in assuming and delegating risk and associated risk mitigation measures.

4-85. Risks to the force describe protection tasks articulated in paragraphs 4-79 through 4-83 and some additional considerations, including—

- Sustainment operations, including convoys, rail movements, and intra-theater lift.
- Dispersion and quality of unit locations.
- Aerial and ground medical evacuation capabilities.
- Lines of communications security.
- Survivability operations.
- Distribution operations.
- Communications.
- Maintenance capabilities and functioning.
- Medical facilities operational capacity.
- Fuel available.
- OCS.

4-86. Units assigned missions during operations to prevent generally have branch plans that describe how they can respond should a situation escalate. Those plans describe conditions that are created or maintained to reduce the risk to the overall mission of the force. The sequencing of forces into theater and their subsequent AOs considers risks to the mission.

4-87. Successful operations to prevent ensure that tensions and conditions do not escalate further. The very nature of these operations implies that the forces assigned to this task have the capacity to conclude the situation with force. While this is a powerful message to an adversary, there is always the potential for misinterpretation of friendly actions and those misinterpretations leading to large-scale combat operations. To reduce the risk of escalation, the theater army considers the initial rules of engagement, force positioning and posture, weapon control status, and operating parameters (for example, limits of advance and AOs.) (See appendix B for additional risk considerations).

Enable Land Forces

4-88. Theater Army staffs face many challenges in supporting the campaign plan. A significant challenge is linking operations planning and outcomes to logistics resources and capacity. Conventional forces (units smaller than a brigade) conducting security cooperation tasks in austere environments, or the staffs planning their deployment, may not have significant existing infrastructure in many AOs. Forces may operate in immature AOs and operate without organic medical evacuation support.

4-89. Theater special operations forces (SOF) activities, National Guard State Partnership Program activities, security assistance missions, and joint exercises take place simultaneously in most theaters at any given time. Theater army staffs coordinate DOD actions to support the chief of mission and the integrated country strategy.

4-90. Concurrently with JTF actions intended to confront and deter an adversary, the theater army commander sets the theater to enable land forces to exert their full capabilities. This includes extending the existing infrastructure to accept the land component and its supporting units, negotiations for use of land and the electromagnetic spectrum (EMS), airspace coordination, and contracted support arrangements. It also includes coordinating for EMS mapping of adversaries to facilitate offensive cyberspace operations (OCO), which is time consuming and resource intensive. There may not be time during a crisis to generate desired effects without prior preparation.

Contingency Command Post

4-91. During operations to prevent, the theater army can provide the GCC with a deployable command post (CP) for contingencies that develop in the AOR. The GCC may use the theater army contingency CP as the nucleus of a small JTF headquarters. The contingency CP receives additional personnel based upon a joint manning document or individual augmentation requests for Army and other Service personnel to accomplish the mission. The contingency CP is a viable option for short notice and limited duration operations because of its established internal staff working relationship and experience base in the AO. This selection capitalizes on the contingency CP's flexibility and rapid deployment.

4-92. If a limited contingency operation is required simultaneously with large-scale combat operations, the contingency CP may deploy as the Army headquarters responsible for exercising mission command over that operation. However, the contingency CP should redeploy as soon as another Army headquarters is operational.

Land Component Headquarters in Large-Scale Combat Operations

4-93. In extreme circumstances, the theater army may start transitioning to a land component command for large-scale combat operations. This could require a theater army headquarters to expand and transform into a joint force land component headquarters for the GCC to exercise command over multiple Army units, multinational corps, or Marine expeditionary forces. To assume this role, the theater army requires extensive augmentation and time to assimilate personnel, integrate additional responsibilities, and train for large-scale combat operations. If designated as a joint or multinational LCC headquarters, the Army headquarters follows the joint doctrine contained in JP 3-31 and JP 3-16.

THE CORPS

4-94. During operations to prevent, the corps headquarters may deploy into an operational area as a tactical headquarters with subordinate divisions and brigades as a show of force. Corps may deploy an early entry CP, comprised of selected personnel from within the headquarters to provide command over arriving forces. In the event conflict escalates, large-scale combat operations may require the corps headquarters to function under the command of a multinational force land component or subordinate to a field army equivalent established as part of a coalition.

PLANNING

4-95. At home-station, the corps headquarters begins planning for all phases of an operation from mobilization to redeployment. The extended battlefield requires planners to consider functions and capabilities required to gain and maintain positions of relative advantage. It is important to understand the planning to execution timelines of certain functions when considering integrating those capabilities into the overall plan. The corps headquarters should have enough information to—
- Identify assembly areas.
- Assign subordinate units areas of operations.
- Identify multiple routes from the points of debarkation through assembly areas to staging areas.
- Identify lines of communications.
- Establish a proposed operational framework.
- Plan for consolidation of gains.
- Coordinate for electronic attack mitigation, including cyberspace distributed denial of service, malicious software, or system intrusion which may damage or alter information in the system.

4-96. Rapid force projection establishes a credible, combat-ready force on the ground before the existing situation worsens. In most places around the world, adversaries are geographically closer to likely trouble spots than U.S. Army units are. Credible deterrence requires that Army forces be ready to deploy and be combat ready when they arrive.

4-97. The situation may require the corps headquarters to plan to deploy a force assigned a security mission that allows the time and space needed for the corps to build combat power. A security force should be capable

Chapter 4

of operating independently from the corps main body. A security force may also be required to maneuver into a position between a potential enemy and the staging areas. There is potential for the security force to be engaged prior to the corps building the minimal combat power required to move into the AO. The security force must have the sustainment capabilities and combat power to sustain combat operations. (See FM 3-90-2 for additional information on the requirements for security missions.)

4-98. Commanders understand how the mix of deploying forces (tailoring) affect their employment options. During planning, the corps commander recommends a mix of forces and the arrival sequence to the theater army commander that best supports future operations. However, if the corps and theater army commander focus only on the land component to the exclusion of complementary joint capabilities, they may not achieve the most effective force mix or sequencing.

4-99. The corps headquarters plans for multiple movement routes to provide division and brigade commanders the flexibility to deconflict forces in time and space. Use of multiple routes reduces adversary targeting of friendly forces, and it reduces the likelihood of disruption along key routes.

4-100. As outlined in both the deployment order and operation order (OPORD), the corps headquarters plans task organization and employment of subordinate units, integration and synchronization of operations, massing of effects, allocation of resources, and determination of priorities. The corps staff provides the primary Army interface to manage the time-phased force and deployment list for all the forces assigned to it in operations to prevent. The corps carefully considers when the corps CPs will arrive at a port of debarkation in relation to the arrival of protection and maneuver forces. The corps tactical CP is mobile, but it has no organic security. Deployment of the corps main CP is lift intensive and requires careful integration into the deployment sequence.

4-101. A corps commander can influence, but not dictate, the tailoring of the corps' subordinate units. The GCC, theater army, supporting combatant commands, USTRANSCOM, and FORSCOM all make decisions concerning the composition and deployment sequence of the corps. A corps commander can, however, organize and prepare the corps CPs for efficient and effective movement throughout the deployment. By selecting personnel with the right skill sets and providing them the right mix of equipment, the commander, chief of staff, and assistant chief of staff, operations (G-3) match corps capabilities with the requirements at home station and in the JOA.

4-102. Assigning initial staging areas, movement routes, and subsequent assembly areas in uncertain threat conditions requires a delicate balance between force protection and building combat power. Staging areas for subordinate division and separate brigade elements should be large enough for the division and separate brigades to disperse elements into company-sized tactical formations while they are performing maintenance checks, loading munitions, ensuring crew readiness, and preparing to move to forward positions.

4-103. The theater army is responsible for coordinating the RSOI of arriving Army forces, and it normally delegates this to the TSC. The expeditionary sustainment command (ESC) or other sustainment headquarters supporting the corps reassembles deploying units and quickly moves them into staging areas from their various ports of debarkation. The corps headquarters may be asked to assist the theater army with coordinating partner-nation support.

4-104. The corps headquarters assumes a command relationship over units as they complete integration of personnel equipment and unit logistics and the TSC releases units from their staging areas for onward movement and integration. Even as the units complete RSOI, the corps headquarters ensures that unit commanders and key leaders are briefed on the situation and their subsequent missions. Effective RSOI establishes a smooth flow of personnel, equipment, and materiel from ports of debarkation through employment as reassembled, mission capable forces. A deploying unit is most vulnerable between its arrival and operational employment, so operations security (OPSEC), protection, deception, and dispersion are vital planning considerations. If the corps is conducting operations while major subordinate units are still arriving, the corps commander may employ the corps tactical CP as the corps' interface with the ESC and arriving forces at the port of debarkation and their various staging areas.

EMPLOYMENT

4-105. The employment of forces for such activities as FDOs or FROs begins when the advanced echelon of the corps headquarters moves to the AO. This movement may be the opening stages of a joint operation or a show of force exercise. Most peer threats have access to space capabilities, unmanned aircraft systems (UASs), and extensive human intelligence capability. Commanders should assume units are under observation by adversaries at all times. OPSEC is critical to protecting friendly plans and unit capabilities.

Deterrence: EXERCISE REFORGER

For over two decades, the annual EXERCISE REFORGER (Return of Forces to Germany) played a crucial role in deterring a conventional attack by the Warsaw Pact against the North Atlantic Treaty Organization (NATO). In 1968, the United States removed two divisions from Europe during the height of the Vietnam War. Beginning in 1969, REFORGER temporarily deployed one or more U.S. divisions to Germany as a chief means of signaling American resolve to support NATO.

REFORGER exercises were not merely a show of force, but they were also an integral part of NATO's general defense plan. Arriving Soldiers drew their vehicles and heavy equipment from facilities called POMCUS (Prepositioning of Materiel Configured in Unit Sets) sites that stored tanks, artillery, armored personnel carriers, support vehicles, and ammunition. Eventually enough modern equipment was on hand to field six combat-ready divisions. During each annual exercise, deploying divisions took part in live-fire training and in large-scale field maneuvers conducted by the U.S. V Corps and VII Corps.

At its height in 1988, REFORGER deployed 125,000 U.S. troops to Europe. After the end of the Cold War, the exercise became less important, and it ended in 1993. As a vital component of deterrence, REFORGER helped keep the peace in Europe for a generation.

Security Operations

He is best secure from dangers who is on his guard even when he seems safe.

Publilius Syrus

4-106. A force assigned a security task (screen, guard, or cover) can provide protection, early warning, and reaction time for friendly forces already on the ground but not yet ready for employment. Constituting and using a force assigned a security mission may be challenging. Forward-stationed forces may be out of position, or the assigned force may not be ideal for the mission. Urbanization, poor roads, and traffic congestion may hamper movement. Further, movement through areas controlled by partner nations is likely to be controlled by partner nation officials requiring convoy clearances, escorts, and reduced force protection levels (including ammunition status levels that may prohibit ammunition from being combat loaded inside vehicles). A security force could be viewed as provocative and may be deliberately constrained in terms of capabilities and actions to reduce the risk of escalation. (See chapters 5 and 6 for a detailed discussion on security operations and the defense.)

Civil Affairs and Civil-Military Operations

4-107. The movement of military forces from ports of debarkation through staging areas and forward into assembly areas requires coordination with the host nation to facilitate their movement. The corps assigned CA brigade organization supports the establishment of a civil-military operations center to coordinate with host-nation and local populations and institutions. Early coordination with unified action partners facilitates movement through indigenous immigration services, land use agreements, and indigenous local security.

Chapter 4

Information Operations

4-108. During operations to prevent, the corps conducts information operations. These operations include—
- Initiating information operations aspects of the theater campaign plan.
- Working in close coordination with unified action partners and host-nation information operations and information-related capability (IRC) forces to ensure unity of effort.
- Developing and controlling the narrative and countering the adversary's narrative.
- Immediately addressing concerns of vulnerable populations to inoculate them against adversary propaganda.
- Increasing message amplification and repetition via multiple conduits to enhance credibility, reach, and persistence.
- Countering adversarial use of propaganda, misinformation, and disinformation.
- Increasing frequency, size, and publicity of exercises and training activities.
- Ensuring that visual information (including combat camera) forces are embedded in forward troop movements, exercises, and training activities.

Military Information Support Operations

4-109. Psychological factors are an integral part of all operations. Military information support operations (MISO) include words and actions specifically planned to reduce the combat effectiveness of enemy armed forces and to influence hostile, neutral, and friendly groups to support U.S. operations. The corps is primarily concerned with MISO designed to produce short-term results. However, some tactical MISO objectives achieve longer-range goals. (See FM 3-53 for additional information on the use of MISO.)

Engineer Operations

4-110. The corps does not have organic engineer echelon-specific units. Each corps has habitual associations established with an engineer brigade headquarters. Each echelon of engineer headquarters units is capable, within their span of control, of serving directly under the corps headquarters in its operational configuration. Force-tailored engineer units supporting the corps security force are capable of providing mobility, counter-mobility, and survivability engineer capabilities. The corps engineer identifies shortages and recommends engineer force allocation for subordinate units. Engineer priorities typically focus on technical engineer reconnaissance in support of mobility and infrastructure development. The assigned engineer force should be robust enough to provide counter-mobility capacity to protect assailable flanks or disrupt enemy offensive operations and support friendly mobility. As a force moves into position, the engineer force transitions to counter-mobility and survivability tasks. (See ATP 3-34.23 for more information on engineer support to the corps headquarters.)

Intelligence Operations

4-111. Lack of adequate intelligence can hamper the initiation of corps operations. Limited knowledge of the enemy may inadvertently turn a security force movement into a situation that exacerbates a delicate political situation and causes an escalation towards open hostilities. Deploying units are generally at a disadvantage. It is imperative during operations to prevent for the intelligence staff to use all intelligence assets available to fill any gaps of information on the enemy, weather, terrain, and civil considerations. These intelligence assets may include national level intelligence, regionally-aligned military intelligence brigades-theater (MIB-Ts), and Army, theater, or corps intelligence staffs. Information developed during operations to shape should be interpreted in preparation for large-scale combat operations.

4-112. Forward deployed forces have the advantage of focusing their information collection toward enemies, adversaries, and specific OEs. This allows them to develop detailed knowledge of adversary and enemy locations, capabilities, dispositions, intentions, and pertinent civil considerations. Constant updates of intelligence products allow these forces to keep current on adversary and enemy capabilities and dispositions. There are several techniques to help ensure continuity of information on intelligence dissemination to include embedding an intelligence liaison officer forward, using home-station mission command facilities, or ensuring adequate connectivity and communication en route.

Fires

4-113. A security force generally does not become decisively engaged, but it may be required to if the enemy attacks. Fires and electronic warfare (EW) capabilities should be allocated to the security force in sufficient quantity to allow the security force to defend itself and the corps assembly areas. The corps designates a headquarters (usually a field artillery brigade) as the force field artillery headquarters. The fires available to the security force should include field artillery systems found in the corps' field artillery brigade and the security force BCTs. The USAF or the U.S. Navy normally provides close air support to the corps. Where the geography allows, naval surface fire support and AMD support may also be available.

Protection

4-114. Host-nation civil and military force capabilities may integrate with U.S. security forces for area and local protection tasks. Military police (MP) units provide a unique set of capabilities that support police operations, detention operations, and security and mobility support. Corps MP support is performed on a command, area, functional, or mission basis. MP capability for the security force should be considered when analyzing the mission variables. (See FM 3-39 for additional information regarding MP operations. ATP 3-39.30 addresses MP conduct of security and mobility support. FM 3-63 addresses Army detainee operations. ATP 4-02.46 addresses Army Health System support to detainee operations.)

Sustainment

4-115. As the corps builds combat power, it must ensure adequate sustainment of the security force, regardless of competing demands. The corps sustainment staff, integrated by the chief of sustainment, requires a firm understanding of the corps mission and security force elements. Accurately establishing support force requirements and ensuring effective support relationships is challenging in a dynamic OE. Sustainment support could be heavily reliant on contract support since 80 percent of sustainment capabilities reside in the reserve component, and there may not be adequate time to mobilize them for use in FDOs or FROs. Furthermore, restrictive force management levels may also drive significant use of contracted support in some operations.

CONSIDERATIONS WHILE BUILDING COMBAT POWER

4-116. Corps defensive-focused tasks are considered in conjunction with the considerations of national interests and enabling the corps to transition rapidly to the offense. The rapid buildup of credible offensive (and defensive) combat capability is what provides credible deterrence against adversaries trying to achieve low-cost objectives by making them too costly to attain.

Operations Security

Secret operations are essential in war; upon them the army relies to make its every move.

Sun Tzu

4-117. Reducing electronic signatures (leaving radio and digital systems off or in radio listening silence mode as much as possible) enhances OPSEC and survivability against adversary or enemy direction-finding capabilities. The proliferation of information technologies and dependence upon them for routine mission command functions creates vulnerabilities. Minimizing those vulnerabilities requires strict adherence to disciplined procedures appropriate for the corps AO. Restrictions on the use of personal electronic devices and social media should be a key consideration for OPSEC and reducing the electronic signatures of units.

Cyberspace Electromagnetic Activities

4-118. The corps coordinates with theater and national cyberspace elements to ensure unity of effort of cyberspace operations. Detailed plans and efforts for information penetration, information isolation attacks, cyberspace raids, and the disruption or neutralization of adversary integrated air defense systems (IADSs) should synchronize with the scheme of maneuver. While the compartmentalization of these cyberspace activities promotes OPSEC, this also increases the likelihood of violating one or more principles of joint operations described in JP 3-0.

Chapter 4

Space Operations

4-119. Peer threats have capabilities that can contest the space domain and attack the on-orbit, link, and terrestrial segments of U.S. satellite communications; positioning, navigation, and timing; missile warning; environmental weather; and space-based intelligence, surveillance, and reconnaissance efforts. These attacks may have significant impact across all warfighting functions, and they may significantly disrupt timelines and resources expended to accomplish the mission. Unified action partners must consider—and be prepared to operate in—denied, degraded, and disrupted space operational environments (D3SOE).

Corps Support Area

4-120. The corps headquarters establishes a support area. The corps commander assigns protection responsibility for the corps support area to a single commander; this may be subordinate to a division assigned for area security in the corps consolidation area. The consolidation area is an AO, assigned to an organization that extends from its higher headquarters boundary to the boundary of the forces in close area operations, where forces have established a level of control and are assigned tasks to consolidate gains. The designated organization may be a division assigned for area security in the corps consolidation area, or a smaller unit designated as the tactical combat force. (See chapter 1 for more information on consolidation areas.) Echelons above corps sustainment forces can operate from within a divisional AO, from ports of debarkation and bases in the joint security area, and from the corps support area. These conditions, associated with the wide dispersal of a sustainment unit, can reduce self-defense capabilities of sustainment forces through the placing of bases and base clusters beyond supporting distance of each other. (See ATP 3-37.10 for more information on base and base cluster interrelationships.)

> A *boundary* is a line that delineates surface areas for the purpose of facilitating coordination and deconfliction of operations between adjacent units, formations, or areas (JP 3-0).

Corps Reserve

4-121. The mission variables may require designating a corps reserve. Unity of command of the reserve is critical. On occasion, the corps reserve is the reserve of one or more of the divisions, with specific restrictions imposed by the corps commander on its employment. The positioning of the reserve facilitates its anticipated employment. The corps reserve is usually committed directly under control of the corps headquarters.

Transition to the Offense

4-122. If required, the corps defeats, destroys, or neutralizes the enemy force. Army commanders seek to mass effects, but not necessarily forces, as they pursue offensive tasks. (See chapter 7 for detailed discussions on the offense.) During operations to prevent, commanders position assets to accomplish objectives during possible follow-on offensive tasks that—

- Defeat, destroy, or neutralize the enemy force.
- Secure decisive terrain.
- Deprive the enemy of resources.
- Gain information.
- Deceive, divert, and fix the enemy in position.
- Disrupt the enemy's attack.
- Set conditions for future operations.
- Consolidate gains.

THE DIVISION

4-123. The division headquarters performs many of the same activities as the corps headquarters. The division headquarters fulfills its primary role as an Army tactical headquarters commanding up to five subordinate BCTs and other subordinate units. Upon deployment into a theater, the division may undergo significant task organization to enable operations. Initially, during operations to prevent, the division conducts defensive, security, and stability tasks supporting joint operations. The primary role of the division is to demonstrate national resolve by presenting a credible coercive force. Divisions should expect to conduct training exercises with multinational partners and perform other activities that demonstrate friendly capabilities. In an immature theater, the division headquarters should be prepared to accommodate the command structure of the next higher echelon until that echelon's systems are in place. (See FM 3-94 and ATP 3-91 for additional information on division operations.)

CONSOLIDATION OF GAINS

4-124. The primary consolidation of gains activities during operations to prevent are those taken to ensure that Army planning accounts for tasks that enable the consolidation of gains. This planning should include considering follow-on forces specifically task-organized to consolidate gains. Planning for the early and effective consolidation of gains enables the achievement of lasting favorable outcomes. The manner in which units execute missions, particularly how positively they interact with local populations and host-nation forces, significantly influences the perceptions of those affected by friendly operations. Early deploying units possess the capability to conduct security and stability tasks as needed. How units are organized and prepared for offensive, defensive, or stability tasks simplifies or complicates future security and stability outcomes.

4-125. Commanders and staffs should also plan for the eventual expansion of the size of consolidation areas, as well as the echelon of command responsible for those areas, as large-scale combat operations may continue over time and space. Conditions could require the commitment of additional forces to conduct consolidate gains activities in multiple division consolidation areas. Planning for such commitments continues as a critical part of operations to prevent. Forces that are identified to conduct consolidation of gains are trained and rehearsed in the conduct of consolidating gains.

4-126. Should operations to prevent transition to large-scale combat operations, forces responsible for consolidation of gains should be positioned to rapidly respond to the needs of the population affected by combat. Liaison and coordination with local civilian authorities should occur prior to the onset of combat, to ensure synchronization of efforts. Finally, any required adjustments to the dimensions of the consolidation area should be made and proper authorities informed of such changes.

This page intentionally left blank.

Chapter 5
Large-Scale Combat Operations

Therefore, the purpose of military operations cannot be simply to avert defeat—but rather it must be to win.

General Donn Starry

This chapter is divided into four sections. Section one provides an overview of large-scale combat operations. Section two addresses tactical enabling tasks that apply to both the defense and the offense. Section three describes forcible entry operations from which Army forces may defend or continue the offense. Section four discusses the transition to consolidation of gains. See chapters 6 and 7 for details on the defense and the offense respectively.

SECTION I – OVERVIEW: LARGE-SCALE COMBAT OPERATIONS

5-1. Section I discusses joint large-scale combat operations in paragraphs 5-2 through 5-13. It discusses Army forces in large-scale combat operations in paragraphs 5-14 through 5-29. It discusses the threat in paragraphs 5-30 through 5-33. It discusses stability in large-scale combat operations in paragraphs 5-34 through 5-54.

JOINT LARGE-SCALE COMBAT OPERATIONS

Jointness implies cross-Service combination wherein the capability of the joint force is understood to be synergistic, with the sum greater than its parts (the capability of individual components).

JP 1, *Doctrine for the Armed Forces of the United States*, 2013

5-2. As a nation, the United States wages war by employing all instruments of national power—diplomatic, informational, military, and economic. The President employs the Armed Forces of the United States to achieve national strategic objectives. The nature and scope of some missions may require joint forces to conduct large-scale combat operations to achieve national strategic objectives or protect national interests. Such combat typically occurs within the framework of a major operation or a campaign.

5-3. When large-scale combat operations commence, the joint force commander (JFC) immediately exploits friendly capabilities across multiple domains and the information environment to gain the initiative. The JFC seeks decisive advantage by using all available elements of combat power to exploit the initiative, deny enemy objectives, defeat enemy capabilities to resist, and compel desired behavior. The JFC coordinates with other U.S. governmental departments and agencies to facilitate coherent use of all instruments of national power in achieving national strategic objectives. Seizing the initiative generally requires force projection. The entry of Army and joint forces into a joint operations area (JOA) or theater of operations may be unopposed or opposed. (See paragraphs 5-110 to 5-113 for a discussion of forcible entry.)

5-4. JFCs strive to achieve air, maritime, space, and cyberspace superiority early to allow the joint force to conduct land operations without prohibitive enemy interference. Previously deployed forward land forces and land forces projected into the theater during large-scale combat operations can enable joint capabilities in the other domains and provide the joint force freedom of action.

5-5. JFCs gain and maintain the initiative by projecting fires, employing forces, and conducting information operations in dynamic combination across multiple domains. Establishing a joint headquarters requires

Chapter 5

detailed planning, active liaison, and coordination throughout the joint force. Such a transition may involve a simple movement of flags and supporting personnel, or it may require a complete change of joint force headquarters. The new joint force headquarters may use personnel and equipment, especially communications equipment, from an existing headquarters, or it may require augmentation from different sources. One technique is to transfer command in several stages. Another technique is for the JFC to use the capabilities of one of the components until the new headquarters is fully prepared. Whichever way the transition is done, staffs must address all of the command and control (C2) requirements and the timing of the transfer of each requirement.

5-6. Single domain success is rarely a solution to any problem, and ceding dominance in any domain to the enemy is not acceptable when there are Army or other joint units with the capability to contest the enemy within that domain. Army aviation, air defense, long range fires, space operations, information operations, electronic warfare (EW), and cyberspace capabilities facilitate operations across multiple domains for the entire joint team in the same way that joint capabilities enable operations on land.

5-7. The JFC, using special operations forces (SOF) independently or integrated with conventional forces, gains an additional and specialized capability to achieve objectives that might not otherwise be attainable. Integration enables the JFC to take fullest advantage of conventional and SOF core competencies. SOF are most effective when special operations are fully integrated into the overall plan and the execution of special operations is through proper SOF C2 elements in a supporting or supported relationship with conventional forces.

5-8. JFCs strive to conserve and increase combat power at the onset of combat operations. Further, joint forces protect host-nation infrastructure and logistics support nodes critical to force projection and sustainment. JFCs counter the enemy's fires and maneuver by making personnel, systems, and units difficult to locate, strike, and destroy. They protect the force from enemy maneuver and fires. Operations security (OPSEC) and military deception (MILDEC) are key elements of this effort. Operations to gain air, space, maritime, and electromagnetic spectrum (EMS) superiority; defensive use of information operations; and protection of airports and seaports, lines of communications, communications networks, and friendly force lodgment areas all contribute significantly to force protection.

5-9. JFCs establish areas of operations (AOs) to decentralize execution of land operations, allow rapid maneuver, and provide the ability to fight at extended ranges. The size, shape, and positioning of land AOs are based on the JFC's concept of operations and the land commanders' requirements to accomplish their missions. Within these AOs, land commanders are designated the supported commander for the integration and synchronization of joint maneuver, fires, and interdiction. Interdicting is an action to divert, disrupt, delay, or destroy the enemy's military surface capability before it can be used effectively against friendly forces, or to achieve enemy objectives. Accordingly, land commanders designate the target priority, effects, and timing of joint interdiction operations within their AOs. Further, in coordination with the land commander, the component commander designated as the supported commander for JOA-wide interdiction—typically the joint force air component commander (JFACC)—has the latitude to plan and execute JFC prioritized missions within the land AO. If JOA-wide interdiction operations would have adverse effects within a land AO, then the commander conducting those operations must either readjust the plan, resolve the issue with the appropriate component commander, or consult with the JFC for resolution.

5-10. The land commander should clearly articulate the vision of maneuver operations to other commanders who may employ interdiction forces within the land AO. The land commander's intent and concept of operations should clearly state how interdiction will enable or enhance land force maneuver in the AO and what is to be accomplished with interdiction (as well as those actions to be avoided, such as the destruction of key transportation nodes or the use of certain munitions in a specific area). Once this is understood, other interdiction-capable commanders can normally plan and execute operations with only that coordination required with the land commander. However, the land commander should provide other commanders as much latitude as possible in the planning and execution of interdiction operations within the AO.

5-11. JFCs pay particular attention and give priority to activities impinging on and supporting the maneuver and interdiction needs of all forces. In addition to normal target nomination procedures, JFCs establish procedures through which land commanders can specifically identify those interdiction targets in multiple domains critical to land maneuver that they are unable to engage with organic assets. These targets may be identified individually, by category, or tied to a desired effect or time period. Interdiction target priorities within land AOs are considered along with JOA-wide interdiction priorities by the JFC and reflected in apportionment decisions. Service or component commanders and supporting agencies use these priorities to plan, coordinate, and execute JOA-wide interdiction efforts in multiple domains.

5-12. The presence of and potential use of weapons of mass destruction (WMD) by enemy forces in large-scale combat operations is a grave concern for the JFC. WMD are chemical, biological, radiological, or nuclear weapons or devices capable of a high order of destruction or that cause mass casualties. Countering weapons of mass destruction (CWMD) entails activities across the U.S. Government to ensure that friendly forces and interests are not attacked or coerced by actors possessing WMD. CWMD activities can be accomplished during any phase (from shaping through enable civil authorities) of a joint operation. During the seize the initiative and dominate phases of large-scale combat operations, CWMD activities control, defeat, disable, and dispose of WMD threats, safeguard the force, and manage consequences in the event of use. (See JP 3-40 for joint doctrine on CWMD.)

5-13. Conditions preceding large-scale combat operations vary depending on the threat. Some adversaries possess significant capability to employ antiaccess (A2) and area denial (AD) strategies. Countering those strategies is the responsibility of the JFC. The land component commander's (LCC's) challenge is deploying significant combat power in an environment where the enemy has an initial advantage.

ARMY FORCES IN LARGE-SCALE COMBAT OPERATIONS

5-14. The Army provides the JFC significant and sustained landpower. *Landpower* is the ability—by threat, force, or occupation—to gain, sustain, and exploit control over land, resources, and people (ADRP 3-0). The Army supports the joint force by providing capability and capacity for the application of land power through maneuver, fires, special operations, cyberspace operations, EW, space operations, sustainment, and area security.

5-15. The JFC applies Army capabilities to neutralize enemy integrated defenses by systematically destroying key nodes and capabilities essential to their coherence. Conventional, special operations, and joint force interdependence is an important part of this effort. Army SOF can engage high value targets to achieve operational and strategic level effects, while Army intelligence capabilities facilitate information collection for all unified action partners and reinforce the idea of interdependence.

5-16. During large-scale combat operations, Army forces defeat the enemy. Defeat of enemy forces in close-combat operations is normally required to achieve campaign objectives and national strategic goals after the commencement of hostilities. Planning for sequels to consolidate gains at higher levels should be informed by combat operations and vice versa. However, the demands of large-scale combat operations consume all available staff capability at the tactical level.

5-17. In large-scale combat operations against a peer threat, commanders conduct decisive action to seize, retain, and exploit the initiative. This involves the orchestration of many simultaneous unit actions in the most demanding of operational environments. Large-scale combat operations introduce levels of complexity, lethality, ambiguity, and speed to military activities not common in other operations. Large-scale combat operations require the execution of multiple tasks synchronized and converged across multiple domains to create opportunities to destroy, dislocate, disintegrate, and isolate enemy forces.

5-18. Army forces defeat enemy organizations, control terrain, protect populations, and preserve joint force and unified action partner freedom of movement and action in the land and other domains. Corps and division commanders are directly concerned with those enemy forces and capabilities that can affect their current and future operations. Accordingly, joint interdiction efforts with a near-term effect on land maneuver normally support land maneuver. Successful corps and division operations may depend on successful joint interdiction operations, including those operations to isolate the battle or weaken the enemy force before battle is fully joined.

5-19. During large-scale combat operations, Army forces enable joint force freedom of action by denying the enemy the ability to operate uncontested in multiple domains. Army leaders synchronize the efforts of multiple unified action partners to ensure unity of effort. Army forces adapt continuously to seize, retain, and exploit the initiative. Army forces use mobility, protection, and firepower to strike the enemy unexpectedly

Chapter 5

from multiple directions, denying the enemy freedom to maneuver and creating multiple dilemmas that the enemy commander cannot effectively address.

5-20. Commanders consider potential future enemy moves and what actions friendly forces may take to counter those moves. Planners anticipate the use of cyberspace operations that can disrupt enemy command and control nodes. Army and joint forces concentrate combat power rapidly from dispersed locations to attack critical enemy assets and exploit opportunities, and then disperse quickly enough to avoid becoming lucrative targets themselves. This provides protection against an enemy with long range fires advantages. The initial Army forces to arrive within a JOA may need to fight outnumbered to buy time for and protect follow-on forces. This requires those initial forces to have tactical mobility, firepower, and resilience.

5-21. Army forces assist in CWMD operations within the JOA and throughout the area of responsibility (AOR) during large-scale combat operations. In accordance with the JFC's concept of operation, Army forces plan and integrate their efforts with the joint force and host nation for CWMD operations. Army forces contribute to the mission success of CWMD through integrated combined arms teams executing tactical tasks that support the activities of CWMD and with an understanding of the special considerations associated with chemical, biological, radiological, and nuclear (CBRN) environments and WMD objectives. For example brigade combat teams (BCTs) plan for site exploitation, use of contaminated routes for movement, casualty evacuation of contaminated casualties, prolonged operations in CBRN environments, and immediate, operational, and thorough decontamination of units. (See ATP 3-90.40 for more information on combined arms CWMD operations.)

5-22. Army forces generally constitute the preponderance of land combat forces, organized into corps and divisions, during large-scale combat operations. Army forces seize the initiative, gain and exploit positions of relative advantage in multiple domains to dominate an enemy force, and consolidate gains. Corps and divisions execute decisive action tasks, where offensive and defensive tasks make up the preponderance of activities. Commanders must explicitly understand the lethality of large-scale combat operations to preserve their combat power and manage risk. Commanders leverage cyberspace operations, space capabilities, and information-related capabilities in a deliberate fashion to support ground maneuver. Commanders also use ground maneuver and other land-based capabilities to enable maneuver in the other domains.

5-23. BCTs and subordinate echelons concentrate on performing offensive and defensive tasks and necessary tactical enabling tasks. During large-scale combat operations they perform only those minimal essential stability tasks necessary to comply with the laws of land warfare. They do not conduct operationally significant consolidate gains activities unless assigned that mission in a consolidation area. BCT commanders orchestrate rapid maneuver to operate inside an enemy's decision cycle and create an increasing cascade of hard choices for the enemy commander.

5-24. Commanders employ the appropriate form of maneuver to close with an enemy to mitigate disadvantages in the capabilities of weapon systems and vehicle protection. This typically requires rapid movement through close or complex terrain during periods of limited visibility. Subordinate unit combat formations move in as dispersed a manner as possible while retaining the capability to mass effects against enemy forces at opportune times and places. Joint enablers become more effective when an enemy has no time to focus on singular friendly capabilities in just one or two domains. Units perform attacks that penetrate enemy defenses or attack them frontally or from a flank. Depending on the situation, they also infiltrate enemy positions, envelop them, or turn enemy forces out of their current positions. Those units then exploit success to render enemy forces incapable of further resistance.

5-25. The consolidation of gains is an integral part of decisive action. Corps and division headquarters assign purposefully task-organized forces designated consolidation areas to begin consolidate gains activities concurrent with large-scale combat operations. Consolidate gains activities provide freedom of action and higher tempo for those forces committed to the close, deep, and support areas. Units begin consolidate gains activities after achieving a minimum level of control and when there are no on-going large-scale combat operations in a specific portion of their AO. Corps and divisions can designate a maneuver force responsible for consolidation areas. Forces assigned the mission of consolidating gains execute area security and stability tasks. This enables freedom of action for units in the other corps and division areas by allowing them to focus on their assigned tasks and expediting the achievement of the overall purpose of the operation. Initially the focus is on combined arms operations against bypassed enemy forces, defeated remnants, and irregular forces to defeat threats against friendly forces in the support and consolidation areas, as well as those short of the rear boundaries of BCT in the close area. Friendly forces may eventually create or reconstitute an indigenous security force through security cooperation activities as the overall focus of operations shifts from large-scale combat operations to consolidating gains. Optimally, a division commander would assign a BCT to secure a consolidation area. A division is the preferred echelon for this mission in a corps AO. The requirement for additional forces to consolidate gains as early as possible should be accounted for early during planning with appropriate force tailoring by the theater army.

5-26. The application of the tenets of unified land operations—simultaneity, depth, synchronization, and flexibility—is foundational to the conduct of operations. These four tenets should inform both plans and the conduct of operations. Army forces are most successful when they synchronize and converge effects across the breadth and depth of their assigned AOs within the scope of their mission. Large-scale combat operations require higher echelon headquarters to empower tactical organizations to make decisions and act, and higher echelon headquarters enable lower echelons within their capabilities. As an example, there are national-level space and cyberspace capabilities that can engage specific target sets to support tactical engagements. Commanders and staffs understand, plan for, and request effects available from these national level assets in much the same way they do joint fires. For tactical commands to successfully orchestrate the various domains and functions, planners must not assume higher-level effects are beyond their scope. However, such requests must generally be made in advance of combat operations, and the execution and results of such requests may not always be known to the commander on the ground.

5-27. It is likely that Army forces will be required to defend against enemy forces with locally superior capabilities at the beginning of large-scale combat operations. Enemy forces are most likely to attack when they have a position of relative advantage and friendly forces are most vulnerable, particularly when conditions make effective employment of joint force capabilities difficult. (See chapter 6 for a discussion of large-scale defensive operations.)

5-28. Army forces may conduct large-scale combat operations in urban areas within the JOA or theater of operations. Currently more than 50 percent of the world's population lives in urban areas, and this is likely to increase to 70 percent by 2050, making large-scale combat operations in cities likely. Commanders may conduct urban operations because they provide a tactical, political, or economic advantage, or when not doing so threatens the joint campaign. Army forces conduct large-scale combat operations in urban areas either as a specific, unique operation, or more typically, as one of a larger series of operations in a joint campaign. Urban operations focus on the threat to or within the urban area and allow other forces to conduct operations elsewhere. Conducting operations in dense urban terrain is complex and resource intensive. Due to the complexity of an urban environment, commanders must carefully arrange their forces and operations according to purpose, time, and space to accomplish the mission. In most urban operations, the terrain, the dense population, military forces, and unified action partners will further complicate this arrangement. (See FM 3-06 for an in depth discussion of operations in urban areas.)

5-29. The performance of offensive tasks is traditionally associated with a favorable combat-power ratio. Combat multipliers often provide positions of relative advantage, even when Army forces are outnumbered. Numerical superiority is not a precondition for performing offensive tasks. Rather, commanders must continuously seek every opportunity to seize the initiative through offensive tasks, even when the force as a whole is on the defense. This requires commanders to perform economy of force measures to adequately resource their decisive operations or main efforts. Mobility, surprise, and aggressive execution are the most effective means for achieving tactical success when performing both offensive and defensive tasks. Bold, aggressive tactics may involve significant risk; however, greater gains normally require greater risks. A numerically inferior force capable of bold and aggressive action can create opportunities to seize and exploit the initiative. (See chapter 7 for a discussion of large-scale offensive operations.)

THREAT

A general in all of his projects should not think so much about what he wishes to do as what his enemy will do; that he should never underestimate this enemy, but he should put himself in his place to appreciate difficulties and hindrances the enemy could interpose; that his plans will be deranged at the slightest event if he has not foreseen everything and if he has not devised means with which to surmount the obstacles.

Frederick the Great

Chapter 5

5-30. The enemy will seek to render U.S. combat power ineffective by systemic and continual attacks in multiple domains and the information environment before and during combat operations. This includes attacks on U.S. sustainment activities and mission command networks in addition to traditional attacks on maneuver forces. The enemy will attempt to isolate and contain friendly maneuver formations and force them into engagements that do not bear on the decisive point. The destruction of high-visibility or unique systems employed by U.S. forces offers exponential value in terms of enemy goals. These actions are not always linked to military objectives. They often seek to maximize effects in the information and psychological arenas to achieve political objectives.

5-31. Friendly high-value systems that could be identified for destruction by the enemy include stealth aircraft, attack helicopters, counterbattery artillery radars, aerial surveillance platforms, or long-range fires systems. Losses of these systems degrade operational capability and inhibit employment of critical weapons systems. Commanders should assume enemy use of precision munitions, massed fires, and area effects munitions. Enemy use of camouflage, deception, decoy, or mockup systems can degrade the effects of friendly sensor systems. Also, enemies can employ global positioning system jamming to disrupt U.S. precision munitions targeting, sensor-to-shooter links, and navigation. An enemy commander may use long-range missiles and rockets, advanced air defense, sensor and EW weapons, unmanned aircraft systems (UASs), precision-guided artillery, area denial munitions, and directed energy weapons to degrade Army and joint capabilities.

5-32. Peer threats will minimize massed formations, patterned echelonment, and linear operations whenever possible. They will hide and disperse forces in areas of sanctuary to limit the ability of friendly forces to apply their full range of capabilities. They will attempt to retain the ability to rapidly mass forces and fires from dispersed locations at the times and places of their choosing. Whenever possible, an enemy will use the physical environment and natural conditions to neutralize or offset U.S. information collection assets.

5-33. Peer threats contest U.S. forces across all domains, including air, maritime, land, space, and cyberspace. They contest U.S. use of the EMS and the information environment. Enemies are likely to employ A2 strategies to prevent the joint force's ability to project and sustain combat power into a region and area denial strategies to constrain U.S. forces' freedom of action within the region. Assured access—the unhindered national use of the global commons and selected sovereign territory, waters, airspace and cyberspace—is achieved by projecting all the instruments of national power. Often U.S. forces require operational access, which is the ability to project military force into an operational area with sufficient freedom of action to accomplish the mission.

STABILITY IN LARGE-SCALE COMBAT OPERATIONS

> *Limited resources and capabilities often make it impossible to accomplish all objectives at once. Therefore, the designation of priorities, allocation of resources, and assignment of tasks require careful consideration. Plans must be appraised realistically in the light of their short and long-range impact on the population and on the benefits which will accrue to the government.*
>
> FM 31-23, *Stability Operations*, 1972

5-34. Generally, the responsibility for providing for the needs of the civilian population rests with the host-nation government or designated civil authorities, agencies, and organizations. Army forces perform minimal essential stability tasks to provide security, food, water, shelter, and medical treatment when there is no legitimate civil authority present. Commanders assess available resources against their missions to determine how best to conduct these minimum essential stability tasks and what risks must be accepted. (See ADRP 3-07 for additional information on minimal essential stability tasks.)

5-35. Large-scale combat operations involve the combination of offense, defense, and stability tasks. The priorities and effort given to stability tasks vary within subordinate unit AOs. Corps and division commanders analyze the situations they face to determine the minimum essential stability tasks and the priority associated with each task. This analysis includes a planned transition to consolidation of gains in operational areas once large-scale combat operations culminate.

STABILITY TASKS

5-36. Stability tasks during large-scale combat operations include restoration of essential services and population control in areas controlled by friendly forces. During the performance of defensive tasks, forces protect civilians from enemy attacks, maintain control, or evacuate civilians from AOs controlled by friendly units. Initially, the performance of stability tasks may be incidental to combat operations. Divisions conducting large-scale combat operations are not task-organized to simultaneously perform stability tasks, but they often include civil affairs (CA) units in their organizations. A corps headquarters retains control over functional support brigades, such as military police (MP) and engineers, so it can reinforce subordinate divisions or direct the performance of stability tasks. A corps headquarters adjusts subordinate division task organizations in anticipation of them performing stability tasks as combat requirements diminish.

5-37. Commanders at all echelons plan for logistic and medical support to indigenous populations affected by combat operations. The land component, corps, and division headquarters staffs typically work directly with any host-nation authorities located within or responsible for portions of their AOs to identify the minimum-essential support that U.S. forces must provide to meet international accords. Corps and division staffs collaborate with the land component, theater army, and joint task force staff to forecast requirements for each successive phase of operations. As required, the theater army requests additional Army units to manage stability tasks in its AO.

5-38. Commanders must often reorganize their forces to perform stability tasks during the course of an operation. This is especially the case as corps and division consolidation areas are established. Corps, division, and BCT commanders, assisted by attached or supporting CA units, identify the initial response, transformation, and fostering sustainability stability requirements of their respective AOs. They plan the performance of stability tasks and their associated transitions from military units to legitimate authorities.

5-39. Corps and division commanders plan the time and method necessary to task-organize forces, re-allocate subordinate AOs and priorities of support, and request additional forces to perform previously identified stability tasks using the above information. These additional forces normally include MP, medical, logistics, general engineering, and explosive ordnance disposal (EOD) units. MP units provide population control and infrastructure security. Medical units provide public health services. Logistics units provide food and water, including water purification and bulk water transport. General engineering units repair infrastructure or provide emergency shelter. EOD units clear unexploded ordnance within their capability according to the commander's priorities. However, the U.S. does not conduct humanitarian demining.

5-40. CA and psychological operations units and other information-related capability (IRC) units support stability tasks during large-scale combat operations engaging the civilian population within the corps or division AO. They also target selected foreign civilian populations outside of those AOs. They publicize curfews, checkpoint procedures, evacuation routes, food and water distribution points, emergency health care, and necessary vector control measures to assist the overarching stability effort, minimize civilian casualties, facilitate freedom of movement, and avert potential humanitarian tragedies.

OPERATIONS IN THE SUPPORT AND CONSOLIDATION AREAS

In the past, combat operations in the rear area have proven to be difficult to defend against and to be very disruptive to forward support.

FM 90-14, *Rear Battle*, 1985

5-41. A corps or division commander typically assigns responsibility for a corps or division support area to a maneuver enhancement brigade (MEB) while assigning responsibility for the consolidation area to a task organized BCT (in the division) or a task-organized division (in the corps). The forces necessary to consolidate gains represent a separate and distinct requirement beyond the BCTs and divisions required to conduct close and deep operations. To properly consolidate gains, the theater army must plan and request the additional required forces through the force tailoring process during the early stages of the conflict buildup.

5-42. For a variety of reasons, echelon-specific support areas may or may not be located within a larger consolidation area assigned to a BCT or division. However, positioning the support area within a consolidation area when possible reduces the area security burden on the MEB significantly. The provision of a BCT headquarters with its associated United States Air Force (USAF) tactical air control party enables

Chapter 5

the rapid employment of joint fires against threats to friendly units in the support area. Likewise, a division headquarters with its associated USAF theater air control systems element and joint air-ground integration center (JAGIC) enables the rapid employment of joint fires in the corps consolidation area. The MEB is specifically organized to execute the nine doctrinal responsibilities inherent to a unit assigned an AO. Division commanders typically assign their task-organized MEB responsibility for base and base cluster defense within the division support area.

Support of Base Security and Defense

5-43. An assigned unit conducts and coordinates base camp and base cluster security and defense. The assigned unit is responsible for area security, base and base cluster security, and defense within its AO. The designated base camp commander or base operating support-integrator within the assigned unit's AO should be under the tactical control (TACON) of the assigned unit's commander. The base operating support-integrator is the designated service component or joint task force commander assigned to synchronize all sustainment functions for a contingency base. The elements operating within individual bases or base camps should be under the operational control (OPCON) or TACON of the base camp commander. The assigned unit tasks these tenet organizations to conduct information collection, security, and local or perimeter defense operations.

5-44. The assigned unit integrates the base and base cluster security and self-defense plans. Either the base camp commander or base operating support-integrator perform this additional responsibility under the oversight of the assigned unit commander. The assigned unit can mass forces, capabilities, or systems from several bases or base clusters to integrate, synchronize, and mass combat power to defeat a level III threat. Paragraphs 5-44 through 5-46 discuss threat levels. (See ATP 3-37.10 for details on base security and defense.)

5-45. Threats in the division support area are categorized by the three levels of defense required to counter them. Any or all threat levels may exist simultaneously in the division support area. Emphasis on base defense and security measures may depend on the anticipated threat level. A *Level I threat* is a small enemy force that can be defeated by those units normally operating in the echelon support area or by the perimeter defenses established by friendly bases and base clusters (ATP 3-91). A Level I threat for a typical base consists of a squad-sized unit or smaller groups of enemy soldiers, agents, or terrorists. Typical objectives for a Level I threat include supplying themselves from friendly supply stocks; disrupting friendly mission command nodes and logistics facilities; and interdicting friendly lines of communication.

5-46. A *Level II threat* is an enemy force or activities that can be defeated by a base or base cluster's defensive capabilities when augmented by a response force (ATP 3-91). A typical response force is an MP platoon (with appropriate supporting fires); however, it can be a combined arms maneuver element. Level II threats consist of enemy special operations teams, long-range reconnaissance units, mounted or dismounted combat reconnaissance teams, and partially attrited small combat units. Typical objectives for a Level II threat include the destruction, as well as the disruption, of friendly mission command nodes and logistics and commercial facilities; and the interdiction of friendly lines of communications.

5-47. A *Level III threat* is an enemy force or activities beyond the defensive capability of both the base and base cluster and any local reserve or response force (ATP 3-91). It consists of mobile enemy combat forces. Possible objectives for a Level III threat include seizing key terrain, interfering with the movement and commitment of reserves and artillery, and destroying friendly combat forces. Its objectives could also include destroying friendly sustainment facilities, supply points, command post facilities, airfields, aviation assembly areas, arming and refueling points, and interdicting lines of communications and major supply routes. The division response to a Level III threat is a tactical combat force (TCF).

5-48. Each designated base and base cluster commander, or base operating support-integrator, is responsible for organizing and preparing either an initial-response force or a mobile security force. An initial-response force comes from the assets of the base's assigned, attached, OPCON, or TACON units. A mobile security force, typically an MP unit, comes from supporting or reinforcing combat forces directed to conduct combat operations in support of the unit owning the AO. The initial-response force operates under control of its base defense operations center to defeat Level I and some Level II threats until the base cluster commander can respond with the mobile security force or TCF.

5-49. A base cluster commander is responsible for organizing and preparing a mobile security force for Level II threats from the assets available within the base cluster. The mobile security force assembles and counterattacks to eliminate Level II threats. The base cluster commander reconstitutes another mobile security force upon commitment of the original response force. The base cluster commander, or base operating support-integrator, notifies the local commander when a Level II threat is detected or being reacted to. The commander typically responds to a Level III threat with a reserve or dedicated TCF. A *tactical combat force* is a rapidly deployable, air-ground mobile combat unit, with appropriate combat support and combat service support assets assigned to and capable of defeating Level III threats including combined arms (JP 3-10).

Tactical Combat Force in the Support Area

5-50. Support area commanders defeat Level I and some Level II threats within their assigned bases and base camps. The commitment of a mobile security force, reserve, or a TCF becomes a significant mission command and potential fratricide problem that rehearsals and standing operating procedures can mitigate. Typically, the support area's commander supervises the operations of the response force, the echelon reserve, and the TCF since these friendly forces may converge with other stationary friendly forces.

Airspace Control In Consolidation and Support Areas

5-51. Both corps and divisions require the capability to control airspace. Airspace control in the consolidation and support areas may be accomplished by a variety of methods. If the corps is primarily focused on shaping for its forward divisions to defeat enemy forces in the close and deep areas, then the corps may delegate airspace control over the consolidation area to the unit (or division) responsible for it. If the corps is primarily conducting operations to consolidate gains, then it may choose to retain control of airspace over the entire corps AO.

5-52. Delegating airspace control to the division responsible for the consolidation area allows the corps to focus on rapidly coordinating airspace for corps fires and for dynamic repositioning of corps aircraft. The nature of corps deep fires requires that the consolidation area division, the field artillery brigade, the corps, and the JFACC control and reporting center establish a rapid airspace coordination network to enable responsive fires.

5-53. During operations to consolidate gains, corps airspace control elements focus on the non-corps airspace users operating over the corps AO. Airspace users over the corps consolidation areas may require positive control beyond the capability of the division assigned the consolidation area. This may require either an air traffic control facility or a JFACC control and reporting center.

5-54. A MEB assigned a support area has the capability to control Army airspace users over its AO. If the corps has delegated airspace control to the consolidation area division, the MEB controlling the corps support area coordinates airspace control requirements with that division. If the corps has retained airspace control over the consolidation area, then the MEB coordinates airspace control requirements with the corps. In a similar manner, the BCT responsible for the division consolidation area coordinates airspace control requirements with that division.

SECTION II – TACTICAL ENABLING TASKS

5-55. Commanders direct tactical enabling tasks to support the performance of all offensive, defensive, and stability tasks. Tactical enabling tasks are usually employed by commanders as part of shaping operations or supporting efforts. The tactical enabling tasks are reconnaissance, security, troop movement, relief in place, passage of lines, encirclement operations, and mobility and countermobility operations.

RECONNAISSANCE

> *The purpose of reconnaissance is to gather information upon which commanders may base plans, decisions, and orders. Reconnaissance includes surveillance; that is, systematic observation by any means.*
>
> FM 17-95, *Cavalry Operations*, 1981

Chapter 5

5-56. *Reconnaissance* is a mission undertaken to obtain, by visual observation or other detection methods, information about the activities and resources of an enemy or adversary, or to secure data concerning the meteorological, hydrographic, or geographic characteristics of a particular area (JP 2-0). Reconnaissance primarily relies on human beings rather than technical means. Reconnaissance is a focused collection effort. It is performed before, during, and after other operations to provide information used in the intelligence preparation of the battlefield process and by commanders in order to formulate, confirm, or modify a course of action.

5-57. Commanders orient their reconnaissance assets by identifying a reconnaissance objective within an AO. The *reconnaissance objective* is a terrain feature, geographic area, enemy force, adversary, or other mission or operational variable, such as specific civil considerations, about which the commander wants to obtain additional information (ADRP 3-90). The reconnaissance objective clarifies the intent of the reconnaissance effort by specifying the most important result to obtain from the reconnaissance effort. Every reconnaissance mission specifies a reconnaissance objective. Commanders assign reconnaissance objectives based on priority information requirements resulting from the intelligence preparation of the battlefield (IPB) process and the reconnaissance asset's capabilities and limitations. A reconnaissance objective can be information about a specific geographic location, such as the cross country trafficability of a specific area, a specific enemy or adversary activity to be confirmed or denied, or a specific enemy or adversary unit to be located and tracked. A reconnaissance unit uses the reconnaissance objective to guide it in setting priorities when it does not have enough time to complete all the tasks associated with a specific form of reconnaissance.

5-58. There are seven fundamentals of successful reconnaissance operations. Commanders—

- Ensure continuous reconnaissance.
- Do not keep reconnaissance assets in reserve.
- Orient on the reconnaissance objective.
- Report information rapidly and accurately.
- Retain freedom of maneuver.
- Gain and maintain enemy contact.
- Develop the situation rapidly.

> There are eight forms of contact—direct, indirect, nonhostile or civilian, obstacle, CBRN, aerial, visual, and electronic. (Electronic includes contact in cyberspace.)

5-59. The responsibility for conducting reconnaissance operations does not reside solely with specifically organized units. Every unit has an implied mission to report information about the terrain, civilian activities, and friendly and enemy dispositions. This is regardless of a unit's location or primary function. Troops in contact with an enemy and reconnaissance patrols of maneuver units at all echelons collect information on enemy units and activities. In the support and consolidation areas, reserve maneuver forces, functional and multifunctional support and sustainment elements, other governmental agencies, and multinational forces observe and report civilian, adversary, and enemy activity and significant changes in terrain trafficability. Although all units conduct reconnaissance, those specifically trained in reconnaissance tasks are ground cavalry, aviation attack reconnaissance units, scouts, Ranger units, CA units, and SOF. Some units, such as engineers, CA, and the chemical corps, have specific reconnaissance tasks to perform that complement the force's overall reconnaissance effort. However, BCT, division, and corps commanders primarily use their organic or attached reconnaissance—ground or air—and intelligence elements to conduct reconnaissance operations.

5-60. There are five forms of reconnaissance operations. They are—

- Route reconnaissance.
- Zone reconnaissance.
- Area reconnaissance.
- Reconnaissance in force.
- Special reconnaissance.

These forms of reconnaissance are discussed in paragraphs 5-61 through 5-64.

ROUTE RECONNAISSANCE

5-61. *Route reconnaissance* is a directed effort to obtain detailed information of a specified route and all terrain from which the enemy could influence movement along that route (ADRP 3-90). That route may be a cross country mobility corridor. Route reconnaissance provides new or updated information on route conditions, such as obstacles and bridge classifications, and enemy, adversary, and civilian activity along the route. Commanders normally assign this mission when wanting to use a specific route for friendly movement.

ZONE RECONNAISSANCE

5-62. *Zone reconnaissance* is a form of reconnaissance that involves a directed effort to obtain detailed information on all routes, obstacles, terrain, and enemy forces within a zone defined by boundaries (ADRP 3-90). Obstacles include existing, reinforcing, and areas with CBRN contamination. Commanders assign a zone reconnaissance mission when they need additional information on a zone before committing other forces. It is appropriate when the enemy situation is vague, existing knowledge of the terrain is limited, or combat operations have altered the terrain. A zone reconnaissance may include several route or area reconnaissance missions assigned to subordinate units.

AREA RECONNAISSANCE

5-63. *Area reconnaissance* is a form of reconnaissance that focuses on obtaining detailed information about the terrain or enemy activity within a prescribed area (ADRP 3-90). This area may include a town, a ridgeline, a forest, an airhead, or any other critical operational feature. The area may consist of a single point, such as a bridge or an installation. The primary difference between an area reconnaissance and a zone reconnaissance is that in an area reconnaissance, units conducting the reconnaissance first move to the area in which the reconnaissance will take place. In a zone reconnaissance the units conducting the reconnaissance start from a line of departure. Areas are normally smaller than zones and are not usually contiguous to other friendly areas targeted for reconnaissance. Because the area is smaller, an area reconnaissance typically takes less time to complete than a zone reconnaissance.

RECONNAISSANCE IN FORCE

5-64. A *reconnaissance in force* is a deliberate combat operation designed to discover or test the enemy's strength, dispositions, and reactions or to obtain other information (ADRP 3-90). Battalion-size task forces or larger organizations usually conduct a reconnaissance in force. A commander assigns a reconnaissance in force when an enemy is operating within an area and the commander cannot obtain adequate intelligence by any other means. A unit may also conduct a reconnaissance in force in restrictive terrain where an enemy is likely to ambush smaller reconnaissance forces. A reconnaissance in force is an aggressive reconnaissance, conducted as an offensive task with clearly stated reconnaissance objectives. The overall goal of a reconnaissance in force is to determine enemy weaknesses that can be exploited. It differs from other reconnaissance operations because it is normally conducted only to gain information about an enemy and not the terrain. The unit commander plans for both the retrograde or reinforcement of the force, in case it encounters superior enemy forces, and for the exploitation of its success.

> **Ambiguous Environment: Fighting for Information**
>
> On 5 April 2003, 1st Brigade, 3d Infantry Division (Mechanized) mounted an operation into western Baghdad in preparation for the division's advance on the Iraqi capital. The mission was planned as a battalion-sized reconnaissance-in-force in order to determine the composition, strength, and disposition of enemy defenses. This marked the first armored foray into a major city since World War II.
>
> The operation, executed by 1st Battalion, 64th Armor, was considered a reasonable and acceptable risk despite the ambiguity of the enemy situation. Command guidance was simply to "conduct a movement to contact north along Highway 8 to determine the enemy's disposition, strength, and will to fight," a statement that offered both a clear intent and purpose. These mission-type orders allowed for a great deal of flexibility in seizing the initiative and reacting to the enemy. Unit commanders believed that the superior quality of their Soldiers would mitigate the inherent risk in the operation. They were right. Despite heavy resistance, the armored column quickly reached its objective. Commanders concluded the reconnaissance in force had taken enemy forces completely by surprise and damaged their physical and mental ability to resist.
>
> The operation demonstrated that U.S. armored forces could penetrate Baghdad while suffering minimal casualties. Moreover, it met the original intent by providing excellent indicators of enemy tactics, strength, and locations. Senior commanders came to view the reconnaissance-in-force as a prelude to additional armored missions in and out of the city that would disrupt the Iraqi defenses in the city and ultimately cause the regime to collapse. Using the lessons learned on 5 April, 3d Infantry Division launched another operation on a larger scale on 7 April, which resulted in the occupation of downtown Baghdad and the final fall of the Baathist government.

SPECIAL RECONNAISSANCE

5-65. *Special reconnaissance* is reconnaissance and surveillance actions conducted as a special operation in hostile, denied, or diplomatically and/or politically sensitive environments to collect or verify information of strategic or operational significance, employing military capabilities not normally found in conventional forces (JP 3-05). These actions provide an additional capability for commanders and supplement other conventional reconnaissance and surveillance actions. Even with long range sensors and overhead platforms, some information can be obtained only by visual observation or other collection methods in the target area. SOF capabilities for gaining access to denied and hostile areas, worldwide communications, and specialized aircraft and sensors enable them to conduct special reconnaissance against targets inaccessible to other forces or assets. Special reconnaissance tasks include—

- Environmental reconnaissance.
- Armed reconnaissance.
- Target and threat assessment.
- Post-strike reconnaissance.

(See JP 3-05 for additional information on these special reconnaissance tasks. See FM 3-90-2 and FM 3-98 for additional information on the other four forms of reconnaissance.)

SECURITY OPERATIONS

> *Security includes all measures taken by a command to protect itself from enemy observation, sabotage, annoyance, or surprise. Its purpose is to preserve secrecy and to gain and maintain freedom of action.*
>
> FM 7-19, *Combat Support Company, Infantry Division Battle Group*, 1960

5-66. *Security operations* are those operations undertaken by a commander to provide early and accurate warning of enemy operations, to provide the force being protected with time and maneuver space within which to react to the enemy, and to develop the situation to allow the commander to effectively use the protected force (ADRP 3-90). The ultimate goal of security operations is to protect the force from surprise and reduce the unknowns in any situation. Security operations may also protect the civilian population, civil institutions, and civilian infrastructure within the unit's AO. Security operations can be either offensive or defensive. A commander may conduct security operations to the front, flanks, or rear of the friendly force. The main difference between security operations and reconnaissance operations is that security operations orient on the force or facility being protected, while reconnaissance is enemy and terrain oriented. Security operations are shaping operations. As a shaping operation, economy of force is often a consideration when planning tactical security operations.

5-67. Security operations encompass five tasks—screen, guard, cover, area security, and local security:
- *Screen* is a security task that primarily provides early warning to the protected force (ADRP 3- 90).
- *Guard* is a security task to protect the main body by fighting to gain time while also observing and reporting information and preventing enemy ground observation of and direct fire against the main body. Units conducting a guard mission cannot operate independently because they rely upon fires and functional and multifunctional support assets of the main body (ADRP 3-90).
- *Cover* is a security task to protect the main body by fighting to gain time while also observing and reporting information and preventing enemy ground observation of and direct fire against the main body (ADRP 3-90). Cover operations also typically operate out of range of the protected force so units must be organized with the enabling capabilities for independent operations.
- *Area security* is a security task conducted to protect friendly forces, installations, routes, and actions within a specific area (ADRP 3-90).
- *Local security* is a security task that includes low level security activities conducted near a unit to prevent surprise by the enemy (ADRP 3-90). Local security is closely associated with unit force protection efforts.

5-68. Screen and guard missions can be forward, on the flank, or in the rear of the main body, and they can be moving or stationary. Moving flank guards must also conduct a zone reconnaissance between the main body and the guard force to prevent bypassing significant enemy forces. The screen, guard, and cover security tasks, respectively, contain increasing levels of combat power and provide increasing levels of security for the main body. However, more combat power in the security force means less for the main body. Also, commanders should incorporate signature reduction techniques into local security plans. Area security preserves a commander's freedom to move reserves, position fire support assets, provide for mission command, and conduct sustaining operations. Local security provides immediate protection to the friendly force. (See ADRP 3-37 for more information on protection.)

5-69. All maneuver forces are capable of conducting security operations. All three types of Army BCTs—armored, infantry, and Stryker—have conduct security operations as part of their mission-essential task lists (METLs). Commanders ensure that subordinate units perform those specific security tasks required by the situation. Habitual support relationships with attachments and standard operating procedures are required to obtain proficiency in the conduct of these tasks.

5-70. Successful security operations depend on properly applying five fundamentals:
- Provide early and accurate warning.
- Provide reaction time and maneuver space.
- Orient on the force or facility to be secured.
- Perform continuous reconnaissance.
- Maintain enemy contact.

(See FM 3-90-2 and FM 3-98 for additional information on the conduct of security operations.)

TROOP MOVEMENT

5-71. The ability of a commander to posture friendly forces for a decisive or shaping operation depends on the commander's ability to move that force. The essence of battlefield agility is the capability to conduct rapid and orderly movement to concentrate combat power at decisive points and times. Successful movement places troops and equipment at their destination at the proper time, ready for combat. Transition from movement to maneuver occurs when enemy contact is expected.

5-72. *Troop movement* is the movement of troops from one place to another by any available means (ADRP 3-90). Troop movements are made by different methods; such as dismounted and mounted marches using organic combat and tactical vehicles; motor transport; and air, rail, and water means in various combinations. The method employed depends on the situation, the size and composition of the moving unit, the distance the unit must cover, the urgency of execution, and the condition of the troops. It also depends on the availability, suitability, and capacity of the different means of transportation. Troop movements over extended distances have extensive sustainment considerations. When necessary, dismounted and mounted marches can be hurried by conducting a forced march.

DISMOUNTED MARCHES

5-73. A *dismounted march* is movement of troops and equipment mainly by foot, with limited support by vehicles (FM 3-90-2). Also called foot marches, dismounted marches increase the commander's maneuver options. Their positive characteristics include combat readiness (all Soldiers can immediately respond to enemy attack without the need to dismount), ease of control, adaptability to terrain, and independence from the existing road network. Their limitations include a slow movement rate and increased personnel fatigue. Soldiers carrying heavy loads over long distances or large changes in elevation get tired. A unit conducts a dismounted march when the situation requires stealth, the distance to travel is short, transport or fuel is limited, or the terrain precludes using a large number of vehicles.

MOUNTED MARCHES

5-74. A mounted march is the movement of troops and equipment by combat and tactical vehicles. Armored and mechanized units routinely conduct mounted marches. The speed of the march and the increased amounts of supplies that can accompany the unit characterize this march method. Armored and Stryker maneuver units are normally self-sufficient to conduct mounted marches over limited distances. Light infantry maneuver units and most functional and multifunctional support and sustainment units are not completely mobile with organic truck assets, and they need assistance from transportation elements to conduct mounted marches. Considerations for mounted marches over extended distances include—

- The ability of the route network to support the numbers, sizes, and weights of the tactical and combat vehicles assigned to or supporting the unit making the move.
- Available refueling, maintenance, and rest areas.
- The need for maintenance recovery and evacuation assets.

ARMY AIR MOVEMENTS

5-75. Army air movements are operations involving the use of utility and cargo rotary-wing assets for other than air assaults. Commanders conduct air movements to move troops and equipment, to emplace systems, and to transport ammunition, fuel, and other high-value supplies. Commanders may employ air movements as a substitute for ground tactical movements. Air movements are generally faster than ground tactical movements, but they can be vulnerable to enemy air defense systems or influenced by bad weather. The same general considerations that apply to air assault operations also apply to Army air movements. (See ATP 3-04.1 for additional information concerning air movement.)

ARMY RAIL AND WATER MOVEMENTS

5-76. Operating forces can use rail and water modes to conduct troop movements, if they are available in an AO. Their use can provide flexibility by freeing other modes of transport for other missions. Their use normally involves a mixture of military and commercial assets, such as defense freight railway interchange railcars pulled by privately owned diesel-electric engines to transport tanks along railroad right-of-ways from one rail terminus to another. Responsibility for coordinating the use of railroads and waterways resides in the ARFOR headquarters in the theater of operations.

Large-Scale Combat Operations

FORCED MARCHES

5-77. In cases of tactical necessity, a unit can accelerate its rate of movement by conducting a forced march so that it arrives at its destination quickly. Armored, motorized, and infantry units can conduct forced marches. Forced marches require speed, exertion, and an increase in the number of hours marched or traveled by vehicles each day beyond normal standards. Soldiers cannot sustain forced marches for more than limited periods since units do not halt as often or for as long as is typically required for maintenance, rest, feeding, and fuel. Immediately following a long forced march, units may suffer a degradation in combat effectiveness and cohesion due to fatigue, supply, and maintenance issues. Plans for forced marches should account for stragglers, greater fuel consumption, and increased maintenance failures.

ADMINISTRATIVE MOVEMENT

5-78. Administrative movement is a movement in which troops and vehicles are arranged to expedite their movement and conserve time and energy when no enemy ground interference is anticipated. The commander only conducts administrative movements in secure areas. Examples of administrative movements include rail and highway movement in the continental United States. Once units deploy into a theater of war, commanders normally do not employ administrative movements. Since these types of moves are non-tactical, the echelon assistant chief of staff, logistics (G-4) or the battalion or brigade logistics staff officer (S-4) usually supervises them. (See ATP 4-16 for a discussion of route synchronization planning.)

RELIEF IN PLACE

5-79. A *relief in place* is an operation in which, by the direction of higher authority, all or part of a unit is replaced in an area by the incoming unit and the responsibilities of the replaced elements for the mission and the assigned zone of operations are transferred to the incoming unit (JP 3-07.3). (See figure 5-1. *Note.* The Army uses AO instead of a zone of operations.) The incoming unit continues the operation as ordered. A commander conducts a relief in place as part of a larger operation, primarily to maintain the combat effectiveness of committed units. The higher echelon headquarters directs when and where to conduct a relief and establishes the appropriate control measures. Normally, during the conduct of large-scale combat operations, the unit being relieved is defending. However, a relief may set the stage for resuming the offense. A relief may also serve to free the relieved unit for other tasks, such as

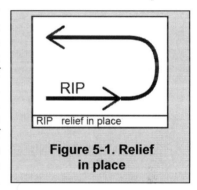

Figure 5-1. Relief in place

decontamination, reconstitution, routine rest, resupply, maintenance, specialized training, or redeployment. Sometimes, as part of a larger operation, a commander wants an enemy force to discover the relief, because that discovery might cause an enemy to do something in response that is contrary to its interests, such as move reserves from an area where the friendly commander wants to conduct a penetration.

5-80. There are three techniques for conducting a relief: sequentially, simultaneously, or staggered. A sequential relief occurs when each element within the relieved unit is relieved in succession, from right to left or left to right, depending on how the unit is deployed. A simultaneous relief occurs when all elements are relieved at the same time. A staggered relief occurs when a commander relieves each element in a sequence determined by the tactical situation, not its geographic orientation. Simultaneous relief takes the least time to execute, but is more easily detected by the enemy. Sequential or staggered reliefs can take place over a significant amount of time. These three relief techniques can occur regardless of the mission and operational environment (OE) in which the unit is participating.

5-81. A relief can be characterized as either deliberate or hasty, depending on the amount of planning and preparations associated with the relief. The major differences are the depth and detail of planning and, potentially, the execution time. Detailed planning generally facilitates shorter execution time by determining exactly what the unit commander believes needs to be done and the resources needed to accomplish the mission. Deliberate planning allows unit commanders and staffs to identify, develop, and coordinate solutions to most potential problems before they occur and to ensure the availability of resources when and where they are needed. (See FM 3-90-2 for additional information on the conduct of a relief in place.)

Chapter 5

PASSAGE OF LINES

The passage of lines is an operation designed to facilitate another tactical operation. The division's task organization supports the primary tactical mission. Centralized planning and execution characterize passage of lines operations.

FM 71-100-3, *Air Assault Division Operations*, 1996

5-82. *Passage of lines* is an operation in which a force moves forward or rearward through another force's combat positions with the intention of moving into or out of contact with the enemy (JP 3-18). A passage may be designated as a forward or rearward passage of lines. A commander conducts a passage of lines to continue an attack or conduct a counterattack, retrograde, or security operation when one unit cannot bypass another unit's position. The conduct of a passage of lines potentially involves close combat. It involves transferring the responsibility for an AO between two commanders. That transfer of authority usually occurs when roughly two thirds of the passing force has moved through the passage point. If not directed by higher authority, the two unit commanders determine, by mutual agreement, the time to transfer command. They disseminate this information to the lowest levels of both organizations.

5-83. There are several reasons for a commander to conduct a passage of lines. They include to—
- Sustain the tempo of the offense.
- Maintain the viability of the defense by transferring responsibility from one unit to another.
- Transition from a delay or security operation by one force to a defense.
- Free a unit for another mission or task.

The headquarters directing the passage of lines is responsible for determining when the passage starts and finishes.

5-84. A passage of lines occurs under two basic conditions. A *forward passage of lines* occurs when a unit passes through another unit's positions while moving toward the enemy (ADRP 3-90). A *rearward passage of lines* occurs when a unit passes through another unit's positions while moving away from the enemy (ADRP 3-90). Ideally, a passage of lines does not interfere with conducting the stationary unit's operations. (See FM 3-90-2 for additional information on passage of lines.)

ENCIRCLEMENT OPERATIONS

East and north of Bastogne, a line roughly three miles from the town that stalwart Soldiers of the 101st Airborne Division and assorted lesser units had established and held through the days of encirclement remained intact.

Center for Military History Publication 7-9-1, *The Last Offensive*, 1993

5-85. *Encirclement operations* are operations where one force loses its freedom of maneuver because an opposing force is able to isolate it by controlling all ground lines of communications and reinforcement (ADRP 3-90). A unit can conduct offensive encirclement operations designed to isolate an enemy force or conduct defensive encirclement operations as a result of the unit's isolation by the actions of an enemy force. Encirclement operations occur because combat operations involving modernized forces are likely to be chaotic, intense, and highly destructive, extending across large areas containing relatively few units as each side maneuvers against the other to obtain positional advantage. (See chapter 6 for a discussion of defending encircled and breakout from an encirclement. See chapter 7 for a discussion of offensive encirclement operations.)

MOBILITY OPERATIONS

5-86. Freedom to move and maneuver within an operational area is essential to the application of combat power and achieving results across the range of military operations. An OE will present numerous challenges to movement and maneuver. These are typically overcome through the integration of combined arms mobility and countermobility in support of mission requirements.

Large-Scale Combat Operations

BREACHING OPERATIONS

5-87. Breaching activities are conducted to allow maneuver despite the presence of enemy reinforcing obstacles that are covered by fire and used to shape engagement areas. Breaching is an inherent part of maneuver, and it is one of the most difficult combat tasks to perform. Breaching activities should be conducted only if bypassing an obstacle is not possible. Breaching activities are characterized by thorough reconnaissance, detailed planning, extensive preparation and rehearsal, and a massing of combat power.

5-88. A *breach* is a synchronized combined arms activity under the control of the maneuver commander conducted to allow maneuver through an obstacle (ATP 3-90.4). Breaching begins when friendly forces detect an obstacle and begin to apply the breaching fundamentals. Breaching ends when battle handover occurs between follow-on forces and the unit conducting the breach. Breaching establishes one or more lanes through a minefield and reduces other obstacles. Generally, breaching requires significant combat engineering support to accomplish. When planning a breach, the staff considers the fundamentals described by the memory aid SOSRA, which stands for—

- Suppress.
- Obscure.
- Secure.
- Reduce.
- Assault.

(See ATP 3-90.4 for a detailed discussion of each fundamental.)

5-89. Most combined arms breaching is conducted by a BCT or a battalion-size task force as a tactical mission, but corps and divisions may also execute combined-arms breaching tasks. Significant engineer augmentation from echelons above brigade is typically required to enable a BCT breach or a battalion task force hasty or deliberate breach.

5-90. Bulling or forcing through is not the preferred breaching technique. Bulling or forcing through is a decision made when a commander (based on a rapid risk assessment) must react immediately to extricate a force from an untenable position within an obstacle and no other breaching assets are available. When a force is in a minefield and it is receiving fires and taking heavy losses, the commander may decide immediately bulling through the minefield is the lesser risk to the force rather than a withdrawal or reducing the obstacle. When breaching, a unit develops a scheme of movement and maneuver specifically designed to neutralize the effects of the obstacle and continue its mission. Maneuver company teams, tasks forces, and BCTs conduct the breach. Normally, a task force executes a breach and the company teams are assigned as support, breach, and assault forces. Higher echelon units will conduct a breach when the force-allocation ratio indicates that a confirmed enemy situation is beyond subordinate unit capabilities. When a subordinate unit cannot successfully conduct a breach, the higher echelon unit will task organize to conduct the breach. It will also conduct a higher-level breach when a subordinate unit has failed in its attempt to breach enemy obstacles.

CLEARING OPERATIONS

5-91. Maneuver commanders may order clearing activities to facilitate mobility within an AO. They may order a critical route or area to be cleared of mines, explosive hazards, or other obstacles. The activity could be conducted as a single mission to open or reopen a route or area, or it may be conducted on a recurring basis in support of efforts to defeat an enemy's reestablishment of obstacles along critical routes.

5-92. *Clearing* is a mobility task that involves the elimination or neutralization of an obstacle that is usually performed by follow-on engineers and is not done under fire (ATP 3-90.4). It is generally accomplished by destroying, altering, or removing obstacles. Clearing of a route or an area is often performed by a combined-arms force built around an engineer-based clearance unit. Clearing is conducted when a commander orders enemy obstacles to be cleared within an assigned area or along a specified route. A route clearance may include a transition to an in-stride combined arms breach, if the encountered obstacles are covered by fire and an effective breach organization is available within the task organization or can be established. Once the in-stride breach is completed and the enemy's ability to interfere with route clearing has been neutralized, the clearing force reverts to the primary mission of route clearance. A clearing mission is not limited to reducing a lane or lanes through the identified obstacles as is the case in a breaching, since a clearing mission is focused on movement along the route or within an area rather than supporting the maneuver of a combat assault force. In route and area clearance, engineer and EOD units destroy or remove explosive obstacles that are a threat to mobility along the route or within the specified area. Routes and areas cleared are considered cleared only if they remain controlled by friendly forces. (See ATP 3-90.4 for additional information on the performance of the clearing task.)

GAP-CROSSING OPERATIONS

5-93. Gap crossing in support of maneuver is similar to a breach in that the force is vulnerable while moving through a lane or across a gap. Maneuver units are forced to break movement formations, concentrate within lanes or at crossing points, and reform on the far side before continuing to maneuver. While much of the terminology and planning associated with gap crossing is the same as that used in a breach, gap crossing and breach differ in scope. The amount and type of assets involved also differ. Combined-arms breaching sometimes includes gap crossings as a reduction method. Since the primary focus of planning and preparation is on the breach, they are typically discussed as a subordinate part of the breach rather than as a separate gap crossing. Wet-gap crossings tend to be more complex and demanding than dry-gap crossings.

5-94. The gap crossing types are deliberate, hasty, and covert. Each gap crossing type has a general list of conditions that help define its category. The planning requirements for each type of gap crossing are similar. However, the required degree of detail and necessary conditions for a high degree of success will vary based on the type and the unique features associated with a crossing mission. In all cases, the ability to conduct any type of crossing begins by providing a crossing force with the necessary gap-crossing means and control elements and identifying those requirements early during planning. (See ATP 3-90.4 for additional information on gap crossing in support of maneuver and movement.)

COMBAT ROADS AND TRAILS

5-95. Combat roads and trails are built primarily as a hasty supplement or bypass to an existing route that is impacted by natural or man-made obstacles. They fulfill immediate requirements to move personnel, equipment, and supplies throughout an AO to achieve the commander's intent despite terrain restrictions.

5-96. A combat road is a traveled way that has been cleared of obstacles and temporarily surfaced with material (usually gravel or crushed rock) or by an expedient means to increase its trafficability. Combat roads usually do not have an asphalt or concrete surface. Combat roads are generally needed to support wheeled vehicles unsuited for rugged terrain. Combat roads require more effort to build than trails, but they support a broader range of vehicles and tend to last longer. Building combat roads (and trails) is a combat engineering task that is conducted in close support to ground maneuver forces that are in close combat. Higher-level roadwork is a general engineer task (see ATP 3-34.40/MCWP 3-17.7) and the specific application of that level of road construction is detailed in TM 3-34.48-1. The primary difference between combat roads and conventional roads is the degree of permanence and the characteristics of the traffic they are designed to support. Combat roads are built to handle low volumes of traffic for a short time.

5-97. A combat trail is a traveled way that has been cleared of obstacles, but has not been temporarily surfaced. A trail may be roughly graded by combat earthmoving equipment to provide a relatively smooth surface. Combat trails are usually adequate for tracked and wheeled combat vehicles.

5-98. A commander's requirement for combat roads or trails is generally determined by the characteristics of the intended traffic, the duration of the requirement, and the inherent trafficability of the existing ground surface based on its soil characteristics and weather conditions. In flat, desert areas with dry conditions, for example, once a pathway is cleared of obstacles and marked, the resulting trail can usually support a low volume of most types of vehicular traffic for a short time without any additional effort. (See ATP 3-90.4 for additional information on the construction and maintenance of combat roads and trails.)

FORWARD AVIATION COMBAT ENGINEERING

5-99. Combat engineering in support of aviation operations includes the combat engineering tasks performed specifically to support the air movement of personnel, equipment, and supplies within an operational area, especially in complex terrain or when ground lines of communications have been disrupted. Forward aviation combat engineering also includes those tasks that support forward aviation operating facilities, including forward arming and refuel points and tactical UAS launch and recovery strips. Forward aviation operating facilities shorten the distance between an aviation unit's projected area of commitment and its sustainment area, reducing turnaround times. Forward aviation combat engineering also enhances the availability and responsiveness of aviation assets.

5-100. Combat engineering in support of aviation operations is a combat engineering mission; however, general engineering units may be required to perform or augment combat engineers that are performing this task, depending on the situation. The BCT organic engineering capabilities require augmentation to meet the full range of aviation support requirements, especially the requirements that will likely be more permanent as the phases of an operation progress. Combat engineering in support of aviation operations includes these tasks—

- Construct landing zones.
- Construct extraction zones.
- Construct, maintain, and repair manned and unmanned landing strips.
- Construct, maintain, and repair forward aviation operating facilities.

5-101. Combat engineering in support of aviation operations includes requirements for the hasty construction or rehabilitation of aviation facilities that support fixed- and rotary-wing aircraft as well as UASs. It includes the requirements to improve drop zones and extraction zones for airborne insertions and aerial (airdrop) delivery. The considerations associated with performing forward aviation combat engineering are similar to those associated with the construction of combat roads and trails. Both are based on expedient horizontal construction techniques and are considered combat engineering missions, although they can be conducted by units that are designed to perform general engineering tasks.

5-102. Landing zones allow aircraft to land and deliver troops, equipment, and logistics. The combat engineering applied to landing zones can range from clearing trees from the rotor zone to placing expedient surface material to facilitate surface stability and to control dust.

5-103. Sustaining ground forces may require aerial resupply by means other than air landing and airdrops. Extraction zones are specifically used to support low-altitude aerial deliveries that use the low-altitude parachute extraction system. Extraction zones require relatively flat, stump-free terrain with measurements similar to those specified for a C-130 landing strip. Since the fixed-wing transport aircraft, typically a C-130, does not land when discharging its cargo, the ground strength requirements are based on the equipment being discharged. Information on ground strength is provided in TM 3-34.48-1. Extraction zone criteria and marking requirements are described in TM 3-34.48-2.

5-104. The landing strip (also referred to as a flight strip) is the portion of an airfield that includes the landing area, the end zones, and the shoulder areas. A landing strip is designed, constructed, and maintained to support the landing and takeoff of specified fixed-wing aircraft and UASs. Landing strips must be relatively flat with a surface capable of supporting fully loaded aircraft. Proper site selection is essential in minimizing the engineering effort needed for meeting the design and construction criteria prescribed in TM 3-34.48-2.

5-105. Combat engineers construct, maintain, and repair forward aviation operating facilities, including landing zones, initial heliports and helipads, and forward arming and refueling points. These aviation facilities are positioned to extend the tactical reach of attack, lift, reconnaissance, and unmanned aviation assets and optimize the time on target. Maintenance includes the activities needed to correct deficiencies resulting from normal use and deterioration due to weather, and repair items with damage due to abnormal use, accidents, severe weather, and threat actions. (See ATP 3-90.4 for additional information on forward aviation combat engineering.)

COUNTERMOBILITY OPERATIONS

5-106. *Countermobility operations* are those combined arms activities that use or enhance the effects of natural and man-made obstacles to deny enemy freedom of movement and maneuver (ATP 3-90.8). The primary purposes of countermobility operations are to shape enemy movement and maneuver and to prevent the enemy from gaining a position of advantage. Countermobility operations include the following tasks:

Chapter 5

- Site construction.
- Construct, emplace, or detonate obstacles.
- Mark, report, and record obstacles.
- Maintain obstacle integration.

5-107. In support of offensive tasks, countermobility operations are conducted to isolate objectives and prevent enemy forces from repositioning, reinforcing, and counterattacking. They are also conducted to enable flank protection as combat operations progress into the depth of enemy defenses or as an integrated economy-of-force effort to provide general flank security. The commander's options for emplacing reinforcing obstacles in the offense are often limited because of challenges in gaining early access to much of an AO. This increases the importance of taking advantage of existing obstacles and the natural restrictiveness of the terrain. It also increases the commander's reliance on rapid obstacle emplacement capabilities, especially remotely delivered obstacles to reinforce the terrain.

5-108. In support of defensive tasks, countermobility operations are conducted to disrupt enemy attack formations and assist friendly forces in defeating an enemy in detail, to channel attacking enemy forces into engagement areas throughout the depth of the defense, and to protect the flanks of friendly counterattack forces. They shape engagements, maximize the effects of fires, and provide close-in protection around defensive positions to help defeat an enemy's final assault and to prevent and warn of intrusion into critical fixed sites such as base camps and sustainment sites. Erecting obstacles is also a critical supportive task to conducting denial operations.

5-109. In addition to the rapid obstacle emplacement capabilities that are often used in the offense, commanders can also use more time and resource-intensive techniques to reinforce the terrain for longer time periods. Such techniques primarily include constructed obstacles and some demolition obstacles. Ensuring that these tasks are conducted efficiently and effectively is critical to success. Additionally, the prioritization of tasks is essential in resolving the competing demands of mobility, countermobility, and survivability tasks and ensuring that the most important tasks are accomplished first.

5-110. The conduct of countermobility operations typically involves engineers and includes proper obstacle integration with the maneuver plan, adherence to obstacle emplacement authority, and positive obstacle control. Combined arms obstacle integration synchronizes countermobility operations into the concept of operations. Commanders must ensure that obstacles are integrated with observation and fires and fully synchronized with the concept of operations to avoid hindering any friendly-force mobility. Because most obstacles have the potential to deny the freedom of movement and maneuver to friendly forces and enemy forces, it is critical that commanders properly weigh the risks and evaluate the trade-offs of employing various types of obstacles. (See ATP 3-90.8 for more information about countermobility.)

SECTION III – FORCIBLE ENTRY

Airborne forces execute parachute or airlanded assaults to seize and hold important objectives until ground linkup or withdrawal can be accomplished, or until reinforced by air or amphibious landing.

FM 57-1, *Airborne Operations*, 1967

5-111. *Forcible entry* is the seizing and holding of a military lodgment in the face of armed opposition or forcing access into a denied area to allow movement and maneuver to accomplish the mission (JP 3-18). A forcible entry operation may be the JFC's opening move to seize the initiative. For example, a JFC might direct friendly forces to seize and hold an airhead or a beachhead to facilitate the continuous landing of troops and materiel and expand the maneuver space needed to conduct follow-on operations. Forcible entry operations during the dominate phase of an operation or campaign may be used for a coup de main, conducting operational movement and maneuver to attain positional advantage, or as a tactical deception.

5-112. Army forces, as part of a joint force, must be capable of deploying and fighting to gain access to geographic areas controlled by forces hostile to U.S. national interests to be credible both as a deterrent and as a viable military option for policy enforcement. Swift and decisive victory in these cases requires forcible entry and the ability to surge follow-on forces.

5-113. Commanders design their forcible entry operations to seize and hold a lodgment against armed opposition. A *lodgment* is a designated area in a hostile or potentially hostile operational area that, when seized and held, makes the continuous landing of troops and materiel possible and provides maneuver space for subsequent operations (JP 3-18). A lodgment may be an airhead, a beachhead, or a combination thereof. A lodgment area is large enough to provide maneuver space for subsequent operations. A force defends the perimeter of a lodgment until it has sufficient forces to break out of the lodgment and conduct offensive tasks. Forcible entry operations are inherently joint. Forcible entry demands careful planning and thorough preparation; synchronized, violent, and rapid execution; and leader initiative at every level to deal with friction, chance, and opportunity. There are four types of forcible entry operations: amphibious, airborne, air assault, and cross border. However, only the first three have specialized doctrine.

> **Forcible Entry: OPERATION JUST CAUSE**
>
> During OPERATION JUST CAUSE, the armed forces of the United States rapidly assembled, deployed, and conducted a forcible entry operation. The well-tailored force involved in this operation simultaneously seized multiple key targets in Panama, virtually eliminating organized resistance in the space of a few hours. The operation demonstrated the capability of the U.S. military to project forces rapidly against opposition while synchronizing multiple elements of combat power.
>
> In the mid-1980s, diplomatic relations began to deteriorate between the leader of Panama, Manuel Noriega, and the U.S. In the late 1980, as relations deteriorated, Noriega appeared to shift his Cold War allegiance from the U.S. towards the Soviet bloc, soliciting and receiving military aid from Cuba, Nicaragua, and Libya. By December 1989, Noriega declared that a state of war existed between the U.S. and Panama and threatened the lives of the approximately 35,000 U.S. citizens living in Panama. There were numerous clashes and confrontations between U.S. and Panamanian forces; one U.S. Marine had been killed. American military planners began preparing contingency plans to invade Panama.
>
> Late on 19 December 1989, a joint force of 7,000 Soldiers, Sailors, Airmen, and Marines deployed from U.S. bases bound for Panama. During the early morning hours of 20 December, this force—supported by United States Southern Command forward-deployed forces in Panama—simultaneously attacked targets at 26 separate locations. The success of the attack against key Panamanian Defense Force strongholds required a sequence of stealthy moves by an assortment of U.S. special operations forces and elements from the 82d Airborne Division, the 5th Mechanized Division, the 7th Infantry Divisions, and U.S. Marine Corps. These were supported by the USAF and U.S. Navy in a variety of ways, including airlift and sealift, suppression of enemy air defenses, and AC-130 gunship strikes. Mission orders, decentralized execution, and individual ingenuity contributed greatly to the ability of the joint force to rapidly paralyze the Panamanian Defense Force's response capability.

5-114. Operational access is the desired condition that the United States seeks to maintain in areas of strategic importance to achieve assured access. Army forces, as part of joint forces, project forces and capabilities into an operational area. They conduct operations to defeat enemy A2 and AD capabilities and establish security conditions and control of territory. This preserves the force's freedom of movement and action for follow-on operations or denies the enemy use of an area. Forcible entry operations are complex and always involve taking prudent risk to gain a position of relative advantage over an enemy. (JP 3-18 establishes joint doctrine for forcible entry. JP 3-02 establishes joint doctrine for amphibious operations. FM 3-99 establishes doctrine for airborne and air assault operations.)

SECTION IV – TRANSITION TO CONSOLIDATE GAINS

In the case of a state that is seeking not conquest but the maintenance of its security, the aim is fulfilled if the threat be removed—if the enemy is led to abandon his purpose.

Sir Basil H. Liddell Hart

5-115. Army forces provide the JFC the ability to capitalize on operational success by consolidating gains. Consolidate gains is an integral part of winning armed conflict and achieving success across the range of military operations. It is essential to retaining the initiative over determined enemies and adversaries. Army forces reinforce and integrate the efforts of all unified action partners when they consolidate gains.

5-116. Army forces consolidate gains in support of a host nation and its civilian population, or as part of the pacification of a hostile state. These gains may include the establishment of public security temporarily by using the military as a transitional force, the relocation of displaced civilians, reestablishment of law and order, performance of humanitarian assistance, and restoration of key infrastructure. Concurrently, corps and divisions must be able to accomplish these activities while sustaining, repositioning, and reorganizing subordinate units to continue operations in the close area.

5-117. Upon successful termination of large-scale combat operations, Army forces in the close area transition rapidly to the conduct of consolidation of gains activities. Alternatively, they may be relieved in place by another unit. Consolidation of gains activities may encompass a lengthy period of post conflict operations prior to redeployment. This transition to consolidation of gains may occur even if large-scale combat operations are occurring in other parts of an AO in order to exploit tactical success. Anticipation and early planning for activities after large-scale combat operations ease the transition process.

5-118. The JFC defines the conditions to which an AO is to be stabilized. The theater army is normally the overseer of the orderly transition of authority to appropriate U.S., international, interagency, or host-nation agencies. The theater army and subordinate commanders emphasize those activities that reduce post-conflict or post-crisis turmoil and help stabilize a situation. Commanders address the decontamination, disposal, and destruction of war materiel. They address the removal and destruction of unexploded ordnance and the responsibility for demining operations. (The consolidation of friendly and available enemy mine field reports is critical to this mission.) Additionally, the theater army must be prepared to provide Army Health System support, emergency restoration of utilities, support to social needs of the indigenous population, and other humanitarian activities as required. (See ADRP 3-07 and FM 3-07 for more information on the performance of stability tasks.)

Chapter 6
Large-Scale Defensive Operations

In war, the defensive exists mainly that the offensive may act more freely.
Alfred Thayer Mahan

This chapter begins with a general discussion of the defense, followed by a discussion of how an enemy may attack. It continues with sections on planning and preparing corps and division defenses. It then addresses the three primary defensive tasks and concludes with a discussion on defending encircled.

OVERVIEW OF LARGE-SCALE DEFENSIVE OPERATIONS

6-1. A *defensive task* is a task conducted to defeat an enemy attack, gain time, economize forces, and develop conditions favorable for offensive or stability tasks (ADRP 3-0). There are three primary defensive tasks: area defense, mobile defense, and retrograde. However, the performance of defensive tasks alone normally cannot achieve a decision. Their purpose is to create conditions for a counteroffensive that allows Army forces to regain the initiative. Other reasons for performing defensive tasks include—
- Retaining decisive terrain or denying a vital area to an enemy.
- Conducting defensive covering force operations to protect forces flowing into theater.
- Attriting or fixing an enemy as a prelude to offensive tasks.
- Countering an enemy attack.
- Increasing enemy vulnerabilities to force the enemy commander to concentrate subordinate forces.

6-2. The defense is what provides time for a commander to build combat power and establish conditions to transition to the offense. The defense is ideally a shield behind which a commander maintains or regains the initiative. Initially, a defending commander is likely to be at relative disadvantage against an attacking enemy, since that enemy can choose when and where to strike. Significant capability gaps in terms of fires, including air and missile defense (AMD), countermobility, protection, and aviation may exist early on in any campaign. Also, joint fires may not be available initially in sufficient quantity, or the enemy may have dominance in one or more of the domains that limits joint capabilities.

6-3. The inherent strengths of the defense include the defender's ability to occupy positions before an attack and use the available time to prepare those defenses. A defending force stops improving its defensive preparations only when it retrogrades or begins to engage an enemy. During combat, a defending force takes the opportunities afforded by lulls in an action to improve its positions and repair combat damage. A defender does not wait passively to be attacked. A defender aggressively seeks ways of attriting and weakening attacking enemy forces before close combat begins. A defender maneuvers to place an enemy in a position of disadvantage and attacks that enemy at every opportunity, using fires, electronic warfare (EW), aviation, information-related capabilities (IRCs), cyberspace operations, and obstacles, as well as joint assets, such as air interdiction, close air support (CAS), and special operations forces (SOF). The static and mobile elements of a defense combine to deprive an enemy of the initiative. A defender contains an enemy while seeking every opportunity to transition to the offense.

6-4. Characteristics of the defense include disruption, flexibility, maneuver, mass and concentration, operations in depth, preparation, and security. (See ADRP 3-90 for a discussion of these seven characteristics.) The performance of effective defensive tasks capitalizes on accurate and timely intelligence and other relevant information regarding enemy forces, weather, terrain, and civilian locations. A defender deliberately seeks opportunities to disrupt and attrit an attacker throughout the depth of the enemy formation to establish the conditions for the decisive operation.

ENEMY ATTACK

No plan survives contact with the enemy.

Field Marshal Helmuth von Moltke the Elder

6-5. Operationally, the defense buys time, economizes forces, and develops conditions favorable for resuming the offense. It prevents enemy forces from achieving their goals and meeting their operational objectives. The British defeat in Malaya is a historical example of a poorly executed defense at both the operational and tactical levels. It also is an example of a peer threat exploiting superior planning, leadership, and aggressive offensive action across multiple domains while overcoming numerical inferiority against the defending British force.

> ### Faulty Assumptions—Defeat Across Multiple Domains: Malaya and Singapore
>
> On 8 December 1941, Imperial Japanese forces launched the invasion of Malaya from their strongholds in French Indochina. Japanese intelligence suggested that LTG Yamashita's 70,000 troops were outnumbered nearly 2 to 1, so he concluded that only a bold, rapidly executed attack with aggressive, well trained troops would ensure victory. The British Empire was decisively engaged in North Africa, the battle of the Atlantic, and the defense of the home islands. While it deployed additional ground, naval, and air units to Singapore, the combination of time, distance, and demands in other theaters made further reinforcement unlikely. Those forces that arrived before the commencement of hostilities had little time to train or prepare for operations against an opponent of unknown quality in an unfamiliar environment.
>
> The allied defense of Malaya rested on several false assumptions. First, allied forces assumed there would be sufficient warning of an attack, which in turn, would allow for adequate air and naval reinforcements. The few major naval reinforcements (the battleship Prince of Wales and battle cruiser Repulse) were sunk, and the aircraft rapidly lost against a qualitatively superior Japanese air arm. The Allies assumed that the dense jungle terrain would impede enemy ground movement and simplify defense of the peninsula; but it actually hindered the defenders more than the Japanese. The Allies assumed the enemy main effort would come from the sea, which was the most heavily fortified part of their defense. It did not. They assumed their air force to be superior to that of the Japanese. It was not. They also assumed that the British and Commonwealth troops were better trained than the Japanese Army. They were not. These beliefs cost the Allies dearly.
>
> The Japanese launched a successful night amphibious assault on the northern coasts of Malaya and advanced south along the eastern coast. They also made simultaneous landings in Thailand, just north of the Thailand-Malayan border, and sent units south into the interior and along the west coast. The Allies attempted to slow the Japanese advance with prepared positions along the roads and by destroying bridges, but the Japanese repeatedly bypassed, turned, or enveloped static positions by moving off the roads. On 10 December, Japan sank the last major allied naval units and achieved air supremacy. The Japanese were then able to land troops at will, allowing them to repeatedly turn the allied defenders out of their battle positions.

Large-Scale Defensive Operations

Faulty Assumptions—Defeat Across Multiple Domains: Malaya and Singapore, continued

By 27 January 1942, allied forces had retreated across the strait onto the island of Singapore. Japan subsequently invaded Singapore on 7 February and completed its conquest on the 15th. (See figure 6-1.) In just over two months, Yamashita's 25th Army of 70,000 troops decisively defeated a defending force of over 140,000 soldiers at the cost of just under

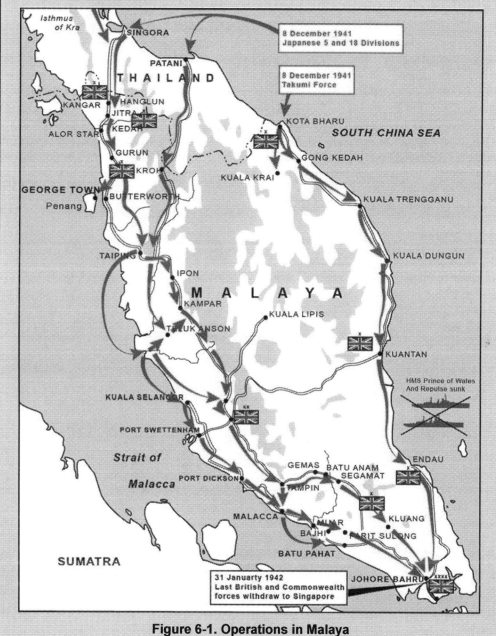

Figure 6-1. Operations in Malaya

> **Faulty Assumptions—Defeat Across Multiple Domains: Malaya and Singapore, continued**
>
> 10,000 casualties. Yamashita's use of speed, maneuver, and surprise allowed him to dictate the operational tempo of the invasion. He ably exploited opportunities provided by the Japanese air and naval forces while generating his own with aggressive maneuver on the ground. The simultaneous presentation of multiple dilemmas across the air, maritime, and land domains led to the largest surrender of British forces in history.

6-6. Enemy tactics, equipment, capabilities, and probable courses of action all inform planning a defense. Defending commanders must see the terrain and their own forces' capabilities and dispositions from the enemy perspective to fully understand weaknesses that an enemy could exploit and act to counter them. Defending commanders identify probable enemy objectives and possible enemy approaches to achieve them. Identifying enemy limitations and constraints determines where and when to exploit positions of relative advantage.

6-7. During offensive operations, enemy forces typically attempt to mask the location of their main effort with multiple fixing attacks on the ground while using fires to disrupt critical friendly nodes (for example command posts [CPs], radars, and fire direction centers) and isolate friendly forward units. Generally, enemy forces seek to reinforce success, massing capabilities at a vulnerable point to achieve large force ratio advantages to enable a rapid penetration of friendly defenses. The enemy uses mobile forces to exploit the penetration rapidly to the maximum possible depth in order to make the overall friendly defensive posture untenable. Threat forces can have advantages in both volume and range of fires, so they can simultaneously mass fires on the point of penetration to enable rapid closure and breakthrough, fix other friendly elements along the forward line of own troops (FLOT), and target key friendly mission command and logistics nodes along the depth of the defense. Threat forces prefer to use fires to move around fixed positions when possible and through destroyed units when necessary. Threat forces seek to maneuver tactically to a depth that achieves operational objectives in support of his overall strategic purpose. Threat forces will employ intelligence, surveillance, and reconnaissance; EW; information warfare; SOF, and all other capabilities at their disposal. These are likely to include chemical weapons. Figure 6-2 shows a typical enemy disruption force in the attack. Figure 6-3 on page 6-6 shows typical enemy fixing and breaching a brigade combat team (BCT) defense. Figure 6-4 on page 6-7 illustrates an enemy exploitation force. Figure 6-5 on page 6-8 illustrates the organization of an army-level integrated fires command.

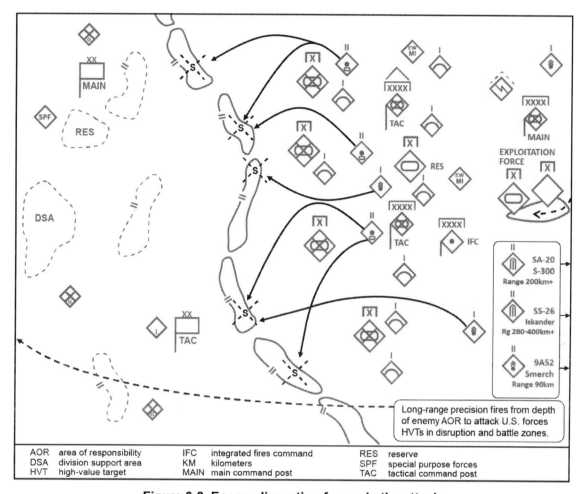

Figure 6-2. Enemy disruption forces in the attack

Chapter 6

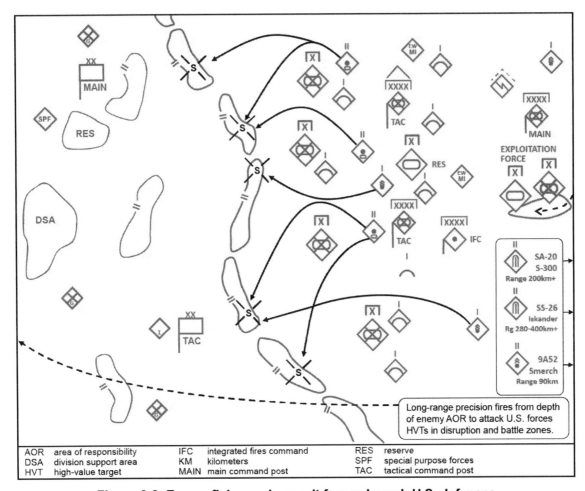

Figure 6-3. Enemy fixing and assault forces breach U.S. defenses

Large-Scale Defensive Operations

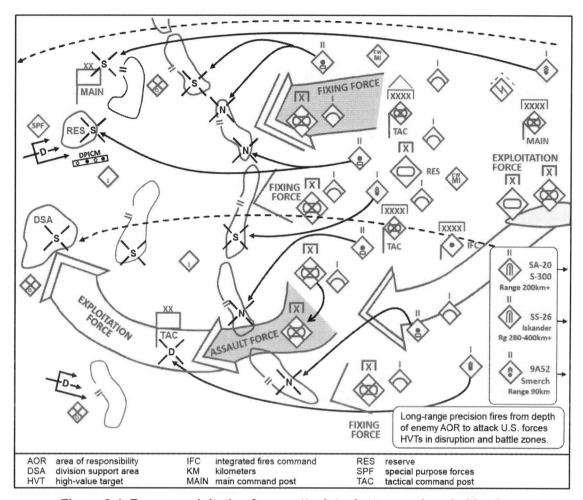

Figure 6-4. Enemy exploitation forces attack to destroy assigned objectives

Chapter 6

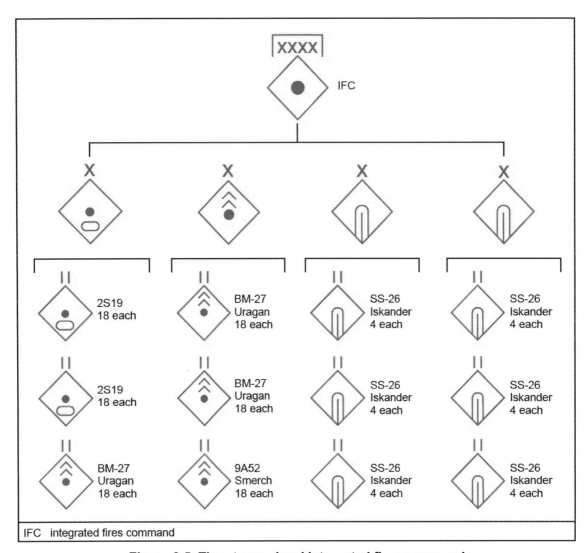

Figure 6-5. Threat army-level integrated fires command

PLANNING CORPS AND DIVISION DEFENSIVE TASKS

In preparing for battle, I have always found that plans are useless, but planning is indispensable.

President Dwight D. Eisenhower

6-8. Defending corps and division commanders seek to push the enemy off-balance in multiple domains and the information environment when that enemy initially has the initiative. The joint task force (JTF), land component, or ARFOR commander sets the stage for planning the conduct of corps and division defensive tasks. One of these operational commanders assigns the corps or division an area of operations (AO) and provides the corps or division commander a clear commander's intent and a concept of operations. Likewise, the corps or division commander assign subordinate AOs and provide the corps or division's attached, operational control (OPCON), or tactical control (TACON) subordinate division or brigade commanders a clear intent and concept of operations. This guidance is how a corps or division commander initiates planning for the conduct of the defense. (See ADRP 5-0 for doctrine on planning and the operations process.)

6-9. The corps or division commander's concept of operations will normally reflect the operational framework used by the higher echelon commander. (Chapter 1 discusses the operational framework.) Corps

Large-Scale Defensive Operations

and division commanders organize their defensive plans based on the orders of their higher echelon commanders and the situation in the AO. They identify and war-game possible enemy reactions for inclusion in those plans. Branch plans to the defensive plan enable subordinate commanders and staffs to remain proactive and ready for possible future situations.

6-10. The key to a successful corps or division defense is the orchestration and synchronization of combat power across all available domains and the information environment to converge effects. Commanders decide where to concentrate combat power and where to accept risk as they establish engagement areas (EAs). (See figure 6-6) Success may require that a defending unit exploit opportunities to seize the initiative that do not risk the integrity of the defense, such as a spoiling attack or counterattack.

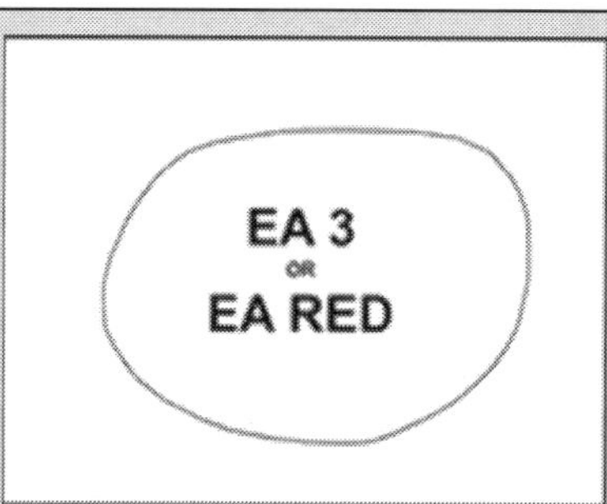

6-11. Time is often the most important resource for defending forces. The enemy chooses the time and location for attack, so the amount of time friendly units have to prepare a defense is often unknown or inadequate. Defending corps and divisions need time to complete their planning, coordinating, information collection, and rehearsals. Their subordinate units need time to develop EAs by preparing battle positions, pre-positioning sustainment assets, and emplacing obstacles. (Figure 6-8 on page 6-11 illustrates obstacle control measures.)

6-12. Defending commanders try to gain adequate time to prepare an effective defense. A corps or division commander may task-organize and resource a forward security force to perform the task of guard or cover to give main battle area (MBA) forces additional preparation time. Lack of time could cause a commander to maintain a larger reserve.

Figure 6-6. Engagement area

An *engagement area* is an area where the commander intends to contain and destroy an enemy force with the massed effects of all available weapons and supporting systems (FM 3-90-1). EAs can be designated by either numbers or names.

SCHEME OF MANEUVER IN THE DEFENSE

6-13. A successful defense requires the integration and synchronization of all available assets. The defending commander assigns missions, allocates forces (including the reserve), and apportions functional and multifunctional support and sustainment resources within the construct of decisive, shaping, and sustaining operations in an operational framework of deep, close, support, and consolidation areas (if applicable). The commander determines where to concentrate defensive efforts and where to take risks based on the results of the intelligence preparation of the battlefield (IPB) process. Commanders strive to defeat enemy attacks across each relevant domain. Commanders can rapidly redirect manned or unmanned attack reconnaissance aviation and artillery systems initially allocated to shaping operations to support the decisive operation at the appropriate time. Commanders organize forces differently for contiguous and noncontiguous areas of operations.

6-14. The *main battle area* is the area where the commander intends to deploy the bulk of the unit's combat power and conduct his decisive operations to defeat an attacking enemy (ADRP 3-90). In the example in figure 6-7 on page 6-10, the MBA is also the division close area. Corps and division defensive planning normally calls for the decisive operation to culminate in the MBA with the attacking enemy's defeat. The plan allows the corps or division to shift and synchronize combat power where necessary to reinforce MBA units. Spoiling attacks and counterattacks designed to disrupt the enemy and to prevent the enemy from massing or exploiting success are part of MBA operations. The headquarters future operations and plans integrating cells conduct contingency planning to counter potential enemy penetrations of forward defenses within the MBA.

Chapter 6

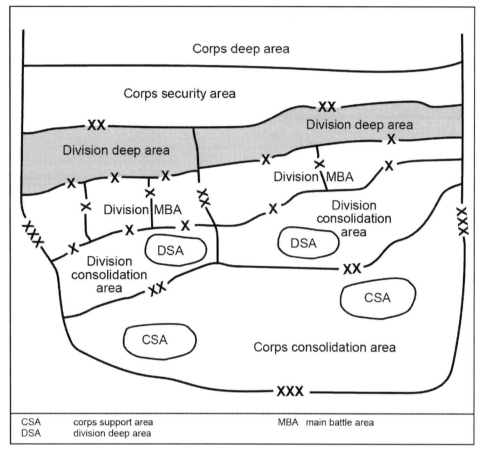

Figure 6-7. Main battle area

6-15. Commanders have several considerations when planning a defense. The keys to a successful defense are—

- Timely detection of the enemy's course of action.
- Concentrating effects at the decisive time and place.
- Depth.
- Security (forward, flank, area, and local security in support areas).
- Ability to shape and exploit terrain advantageously.
- Flexibility.
- Designation, composition, location, and employment of the reserve.
- Timely resumption of the offense.

Large-Scale Defensive Operations

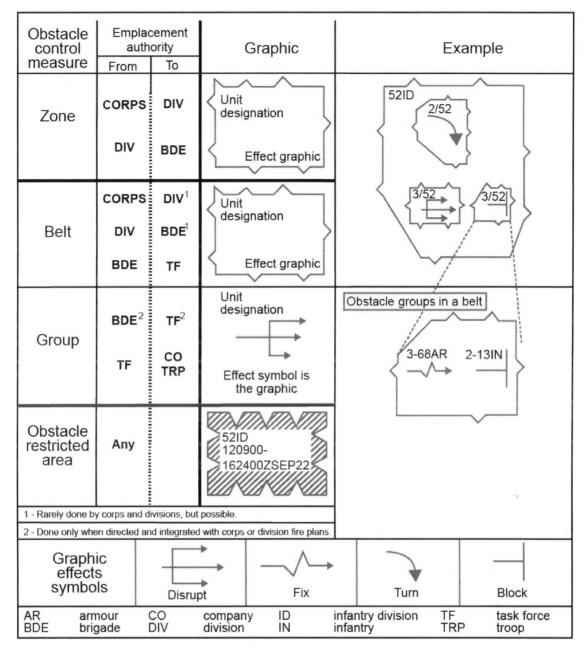

Figure 6-8. Obstacle control measures

6-16. The defense concept requires that units in defensive positions accomplish their mission independently or in combination by defeating the enemy by fire, absorbing the strength of the attack within the position, or destroying the enemy with a local or major counterattack. Commanders combine the advantages of fighting from prepared positions, obstacles, planned fires, and counterattacks to isolate and overwhelm selected enemy formations. Commanders must be prepared to rapidly shift the nature and location of their main efforts, repositioning units to mass fires against the attacker to prevent breakthroughs or preserve the integrity of the defense.

6-17. A defensive plan designates axes of advance and routes for the commitment or movement of forces, or the forward or rearward passage of one unit through another. It should identify air movement corridors and other airspace coordinating measures to enhance the aerial maneuver of attack reconnaissance helicopters, unmanned aircraft systems (UASs), air assault units, or fixed-wing aircraft. The operations process identifies decision points associated with the initiation of counterattacks, repositioning of forces, commitment of the reserve, execution of situational obstacles, and other actions. The ability to dynamically reposition is dependent on the defending force having superior tactical mobility. Without tactical mobility, defending forces may be forced to remain in prepared positions and accept the possibility of being fixed and bypassed or encircled.

6-18. A commander assigning the defensive mission defines the area to defend. A commander defending on a broad front may be forced to accept gaps and conduct noncontiguous operations. The FLOT will not be contiguous. (See figure 6-9.) Defending shallow AOs reduces flexibility and requires the commander to fight well forward. Narrow frontages and deep AOs increase the elasticity of a defense by increasing the commander's maneuver options.

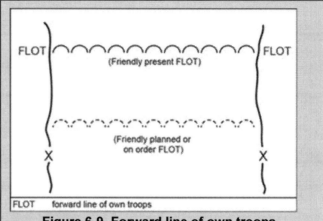

Figure 6-9. Forward line of own troops

The *forward line of own troops* is a line that indicates the most forward positions of friendly forces in any kind of military operation at a specific time (JP 3-03).

6-19. The ideal defense is one where effective mutual support exists throughout the width and depth of the defender's tactical positions across multiple domains. The commander considers physical, temporal, virtual, and cognitive factors when assigning positions to subordinate units. This occurs whether the defending commander employs a defense in an AO, defends by battle position or employs a combination of both. (See figure 6-10.) The defending unit maintains tactical integrity within each type defensive area. A unit conducting an area defense normally addresses the security requirements of each flank by assigning responsibility to a subordinate element or organizing a separate security force to specifically perform flank guard or screen tasks to provide early warning, reaction time, and maneuver space to the defending commander.

6-20. Commanders take transitions into consideration when considering how to shape their AO. For example, targeting a bridge for destruction in the defense should be weighed against the need for the bridge in future offensive operations or during the consolidation of gains.

Planning Operations in Depth in the Defense

6-21. A corps or division commander employs fires to neutralize, suppress, or destroy enemy forces. Fires delay or disrupt a targeted enemy force's capability to execute a given course of action. A corps or division commander directs the establishment of permissive and restrictive fire support coordination measures (FSCMs) that support the course of action.

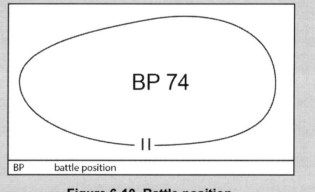

Figure 6-10. Battle position

A *battle position* is a defensive location oriented on a likely enemy avenue of approach (ADRP 3-90). There are five types of battle positions—primary, alternate, supplementary, subsequent and strong point.

The effective planning and use of FSCMs enables rapid clearance of fires and strikes with the required simultaneity to mass combined lethal and nonlethal effects.

6-22. Commanders determine what cross-domain effects would best support the defense and request those capabilities that are not organic from higher echelons. They anticipate which capabilities are likely to be

Large-Scale Defensive Operations

available and ensure that the viability of the defense does not depend entirely upon capabilities they do not control.

6-23. The headquarters fires cell plans the employment of available Army and joint fires to achieve depth and simultaneity and secure advantages for future operations. This is one of the primary means by which a commander shapes the AO for subordinate units. The headquarters staff helps the commander integrate indirect fires, electronic attacks, aviation maneuver, and joint fires into the defensive plan. This allows the defending commander to degrade the combat capabilities of advancing enemy forces before they enter close combat operations. This includes disrupting the enemy's approach to and movement within the MBA, destroying high-payoff targets (HPTs), denying or interrupting vital components of enemy operating systems, and using obscurants to conceal friendly movement. Some key HPTs are the enemy's trailing or reserve units, air defense sites, fire direction centers, key enemy command and control (C2) nodes, and key infrastructure, such as bridges over major rivers. Attacks against infrastructure must be deliberate and they are essential to the success of the defensive plan because of infrastructure's importance to consolidation of gains.

6-24. The commander's intent and concept of operations may be to use available fires to defeat, deter, or delay an enemy before major enemy forces come into direct fire range of the BCTs located within the MBA. Alternatively, the commander's concept may be for field artillery cannon and rockets to conduct suppression of enemy air defenses (suppression of enemy air defenses) to enable combat aviation brigade (CAB) attack reconnaissance assets and joint fires to delay or disrupt the approach of enemy second echelon or reserve forces. This allows those MBA BCTs to complete their defeat of the enemy's initial attack with their organic assets before enemy second echelon or reserve forces join close combat operations. The defending commander directs the delivery of effects in multiple domains to establish positions of relative advantage necessary for a successful counterattack. In either case, the controlling commander coordinates for the fire support coordination line (FSCL) to be closer to the forward edge of the battle area (FEBA) to better facilitate the employment of joint fires. BCT commanders establish coordinated fire lines to facilitate the employment of surface-to-surface fires. (See figure 6-11.)

6-25. Part of the targeting process in planning is deciding which enemy systems and capabilities to attack. This decide function begins the "decide, detect, deliver, and assess" targeting cycle. The decide step provides the overall focus and sets priorities and criteria for intelligence collection and engagement planning. The decide function draws heavily on knowledge of the enemy, a detailed IPB, and continuous assessment of the situation. Each phase or critical event within the defense has targeting priorities, reflected in products that include a HPT target list and attack guidance. The staff designates target areas of interest (TAIs) where weapon systems can best attack HPTs. The shape of a TAI reflects the type of target and the weapon system intended to engage that target. (See figure 6-12 on page 6-14.)

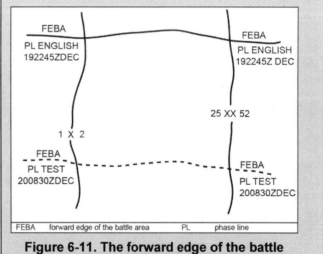

Figure 6-11. The forward edge of the battle area

The *forward edge of the battle area* is the foremost limit of a series of areas in which ground combat units are deployed, excluding the areas in which the covering or screening forces are operating, designated to coordinate fire support, the positioning of forces, or the maneuver of units (JP 3-09.3).

6-26. In the fire support context, a *target* is an area designated and numbered for future firing (JP 3-60). There are control measures for point targets, circular targets, rectangular targets, and linear targets. Figure 6-13 on page 6-15 depicts these symbols. Commanders may group two or more targets for simultaneous engagement. This is called a *group of targets*. Commanders may also attack individual targets and groups of targets in series or in a predetermined sequence. When this occurs, it is referred to as a *series of targets*. The fact that a series or group of targets has been designated does not preclude the attack of individual targets within the series or group. It also does not preclude the attack of one or more groups of targets within the series.

6-27. The plan for a mobile defense might involve the striking force attacking beyond conventional artillery range. Commanders address the forward displacement of cannon and rocket artillery and CAB assets separately or incorporate them into the striking force's movement columns. Army and joint fire assets, such as offensive cyberspace operations (OCO) and EW, add weight to the combat power of the striking force. Commanders plan and execute (with the proper authority) effects on, in, and through cyberspace and the electromagnetic spectrum (EMS). Additional planning considerations for fires in the defense can be found in FM 3-90-1. (See FM 3-09 for a detailed discussion of planning considerations for lethal fires and nonlethal effects. See JP 3-09 for a discussion of the employment of joint fire support.)

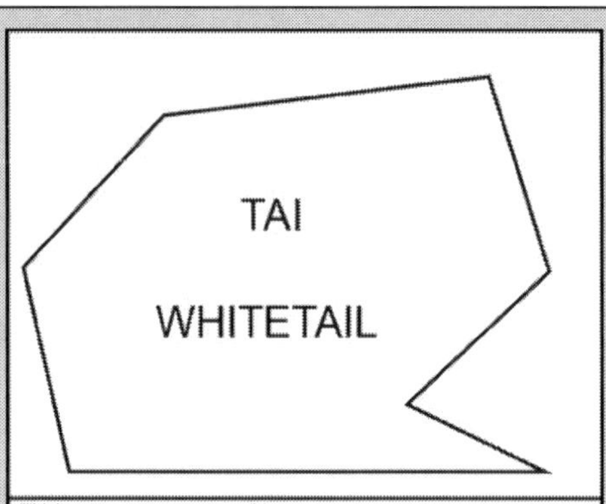

Figure 6-12. Target area of interest

A *target area of interest* is the geographical area where high-value targets can be acquired and engaged by friendly forces (JP 2-01.3).

6-28. Complementary and reinforcing joint and multinational fire capabilities provide redundancy to mitigate organic fires shortcomings. Supporting the scheme of fires during the defense involves acquiring, discriminating, and engaging targets throughout an AO with massed and precision fires including joint, EW, OCO, and IRCs. The essential element of successful fires execution is the ability to detect targets through the depth of an AO. The reconnaissance and surveillance plan must be redundant and in depth with primary and alternate observers across all intelligence disciplines—counterintelligence, geospatial intelligence, human intelligence, measurement and signature intelligence, signals intelligence (SIGINT), and technical intelligence. Planning considerations for supporting the scheme of maneuver during the defense include—

- Weight the main effort.
- Consider positioning fires and reconnaissance assets to exploit weapons ranges and preclude untimely displacement when fires are needed the most.
- Provide suppression of enemy air defense systems.
- Provide counterfire.
- Provide early warning and dissemination.
- Provide wide area surveillance.
- Provide fires to protect forces preparing for and assets critical to the unit's defensive actions.
- Disrupt enemy attacks by attacking enemy forces massing to attack.
- Interdict enemy indirect fires and sustainment efforts.
- Plan observation and fires to defeat or disrupt enemy attempts to breach friendly obstacles.
- Plan obscurant fires to deny enemy observation or screen friendly movements between defensive positions.
- Allocate responsive fires to support the decisive operation.
- Allocate fires to neutralize bypassed enemy combat forces during the conduct of counterattacks.
- Plan for target acquisition and sensors to provide coverage of named areas of interest (NAIs), TAIs, and critical assets.

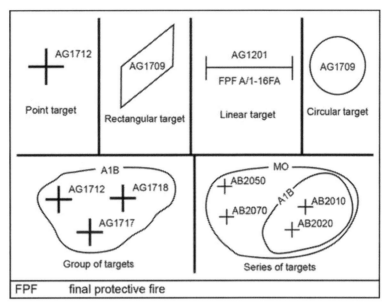

Figure 6-13. Fire support target symbol

PLANNING RECONNAISSANCE AND SECURITY OPERATIONS IN THE DEFENSE

6-29. Commanders use reconnaissance and security operations to confuse the enemy about the location of the friendly MBA positions, prevent enemy observation of friendly preparations and positions, and keep the enemy from delivering observed fire on friendly positions. Commanders plan to use their security forces to force the attacking enemy to deploy prematurely. Security operations can offset the attacker's inherent advantage of initiative regarding the time, place, plan, direction, strength, and composition of the attack by forcing the enemy to attack blind into prepared defenses.

6-30. Commanders counter enemy ground reconnaissance activities through both active and passive measures. Commanders prevent the enemy from determining the precise location and strength of defensive positions, obstacles, EAs, and reserves in two ways. First, the defending force conducts reconnaissance to gain and maintain contact with the enemy. Second, each echelon normally establishes a security area forward of its MBA to deny the enemy freedom of action. The *security area* is that area that begins at the forward edge of the battle area and extends as far to the front and flanks as security forces are deployed. Forces in the security area furnish information on the enemy and delay, deceive, and disrupt the enemy and conduct counterreconnaissance (ADRP 3-90).

6-31. All units conduct aggressive security operations within their AOs, including the support area, to identify and destroy enemy reconnaissance forces. Counterreconnaissance includes defeating enemy SOF and irregular forces, UAS, conventional reconnaissance forces, and aircraft. It also includes denying enemy network intrusions. Units minimize EMS and light signatures; use effective camouflage, cover and concealment; emplace decoy positions; and implement operations security (OPSEC) and other information protection measures to deny or deceive the enemy information about friendly dispositions and locations. (See FM 3-90-2 for more information on the tactics associated with the conduct of security tasks.)

PLANNING RESERVE OPERATIONS IN THE DEFENSE

> *A sudden powerful transition to the offensive-the flashing sword of vengeance-is the greatest moment for the defense.*
>
> Carl von Clausewitz

6-32. The defense plan retains a reserve regardless of the defensive task assigned. The reserve is an uncommitted force available for commitment at the decisive moment. It provides flexibility for the commander through offensive action. The reserve is more difficult to resource in the mobile defense because so much of the available combat power is allocated to the striking force. The division or BCT tasked to provide the defensive forward security force—conducting either a cover or a guard—should not also be designated as the corps or division reserve on completion of its security operations due to the likelihood of requiring extensive reconstitution following completion of the security mission.

6-33. Commanders specify whether the reserve is to counterattack by fire or assault the objective or enemy force. The reserve must remain agile to respond to any enemy penetration of the MBA that occurred earlier than, or at a different location than, that visualized by the commander.

6-34. Commanders designate planning priorities with supporting, on-order graphic control measures for commitment of the reserve. Commanders plan the movement of the reserve from its assembly area to likely commitment locations to deconflict the reserve's movement with other movements, including sustainment convoys, artillery repositioning, and the movement of BCTs within their AOs.

6-35. The reserve can be as large as one-third of the commander's available combat power, but it is normally not smaller than a BCT-sized element for a corps. Uncertainty about enemy intentions and strength generally drives a requirement for larger reserve. The reserve is the commander's principal means of influencing close combat operations. Upon commitment it is supported with additional assets such as artillery; CAS; attack reconnaissance helicopters; EW; engineer; chemical, biological, radiological, and nuclear (CBRN) defense; and sustainment.

PLANNING RETROGRADE OPERATIONS IN THE DEFENSE

6-36. A defensive situation may require the headquarters staff to plan for retrograde tasks in combination, sequentially, or subsequent to the defensive mission. Planning and rehearsing a rearward passage of lines of security forces forward of the MBA is critical to ensuring proper execution while under enemy pressure. The complexity and fluidity of retrograde operations and the need to synchronize the entire operation dictate the need for detailed, centralized planning and coordination with decentralized execution. A retrograde may be forced or voluntary. In either event, a higher echelon commander must approve it.

6-37. Defensive plans address performing more than one retrograde task. A combination of these tasks—delay, withdrawal, or retirement—is performed either simultaneously by adjacent units or by one retrograde task developing into another. For instance, a withdrawal from action may precede a retirement, or a BCT may execute delaying actions to cover the retirement of other forces. The defensive plan addresses the constitution and location of the reserve throughout the performance of all three retrograde tasks. If subordinate units' retrograde movements take them outside of the corps or division AO, this defensive plan addresses their command and support relationships.

6-38. As in other operations, the commander's concept of operations and intent drive planning for the retrograde. Retrogrades inherently put unit cohesion at risk. Commanders at every level minimize this risk by—
- Thorough planning, efficient control, and aggressive leadership.
- Maintaining an aggressive attitude throughout the force.
- Ensuring the purpose of the operation is clearly understood.
- Ensuring good situational awareness and a common operational picture among attached, OPCON, TACON, and supporting unit headquarters.

6-39. The larger the difference in tactical movement capabilities between a unit conducting a retrograde and an advancing enemy force, the greater the unit's chance to conduct the retrograde successfully. The corps or division commander and subordinate commanders enhance the ability of their units to move by—
- Conducting key leader reconnaissance of potential routes and battle positions.
- Improving existing road networks and controlling the flow of traffic along those routes.
- Executing well-rehearsed unit movement and battle drills.
- Positioning security forces at critical choke points.
- Evacuating civilian personnel or restricting their movements to nonmilitary routes.

- Evacuating casualties, recoverable supplies, and unnecessary stocks early.
- Displacing nonessential CPs and sustainment activities early.

PLANNING FOR SUPPORT AND CONSOLIDATION AREA SECURITY IN THE DEFENSE

6-40. The success of the defense depends heavily on protecting the corps or division support area from enemy attacks. Defensive planning addresses the early detection and immediate destruction of enemy forces attempting to operate in or near unit support areas. The wide range of threats to key nodes in the support area include, but are not limited to, individual saboteurs, airborne or air assault insertions, long range artillery and missile strikes, and ground maneuver penetrations. Attacks against friendly forces in the support area, especially from small units, may precede the onset of large-scale combat operations and be indistinguishable from terrorist acts.

6-41. The commander defines responsibilities for the security of units within the support area. The maneuver enhancement brigade (MEB) commander's AO will be a support area. The MEB commander is responsible for area security operations within that AO. The MEB commander has the authority to designate the commanders of tenant units within the support area (except Army Medical Department officers) as base and base cluster commanders responsible for the local security of their respective bases and base clusters. The MEB commander designates protection standards and defensive readiness conditions for tenant units and units transiting through the support area. The MEB commander coordinates with the corps or division main CP or support area command post (SACP) to mitigate the effects of security operations on the primary functions of units located within the support area.

6-42. Corps and division commanders carefully balance available assets against requirements to determine what constitutes acceptable risk during defensive operations. They must decide between simultaneous competing demands in the support area, consolidation area, and MBA. Balancing effectiveness and efficiency requires accurate information, tolerance for risk, and acute situational awareness. Commanders regularly reexamine their assumptions in terms of friendly and threat capabilities in the context of current and future operations to ensure risk calculations are informed by the best available information.

Note. In the defense, commanders typically only designate consolidation areas when the defense follows an offense that successfully seized new areas.

6-43. Corps and division commanders disperse mission command nodes and support capabilities in the support and consolidation areas to avoid providing lucrative targets for the enemy. They establish redundancy to the maximum extent possible, but they must balance both dispersion and redundancy against potential losses in effectiveness. While dispersion may provide protection against long range fires, it potentially makes defending against other threats, like irregular forces or SOF, more difficult.

6-44. The MEB uses its military police (MP) and other assigned security forces to provide area security around friendly CP facilities and critical sites. The corps or division commander assigns the MEB the mission of developing security plans for bases, base clusters, and designated routes within the support area. These plans address unit, base, and convoy defense against Level I threats. These plans address MEB response force operations directed against Level II threats. Areas previously occupied by enemy forces should be cleared if sustainment assets must transit them and other maneuver forces have not done so.

6-45. The commander augments the MEB with a tactical combat force (TCF) to counter any projected Level III threats. The TCF requires tactical mobility and enough combined arms capability to deal with the threats it is likely to face. It should include or be supported by fires, aviation, and engineer assets. The most likely Level III threat to the support area and its mobility corridors during the conduct of defensive tasks are mobile enemy forces intent on—

- Severing or disrupting corps or division's mission command systems and nodes (such as CPs).
- Disrupting or destroying sustainment elements and stocks being brought forward to committed units and sustainment sites.
- Interdicting main supply routes and supply points.

Chapter 6

- Destroying CP facilities, airfields, aviation assembly areas, and arming and refueling points.
- Interfering with the commitment of the reserve.

6-46. When a consolidation area is designated, BCTs assigned responsibility for a consolidation area provide significant area security capability that reduces the burden on the MEB by reducing the size of the support area. Corps and division commanders coordinate AMD coverage of critical friendly capabilities in both the support and consolidation areas. Reserve forces must be ready to perform counterattacks in these areas as well.

6-47. Defending corps and division headquarters develop plans to find, fix, and destroy enemy forces conducting penetrations into the support areas. A defending commander may take the following actions (in order of priority):

- Assign priority of collection and fires to the most appropriate unit positioned to respond to this enemy threat, including attack aviation, armed UAS, or CAS.
- Coordinate for reinforcing fires from higher echelon or adjacent commands to support threatened subordinate units.
- Direct the repositioning of adjacent units to engage enemy forces, when possible.
- Commit the local TCF to defeat a Level III threat (commitment criteria must be clearly described in orders).
- Commit the reserve to reinforce the threatened unit (commitment criteria must be clearly described in orders).

6-48. The key consideration before diverting any assets from the corps or division's decisive operation is if the threat to its sustainment capabilities jeopardizes mission accomplishment. Although a corps or division may sustain the temporary loss of sustainment from its support area, it loses the battle if defeated in the MBA. (See chapter 9 of ATP 3-91 for additional discussion of division support area security.)

PREPARING CORPS AND DIVISIONS FOR THE DEFENSE

6-49. *Preparation* consists of those activities performed by units and Soldiers to improve their ability to execute an operation (ADP 5-0). Corps and divisions take full advantage of all the time available to prepare the defense. Time is of the essence in getting the force prepared to fight. The division commander and subordinate commanders and staffs conduct simultaneous preparations. FM 3-90-1 discusses generic defensive preparation considerations for all three types of defensive tasks. Paragraphs 6-50 through 6-108 discuss areas of special interest to corps and divisions preparing to conduct a defense.

CONSIDERATIONS WHEN PREPARING AN AREA DEFENSE

The commander adopts a plan which takes advantage of the natural defensive strength of the terrain.

FM 7-100, *Infantry Division*, 1960

6-50. An area defense is typically conducted when the terrain favors the defender, or there is insufficient depth to conduct a mobile defense forward of the no penetration line. Commanders planning an area defense focus their preparations on those additional information collection operations required to answer critical information requirements, refining the plan, increasing coordination and synchronization, and conducting shaping actions. Corps and division commanders may have to commit substantial forces to security operations or conduct a spoiling attack, if they know that the enemy will attack before their subordinate units finish their defensive preparations. The spoiling attack or security operation buys reaction time and maneuver space for subordinates to complete their defensive preparations.

6-51. A unit normally transitions to the defense after it completes the deployment process, completes its offensive actions, or is in an assembly area. The unit commander issues a warning order stating the mission and identifying any special considerations. The headquarters staff conducts detailed planning while the rest of the unit completes its current mission. The staff coordinates for the pre-positioning of ammunition and barrier material in a secure area near subordinate unit defensive positions before starting the operation.

6-52. Leaders at all echelons conduct a reconnaissance before occupying a defensive position. This reconnaissance effort is as detailed as the situation allows; time and the presence of the enemy are normally the controlling variables. Reconnaissance may consist of a simple map reconnaissance or a more detailed leaders' reconnaissance that determines the initial layout of the new position. Leaders also take advantage of digital enablers, such as the Distributed Common Ground System-Army and geospatial intelligence, to increase their understanding of their AOs.

6-53. Defending subordinate divisions and BCTs occupy their defensive positions and begin preparation as soon as practical after receiving the mission. Commanders establish a forward security area before subordinate divisions and BCTs begin occupying their defensive positions.

6-54. Preparation of defensive positions requires the construction of fighting and protective positions and obstacles. The obstacle effects will block, turn, fix, or disrupt different portions of the advancing enemy force in order to allow its defeat in detail. At the BCT level, the engineer effort associated with these survivability and countermobility tasks greatly exceeds the capacity of the BCT's engineer battalion. Constructing survivability positions requires considerable time or the augmentation of the brigade engineer battalion with engineer capacity from higher echelons. A unit may pre-position supplies such as ammunition and barrier materiel once it establishes security for those positions. The approaching enemy, troops and support available, and time available are the deciding considerations in establishing priorities of work; with planning, units can accomplish multiple defensive tasks simultaneously.

6-55. Defending units use camouflage and concealment to hide defensive preparations. They limit EMS emissions and protect mission command systems and the integrity of data on those systems. They use terrain to mask emitters so as to reduce the probability of enemy SIGINT and EW detection, and offset emitters from CPs to increase survivability if they are detected. Information operations focus on encouraging the local civilian population to avoid interfering with unit operations in the close, support, and consolidation areas. (See FM 3-90-1 for additional information on the performance of preparation tasks associated with area defense operations.)

CONSIDERATIONS WHEN PREPARING A MOBILE DEFENSE

6-56. A mobile defense is typically conducted when the terrain favors the attacker and there is sufficient depth to employ a striking force forward of the no penetration line. Preparations for conducting a mobile defense include developing fixing force defensive positions and EAs. Commanders aggressively use information collection assets to track enemy units as they approach. Engineers participate in conducting route and area reconnaissance to locate and classify counterattack routes. They improve existing routes and open new routes for defending forces to use during combat operations. They construct obstacle effects to block, turn, fix, and disrupt different portions of the advancing enemy force and that enemy force's combined arms team to allow its defeat in detail by the striking force.

6-57. The striking force assembles in one or more areas, depending on the width of the AO, the terrain, enemy capabilities, and the commander's intent. Before the enemy attack begins, the striking force may deploy some or all of its elements forward in the MBA to—
- Deceive the enemy regarding the purpose of the force.
- Occupy dummy battle positions.
- Create a false impression of unit boundaries, which is important when operating with a mix of armored, Stryker, and infantry forces or multinational forces.
- Conduct reconnaissance of routes between the striking force's axis of advance and potential EAs.

6-58. Enemy forces attempt to discover the strength, composition, and location of the units that constitute the fixing force and the striking force. Defending commanders use protective measures, such as security forces and OPSEC, to deny the enemy this information and degrade the collection capabilities of enemy reconnaissance and surveillance assets. The corps or division commander repositions subordinate formations to mislead the enemy, protect the force, and portray an area defense to the enemy by hiding the striking force.

Preparing for Mission Command in the Defense

6-59. Corps and division commanders ensure that subordinate and supporting commanders understand their defensive concept of operations by conducting rehearsals. If possible, the commander takes subordinate and supporting commanders to a physical vantage point that facilitates shared visualization to give the commander's intent, to describe common control measures, and to conduct commander-to-commander coordination. When use of a physical vantage point is impractical, commanders use all other resources at their disposal, such as a computer-generated depiction of the terrain, sand tables, or map overlays.

6-60. Corps and division main CPs locate where they can best perform their synchronization and planning activities while denying the enemy intelligence information. Main CPs should be out of the range of enemy cannon artillery and most enemy multiple rocket launcher systems. Locating the main CP in or near other bases and base clusters increases security against Level I threats.

6-61. If the plan calls for the tactical CP to control operations in the forward security area or the commitment of the striking force, the tactical CP displaces to the appropriate location where it can best control those operations. When the enemy has SIGINT capability, the tactical CP should locate away from the unit scheduled to conduct the main effort or decisive operation. The SACP plans and coordinates for sustainment support, terrain management, movement control, and security within the support area.

Preparing for Movement and Maneuver in the Defense

> *Troops carry out the organization of the position in accordance with a plan of construction expressed in orders in the form of priorities.*
>
> FM 100-5, *Field Service Regulations Operations*, 1944

6-62. A detailed reconnaissance of an AO helps commanders and staffs refine the defensive plan and determine the most effective way to use the terrain and available resources. These refinements help security forces, units preparing in the MBA, the reserve, and operations in the support and consolidation areas.

6-63. During the preparatory phase, corps and division subordinate and supporting units move into their assigned locations and continuously improve their positions. They complete task organization and exchange of liaison officers. Adjacent units make contact with each other and complete coordination. The striking force withdraws to attack positions and prepares to conduct the decisive attack. (See FM 3-90-1 for a discussion on considerations for occupying defensive positions in both the security area and the MBA.)

6-64. MP teams from the MEB and MP brigades augment maneuver control teams, assisting the commander, assistant chief of staff, operations (G-3), and transportation officer control troop movements within the corps or division AO during the preparation and execution phases of the defense.

6-65. Units improve the natural defensive qualities of available terrain. They augment defensive and blocking positions by emplacing minefields, preparing obstacles, camouflaging fighting and survivability positions, and improving observation and fields of fire. They conduct the detailed coordination necessary to integrate fire plans, EAs, and maneuver forces, particularly the reserve or striking force. (See FM 3-90-1 for a discussion of priorities of work when occupying an area.)

6-66. Obstacles fix the enemy under fire or divert the enemy into areas to be destroyed by maneuver. The placement of obstacle zones and belts compliments and exploits the advantages of existing natural obstacles. BCTs shape terrain using wire, vehicle ditches, abatis, log obstacles, and minefields to block defiles, impede enemy movement, and canalize the enemy into EAs.

6-67. Infantry task forces construct strong points to retain key terrain necessary for the integrity of the defense. Infantry brigade combat teams (IBCTs) anticipating contact with enemy armor require positions that incorporate enough obstacles covered by anti-tank guided missile systems to prevent enemy armor from closing with and penetrating their defenses.

6-68. When preparing an area defense, the BCT commander determines the priority of effort for engineer units to support the concept of operations and mitigate the primary threat. Normally, the priority is countermobility and then survivability. When necessary, the commander tasks the supporting sustainment brigade to help with the forward staging of obstacle material, mines, and demolitions. The engineer staff officer requests additional engineer assets from the MEB or supporting functional engineer brigade to provide any additional mobility support in the support area once the corps or division occupies its AO.

6-69. In a mobile defense, the priority of effort for engineer units is normally to ensure the striking force's mobility and then to the survivability and countermobility efforts of the BCTs in the fixing force. The commander may restrict obstacle emplacement by subordinate units in the fixing force so as to not hinder the striking force when it is committed. BCT commanders within the striking force task-organize maneuver units with engineers capable of conducting in-stride obstacle and gap breaches, ensuring assets are well-forward and integrated into lead maneuver formations.

6-70. When preparing to conduct a retrograde, the priority of effort for engineer units within each of the brigades is normally the mobility of the main body. A secondary priority is to the mobility of any detachments left in contact, and then to the countermobility effort designed to impede the advance of enemy forces.

6-71. The corps or division commander uses attached, OPCON, or TACON information collection assets to gain information about the enemy, terrain, weather, and civil considerations. The commander focuses reconnaissance forces and surveillance capabilities along probable enemy avenues of approach. The corps G-3 tasks subordinate units for information collection when the corps expeditionary military intelligence brigade (E-MIB) cannot meet all of the corps' information collection requirements.

6-72. BCTs use their cavalry squadrons with organic brigade UAS or augmenting CAB assets to collect information. Combining air and ground assets can provide greater fidelity and timeliness than either capability used alone. Aerial assets can cue ground assets to possible enemy locations either to avoid them or get into a better position to observe them. Ground assets can also cue aerial assets to enemy locations so they can track enemy forces when they move. This information allows BCT direct support artillery battalions, the corps field artillery brigade, and the CAB to coordinate strike coordination and reconnaissance operations, Army aviation attack missions, and joint fires. They can also act as observers during strikes and provide battle damage assessments.

6-73. Ground maneuver and reconnaissance units, such as a cavalry squadron, provide local security for the E-MIB SIGINT systems located within a BCT AO. This enables the E-MIB to position these SIGINT systems forward to gather signals intelligence originating beyond the FLOT.

6-74. Counterintelligence and human intelligence teams from the E-MIB may be under the TACON of BCT commanders. They provide the BCTs with a greater capability to conduct stability and defensive tasks within their AOs, collecting information from detainees and the local population to provide commanders better situational understanding.

6-75. Obscuration considerations include man-made obscurants, dust from moving vehicles, and fog, rain, or snow. Obtaining sufficient man-made obscurants is critical to counter enemy electro-optical sensors.

PREPARING FOR INTELLIGENCE IN THE DEFENSE

6-76. Before a battle, commanders at all echelons require information. Specifically, they need to know—
- The composition, equipment, intent, strengths, weaknesses, and scheme of maneuver of the attacking enemy force.
- The location, direction, and speed of enemy reconnaissance elements.
- The location and activities of enemy units and reserves.
- Enemy C2 and communications facilities.
- The location of enemy fire support and air defense systems with associated command and control nets.

6-77. The assistant chief of staff, intelligence (G-2) uses the preparation phase to complete information collection integration and synchronization. Corps and divisions rely on joint and national systems to detect and track targets beyond their organic capabilities. The headquarters employs available information collection assets to refine its knowledge of the terrain and civil considerations within its AO. Information collection assets identify friendly vulnerabilities and key defensible terrain. The division headquarters conducts periodic information collection of any unassigned areas to prevent the enemy from exploiting these areas to achieve surprise.

6-78. Commanders continuously refine the intelligence picture of enemy forces throughout their areas of interest as part of deep operations. They focus their information collection efforts on key geographical areas and enemy capabilities of particular concern using the NAI control measure. (See figure 6-14.)

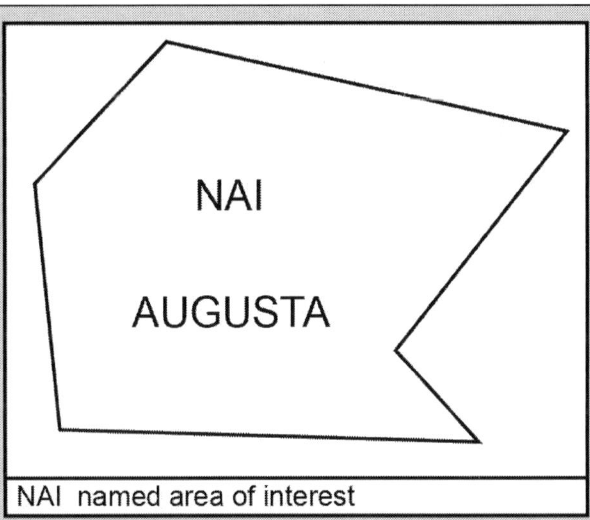

6-79. Constant surveillance of the AO and effective reconnaissance are necessary to acquire targets and to verify and evaluate potential enemy courses of action and capabilities. Information collection efforts focus on identifying when, where, and with what strength the enemy will attack. This allows the commander to identify opportune times to conduct spoiling attacks and reposition forces.

Figure 6-14. Named area of interest

NAI   named area of interest

A *named area of interest* is the geospatial area or systems node or link against which information that will satisfy a specific information requirement can be collected, usually to capture indications of adversary courses of action (JP 2-01.3).

PREPARING FOR FIRES IN THE DEFENSE

6-80. Effective fires planning is essential to the success of any defense. Fire rehearsals help staffs, units, and individual fire support personnel to better understand their specific role in upcoming operations, synchronize execution of the scheme of fires, and ensure equipment and weapons are properly functioning.

6-81. Fires preparation begins before deployment and continues during planning and throughout an operation. Uncommitted firing units prepare for contingencies and anticipate the next phase or branch of an operation. Fire support in the defense is generally more centralized than in the offense in order to quickly mass effects against a superior enemy force. Since a mobile defense includes characteristics of both defense and offense, it requires more flexible fire support planning.

6-82. Defensive fires preparation begins at receipt of mission. Defensive fires preparations initially focus on—
- Detecting HPTs.
- Deep operations against approaching enemy forces.
- Supporting the actions of friendly forces, especially those of forward security forces.
- Employing rapidly emplaced, remotely controlled munitions to selectively deny areas or interdict movement choke points.
- Suppression of enemy air defenses.
- Counterfire.

6-83. Whether in an area defense or mobile defense, fires weight the commander's decisive operation. In an area defense, the decisive operation is where the defense covers the enemy's main avenue of approach. The attack of the striking force is the decisive operation in a mobile defense. The commander seeks positions of relative advantage for Army fire support assets, Army aviation, joint fires, and supporting nonlethal capabilities, such as EW and OCO. These positions are where these capabilities can best survive enemy attack while providing continuous effects massed in support of the decisive operation.

6-84. Firing units plan and prepare primary and alternate firing positions based on the enemy's counterfire capabilities. Firing units rehearse movement to and occupation of alternative positions.

6-85. The division joint air-ground integration center (JAGIC) synchronizes the use of joint air and ground fires to engage enemy air and ground systems while preventing fratricide. The corps works with its corps

Large-Scale Defensive Operations

tactical air control party and the LCC battlefield coordination detachment (BCD) to synchronize the use of joint fires.

Detecting High-Payoff Targets

6-86. Corps and divisions focus on the timely, accurate identification of HPTs during defensive preparations. Reconnaissance units and surveillance systems receive tasks commensurate with their capabilities. Mobile HPTs require continuous tracking after detection. Surveillance assets—including UAS, EW, space-based, and SOF—focused on a TAI normally associated with the acquisition of a HPT. Tracking priorities are based on the concept of the operation and targeting priorities. The fires cell tells the G-2 the degree of accuracy and dwell time required to engage specific targets. The G-2 must match accuracy requirements with the capabilities of the collection systems, while the G-3 directs the appropriate collection assets against target types based upon their electronic, visual, or thermal signatures and estimates of when and where the enemy target will be located.

6-87. As assets collect information for target development, analysis and control element intelligence analysts use the information for target development and situational awareness. They pass target information to the fires cell once they identify a target as an HPT. The fires cell executes the attack guidance against the target. The presence of a field artillery intelligence officer in the analysis and control element facilitates essential close coordination between the intelligence and the fires cells.

6-88. The targeting section helps the commander continuously assess enemy activities that can influence friendly operations. Commanders adjust the attack guidance matrix to exploit opportunities and target those capabilities the enemy commander requires most.

Deep Operations as Part of Defensive Preparations

6-89. In general, corps deep operations occur beyond the area a division can effectively employ its combat power. The extent to which a division conducts deep operations is limited in two ways: through the use of control measures and by the ranges of the capabilities it controls. Information collection, fires, and the ability to manage airspace are critical capabilities during deep operations. Orchestration of those capabilities in the most effective manner requires significant planning and coordination, both of which are most efficient when responsibilities are clearly delineated between echelons. Corps deep operations should facilitate division freedom of action in the close and deep areas. Enemy forces in the deep area are not out of contact in a multi-domain environment, since space, cyberspace, and information operations have few range constraints.

6-90. Corps and division commanders monitor the approach of enemy formations and target HPTs with organic, supporting, and joint systems as far from friendly forces as possible. Air and ground-based fires, EW, space, cyberspace, military deception (MILDEC) and other information-related capabilities, terrain shaping activities, and SOF can all contribute to corps and division deep operations. Commanders may employ subordinate maneuver units to conduct deep air or ground maneuver operations when the desired effects outweigh the inherent risks.

6-91. Deep operations disrupt the enemy's movement in depth, destroy HPTs, and disrupt enemy C2 at critical times. They can deny the enemy the initiative early and limit enemy commander options. It is important to clearly delineate corps and division responsibilities and focus in terms of time, space, and domains. Failure to do so leads to inefficient application of friendly capabilities. Commanders use control measures, such as boundaries, FSCMs, and airspace coordinating measures, to assign deep operations responsibilities to the appropriate echelons.

6-92. An important permissive FSCM is the kill box. Kill boxes allow lethal attack against surface targets without further coordination with the establishing commander and without the requirement for terminal attack control. A *kill box* is a three-dimensional permissive fire support coordination measure with an associated airspace coordinating measure used to facilitate the integration of fires (JP 3-09). Kill boxes are designated as blue or purple. (See figure 6-15 on page 6-24.) The purpose of a blue kill box is to facilitate the attack of surface targets with air-to-surface munitions. The purpose of a purple kill box is to facilitate the attack of surface targets with surface-to-surface and air-to-surface munitions. Surface-to-surface direct fires are not restricted by the establishment of either type of kill box. ATP 3-09.34 discusses blue and purple kill

Chapter 6

boxes. Figure 6-15 does not show the strike coordination and reconnaissance or the kill box coordinator. The use of a kill box requires the designation of a strike coordination and reconnaissance or a kill box coordinator.

6-93. The corps commander determines those capabilities and authorities allocated to the divisions to conduct deep operations while retaining others under corps control. Retaining capabilities at corps level that cannot be effectively employed beyond the division deep area may not use resources to best effect. Conversely, pushing capabilities down to divisions or brigades that lack the capacity to employ them effectively wastes scarce resources. The allocation of resources to the most effective echelon is informed by many factors, including the scheme of fires.

6-94. Corps and division commander areas of interest and influence extend far enough forward of the FLOT to give them time to react to approaching enemy forces, assess options, and execute operations accordingly. Corps and division deep operations begin in the cyberspace and space domains and the information environment long before the enemy physically closes with friendly forces. They are ongoing and continuous. Commanders employ air capabilities, both joint and Army, to achieve a position of relative advantage for subordinate echelons as the battle develops in the forward security area and the MBA.

6-95. The fire support coordinator, supporting United States Air Force (USAF) elements, G-3, and G-2 coordinate to ensure that deep operations support the corps or division commander's defensive concept and successfully support the overall defense. Deep operations that mass effects against enemy forces across multiple domains are the best means of gaining and maintaining the initiative during the performance of defensive tasks. Deep operations begun during the preparation phase continue during the execution phase of the defense. (See ATP 3-94.2 for additional information on deep operations.)

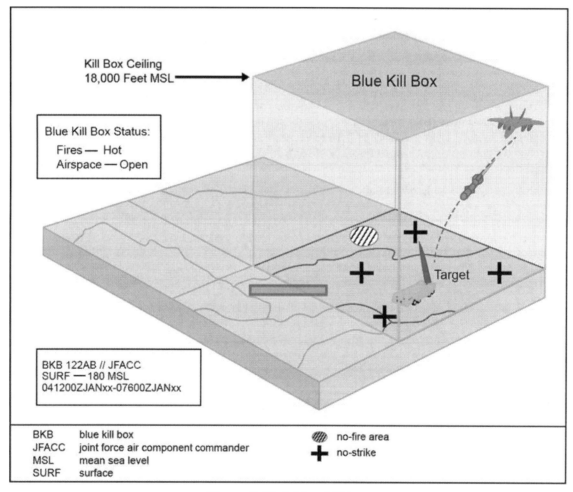

Figure 6-15. A blue kill box

Large-Scale Defensive Operations

PREPARING FOR SUSTAINMENT IN THE DEFENSE

> *The organization of supply and evacuation is controlled by the consideration that it must be capable of adapting itself to the constant and often rapidly changing conditions of military operations. Flexibility, elasticity, mobility, and simplicity are therefore the cardinal principles of the system of supply and evacuation.*
>
> Field Service Regulations United States Army, 1923

6-96. Subordinate brigades preparing their defense require significant quantities of barrier material and ammunition. The assistant chief of staff, logistics (G-4) ensures that the supporting sustainment brigade provides materiel in times and locations that expedite defensive preparations. Consumption of Class IV and V supplies in the defense is typically higher than in the offense. The division or BCT constituting the command's forward security force or striking force require a greater amount of Class III and maintenance support than their MBA counterparts. In both cases, the more distance the forward security force and striking force must cover to accomplish their missions, the greater sustainment they require. When these two forces are located at a significant distance from the support area, establishing a forward logistics area to support the planned employment of these two forces may be necessary. The preparation phase is also a good time to provide unit replacements.

6-97. Army aviation units from the CAB conduct air movement of personnel, leaders, critical supplies, equipment, and systems during the defense. Air movement operations can support a variety of operations, including the emergency resupply of combat units. Aviation units can move barrier materials, munitions, fuel, and personnel. This can reduce risk to ground logistics units while sustaining the tempo of operations and extending the tactical reach of the combined arms team.

6-98. Medical evacuation from the forward security or striking force area poses significant challenges because of distance, limited situational awareness, and rapid changes in the tactical situation. These forces may be conducting noncontiguous combat operations and there is a high probability that movement routes may be temporarily interdicted. This makes it imperative that the common operational picture (COP) available to ground and air ambulances is accurate and updated as frequently as possible. The corps or division surgeon develops the unit's overall medical evacuation plan working closely with supporting aerial medical evacuation unit and supporting medical units. The unit's medical evacuation plans reflect threat artillery and air defense capabilities.

6-99. Engineer units task organized to MEBs or functional engineer brigades initiate any required general engineering tasks in support areas during defensive preparations. These tasks include establishing temporary sites designed to hold displaced and detained persons.

6-100. During the preparation phase the corps headquarters coordinates with the theater sustainment command, expeditionary sustainment command (ESC) headquarters, and other sustainment assets outside the MBA to protect them against enemy unconventional forces and fires.

6-101. Divisions coordinate the positioning of their supporting sustainment brigade, combat sustainment support battalions (CSSBs), and any supporting functional sustainment battalion elements in a similar manner. Sustainment brigades, CSSBs and functional units providing general support or area support to the division coordinate their positioning where they can best provide support. This may be within the division support area or within the division MBA.

PREPARING FOR PROTECTION IN THE DEFENSE

6-102. The defending force's BCTs occupy their respective AOs as soon as possible to maximize preparation time for defensive positions and obstacles. This includes the construction of fighting and survivability positions. Those protective activities are outlined in FM 3-90-1.

6-103. Security activities during the preparatory phase are similar to those discussed in ADRP 3-90. The protection cell coordinates with the MEB as it organizes the support area, prepares units located within the support area to defend themselves, and provides route and communications security. BCTs and support brigades conduct local security activities in their defensive positions, assembly areas, and attack positions.

6-104. MP units from the MEB or a functional MP brigade enhance force protection capabilities by conducting reconnaissance within support areas. They perform response-force operations to defeat Level II threats against bases and base clusters, and maintain contact with Level III threats until the TCF can respond.

6-105. Explosive ordnance disposal (EOD) units may be in a direct support or general support relationship to the MEB. Maneuver units may breach obstacles, but only EOD units have the capability to reduce minefields and render improvised explosive devices and explosive ordnance safe.

6-106. All units have an inherent responsibility to improve the survivability of their own fighting positions, bases, or base camps. This includes preparation for operations in a CBRN environment, which requires planning for immediate or operational decontamination. Survivability operations enhance the ability to avoid or withstand hostile actions by altering the physical environment. They accomplish this through four tasks: constructing fighting positions, constructing protective positions, hardening facilities, and employing camouflage and concealment. The first three tasks focus on providing cover while the fourth task focuses on providing concealment from observation and surveillance.

6-107. CBRN personnel contribute to unit protection by performing vulnerability assessments. These assessments provide a list of recommended activities actions ranging from CBRN protection to contamination mitigation for commanders to consider.

6-108. All units employ a mix of passive and active AMD measures to protect defensive preparations from enemy aerial observation and attack. Most enemy forces can employ UASs, even if they lack capable fixed- or rotary-wing forces. The protection cell in coordination with the fires cell refines division plans to take advantage of available AMD coverage. When preparing to conduct an area defense, commanders take advantage of the AMD air defense umbrella to the maximum extent possible by placing critical nodes and activities within that coverage. When preparing to conduct a mobile defense, the movement of the striking force complicates coverage by AMD assets. Commanders prioritize between protecting both the fixing force and the striking force when submitting the defended asset list through the chain of command to the area air defense commander (AADC.)

DEFENSIVE TASKS

"[T]he best protection against the enemy's fire is a well-directed fire from our own guns.

Flag Officer David G. Farragut

6-109. There are three basic defensive tasks—area defense, mobile defense, and retrograde. These apply to both the tactical and operational levels of war, although the mobile defense is more often associated with the operational level. The three tasks are significantly different concepts and pose significantly different challenges in planning and execution. Although the names of these defensive tasks convey the overall aim of a selected defense, each typically contains static and mobile elements.

6-110. All three basic defensive tasks use terrain, depth, and mutual supporting fires as force multipliers. Proper use of terrain provides economy of force and helps to mass the effects of combat power at decisive points. Terrain influences the tempo of enemy attacks and provides the defender with cover and concealment. The depth of an AO provides commanders with operational flexibility, allows subordinate units to disperse and maneuver, and reduces risk. Ideally, committed divisions and BCTs should have enough depth to provide security through the employment of either a covering or a guard force. Mutual support results from the defending commander's integration of the fires and movement between subordinate units. This mutual support allows commanders to focus effects across multiple domains at decisive points to defeat attacking enemy forces.

6-111. Normally a successful defensive battle culminates in the MBA through the orchestrated application of combat power that defeats the attacking enemy force. While current combat operations unfold in the MBA, the staffs focus on shaping operations or supporting efforts that set the conditions necessary for conduct of the decisive operations or main effort in the next phase of the joint campaign or major operation. The focus of MBA units is on conducting operations according to the commander's intent within the current phase of the campaign or major operation.

6-112. Corps and division commanders combine area and mobile defense tasks based upon availability of assets, terrain, their higher echelon commander's concept of operations, and enemy capabilities. Corps and divisions may also conduct retrograde operations. Both the area and mobile defenses contain static and dynamic elements. The dynamic elements involve the maneuver of combat forces and the movement of their supporting combat multipliers. Using only purely static defensive elements puts friendly forces at great disadvantage when enemy fire support systems outrange those available to the defender or when the enemy has many more such systems available.

6-113. Corps and division commanders organize their subordinate units and assign them to their respective AOs based on their concept of operations for defeating the attacking enemy's main effort. Commanders accept risk to mass combat power in depth along an enemy's main axis of advance. Corps and division headquarters reinforce the subordinate unit or units conducting their decisive operation or main effort with additional fires, information collection, engineer, attack reconnaissance aviation, and AMD support.

6-114. Corps and division commanders organize their defense in depth. They allow their subordinate divisions or BCTs maneuver room to conduct their own defense. The area selected by the corps or division commander for defense becomes that echelon's MBA.

6-115. Corps and division commanders shape their defensive battles through information collection, joint fires, and cyberspace electromagnetic activities (CEMA) before an enemy reaches the MBA. Defensive cyberspace operations (DCO) are threat-specific and mission prioritized to retain the ability to use the Department of Defense Information Network (DODIN). Corps and division headquarters focus on disrupting attacking enemy forces before they encounter the MBA and breaking up the integrity of their attack. Corps commanders pay particular attention to disrupting an enemy's C2 elements through the employment of lethal and nonlethal effects, with the intent of making enemy forces more vulnerable to corps and division counterattacks and spoiling attacks.

6-116. Aggressive action by Army forces in the land, air, maritime, space, and cyberspace domains and the information environment are necessary for the successful performance of defensive tasks. Army forces attack the enemy's integrated air defense system (IADS) with artillery cannon and rocket systems. Army forces destroy enemy target acquisition and air defense fire control nodes identified by Army and joint information collection efforts. Army maneuver forces destroy enemy IADSs wherever they are encountered. These attacks open windows of vulnerability for Army and joint aircraft to attack other components of the enemy IADS and enemy maneuver units. The corps coordinates for EW, space, and cyberspace capabilities to degrade enemy IADS target acquisition and C2 capabilities. These attacks systematically cause the collapse of the enemy IADS over time. Success against the enemy IADSs allows the JTF commander to allocate additional CAS and air interdiction sorties to destroy enemy long range fires systems and maneuver systems beyond the FLOT. Divisions and corps also employ Army aviation attack reconnaissance units using terrain masking and supported by suppression of enemy air defenses to defeat enemy IADS and maneuver forces. Army aviation attack and reconnaissance units can be highly effective in the defense when massed at the decisive point or as a counter-penetration force. Army aviation assault units can effectively reposition anti-armor teams to add greater depth and agility to the defense or as a tactical combat force to counter penetrations of the MBA.

6-117. During the performance of defensive tasks there may be enemy penetrations of a corps or division MBA into the support area. Defending units maintain contact with penetrating enemy forces to delay them while determining their size, composition, direction of attack, probable objective, and rate of movement. The force in contact reports this information to the higher echelon headquarters as rapidly as possible.

6-118. A defending commander takes immediate action to block further enemy advances in depth or to block expansion of an enemy penetration. This may require changing task organization, adjusting subordinate unit boundaries and tasks, executing situational or reserve obstacles, shifting priority of fires, and assuming risk.

6-119. Defending maneuver and other units may need to move away from a penetration based on the direction of an enemy attack. Commanders control these movements to ensure they do not interfere with counterattack plans or the maneuver of subordinate maneuver forces. A defending commander balances moving sustainment activities and supply points away from the area of enemy penetration and the possible temporary loss of support from those locations with the probability of losing sustainment units and supply stocks permanently, if enemy forces overrun friendly positions.

6-120. Based on the size, composition, and direction of the enemy attack, commanders select the best location to defeat or destroy the penetration. The reserve may counterattack into the enemy flank, or it may establish a defensive position in depth to defeat or block further enemy advances. The corps and division staffs establish control measures for counterattacks by the reserve, develop a supporting scheme of fires, and consider required adjustments to FSCM, such as target reference points and TAIs. They also decide on the employment of obstacles to support the counterattack. Traffic control is especially critical. Sufficient routes must be designated and kept clear for use by the reserve as it maneuvers.

AREA DEFENSE

6-121. The *area defense* is a defensive task that concentrates on denying enemy forces access to designated terrain for a specific time rather than destroying the enemy outright (ADRP 3-90). The focus of an area defense is on retaining terrain where the bulk of a defending force positions itself in mutually supporting, prepared positions. Units maintain their positions and control the terrain between these positions. The decisive operation focuses on fires into EAs, possibly supplemented by a counterattack. The reserve may or may not take part in the decisive operation. Commanders can use their reserve to reinforce fires, add depth, block, restore a position by counterattack, seize the initiative, and destroy enemy forces. Units at all echelons can conduct an area defense. (See FM 3-90-1 for a discussion of the advantages and disadvantages of a defense in depth and a forward defense during the conduct of an area defense.)

6-122. Defending commanders in an area defense deny the enemy designated terrain for a specified time. The corps or division commander allocates sufficient combat power against enemy avenues of approach to achieve a reasonable chance of success, even without the commitment of the reserve. Commanders assume risk in less threatened areas by allocating less force in these areas. Maneuver within an area defense usually consists of repositioning between battle positions and local counterattacks.

6-123. Commanders build the decisive operation around identified decisive points, such as key terrain and enemy HPTs. The commander's decisive operation in an area defense focuses on retaining terrain by using fires from mutually supporting, prepared positions supplemented by one or more counterattacks and the repositioning of forces from one location to another. The commander's decisive operation normally involves close combat since an area defense emphasizes terrain retention.

6-124. Ultimately, the mission of BCTs located within the MBA is to defeat the enemy attack or to destroy the attacking enemy force. These BCTs perform many different tasks—defending, delaying, attacking, or performing in an economy of force role—to accomplish this mission. BCTs plan to conduct forward and rearward passages of lines. They normally plan to avoid being bypassed by enemy forces during combat operations unless it fits within the corps or division commander's intent.

Contiguous Framework

6-125. The presence of a significant linear obstacle along the FEBA, such as a river, favors the use of a contiguous operational framework and the retention of terrain. Figure 6-16 illustrates contiguous AOs. A river or other natural linear obstacle adds to the relative combat power of the defender by making enemy forward movement difficult. Reserves at all levels destroy forces which penetrate defensive obstacles or establish bridgeheads on the far side of linear obstacles. Enemy forces attempting to establish such a bridgehead must be destroyed while the bridgehead is small, before they are reinforced. If they are not, they can enable enough enemy combat power to cross and rupture the coherence of the defense.

Large-Scale Defensive Operations

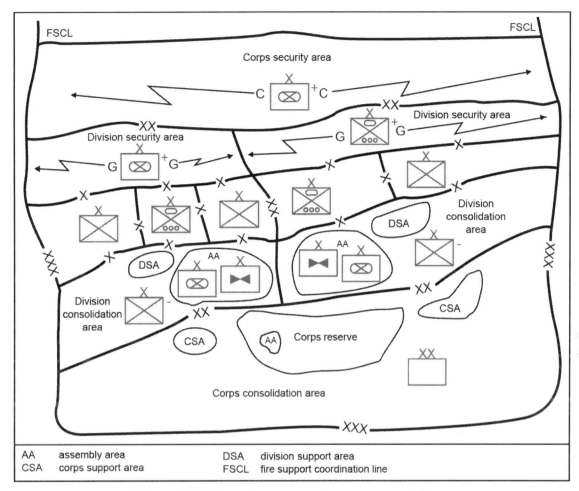

Figure 6-16. Example corps contiguous area defense

6-126. Large enemy forces will attempt to penetrate one or more MBA locations. This is more likely to occur when the enemy is able to use long range fires in conjunction with attack aviation and CAS to target selected friendly defending forces for isolation and destruction. Enemy forces will employ large-scale OCO and EMS denial operations in support of this effort. They may also employ nuclear or chemical munitions. Enemy commanders will attempt to exploit initial success by widening the gap or gaps created by the isolation or destruction of friendly units to penetrate into the depth of the corps or division defense. Corps and division commanders can direct BCTs in the MBA away from the enemy points of penetration to counterattack into the flanks of successful enemy attacks. They can also employ their reserve to block penetrations of the MBA and prevent enemy forces from exploiting initial success.

Noncontiguous Framework

6-127. Corps and division commanders may also structure a noncontiguous defense with elements deployed in depth along major avenues of approach. Figure 6-17 on page 6-30 illustrates noncontiguous AOs with the green arrows illustrating movement corridors. Forward portions of the MBA are occupied by the minimum forces necessary to warn of impending attack, canalize the attacking enemy forces into less favorable terrain, and block or impede the attacking enemy force. These forward defensive positions consist of a mixture of battle positions organized for all-around defense, strong points, and combat outposts. These positions may or may not be mutually supporting. In this case the friendly commander enhances the defense by retaining out a large mobile reserve and by committing fewer elements to the initial MBA defense. Committed elements in such a defense contain initial enemy penetrations and then counterattack to eliminate those penetrations.

Chapter 6

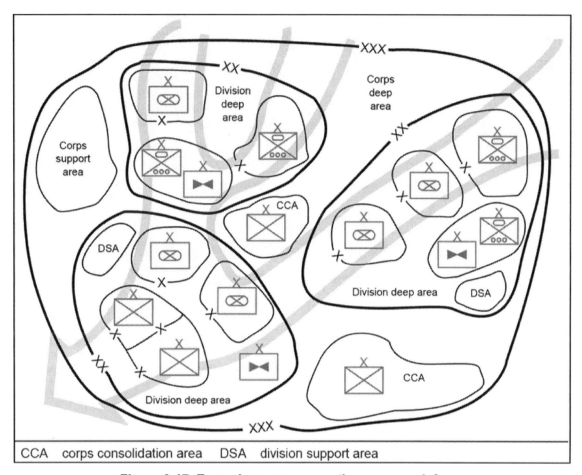

Figure 6-17. Example corps noncontiguous area defense

Organization of Forces for an Area Defense

6-128. Commanders organize a defending force to accomplish information collection, security, MBA, reserve, and sustainment functions. Corps and division commanders normally have few resources to devote to the performance of consolidation of gains tasks during the defense. Commanders may defend forward or in depth. When a commander defends forward, the defending force is organized so that most of the available combat power is committed early in the defensive effort. To accomplish this, a defending commander may deploy forces forward and plan counterattacks well forward in the MBA or even beyond the MBA. When defending in depth, commanders employ security forces and MBA elements to identify, define, and defeat the enemy main effort within the MBA, while using the minimum necessary combat power to defeat supporting attacks. This allows commanders to conserve combat power, strengthen the reserve, and better resource the counterattack.

Information Collection in the Area Defense

6-129. Commanders direct information collection assets to continuously determine the locations, strengths, and probable intentions of the attacking enemy force. Commanders seek early identification of the enemy's main effort. Commanders may need to complement surveillance with offensive action that tests enemy intentions. Fighting for information can have two benefits: it can force the enemy to reveal intentions and disrupt enemy preparations.

6-130. In an area defense, reconnaissance and surveillance operations overlap the defending unit's planning and preparing phases. Leaders performing reconnaissance and surveillance tasks must understand that they often deploy before the commander fully develops the plan. These leaders must be responsive to changes in

Large-Scale Defensive Operations

orientation and mission. The commander ensures that the staff fully plans, prepares, and assesses the execution of the intelligence portion of the overall plan.

Security Forces in the Area Defense

6-131. Defending corps and divisions need to perform forward and flank security tasks. There are also area and local security tasks to perform in the corps and division support and consolidation areas. A corps or division commander must plan and, if time allows, rehearse a rearward passage of lines and battle handover by forward security forces, such as a covering force, through units in the MBA.

6-132. Corps commanders employ security forces to their front and flanks to provide additional security for divisions and BCTs in the MBA. These security forces perform a cover, guard, or screen task depending on the mission variables. Usually the corps employs a defensive covering force in front of MBA divisions. This covering force is a combined arms force task organized to operate forward out of supporting range of the corps' MBA divisions (the protected force). The corps commander assigns the covering force an AO with enough depth to allow sufficient room to maneuver and to force the enemy to reposition its artillery and air defense systems prior to attacking into the MBA. Enemy repositioning of artillery and air defense artillery (ADA) indicates the enemy main effort, makes those systems vulnerable to friendly attacks in depth, and limits the effectiveness of enemy massed artillery fires.

6-133. The covering force battle in the forward security force area is directly tied to the future battle in the MBA as part of the overall defensive scheme of maneuver. Corps and division commanders must understand how the covering force influences operations in the close area. Security force operations normally support the commander's intent regarding where to defeat or destroy the enemy. For example, a commander may direct the covering force to avoid inflicting attrition that causes the enemy to culminate forward of the MBA because the corps plan is to destroy the majority of enemy forces with counterattacks in the MBA to set conditions for future offensive operations.

6-134. The defensive covering force's mission may include shaping the location of the enemy's penetration, identifying the enemy main effort, and attriting lead enemy echelons. Security tasks are similar for both the mobile and area defense. However, the corps commander normally retains control of the covering force to ensure unity of effort in shaping the enemy penetration during a mobile defense.

6-135. A division commander normally assigns a reinforced BCT to conduct a division covering force battle in the absence of a corps covering force. A division-controlled covering force allows a division to seize the initiative from an attacking enemy force. The size and composition of the covering force depends on its mission, the enemy, the terrain, and forces available. These mission variables take on added significance and complexity, depending on the enemy course of action. The depth and width of the AO comprising the forward security area and the time required to prepare the MBA greatly influence covering force operations as well.

6-136. Defending corps and division commanders employ security forces on their exposed flanks. Typically, these security forces perform the security task of guard or screen. The resources required to establish a covering force tend to preclude being able to simultaneously establish a flank security force capable of performing the cover task.

6-137. Corps and divisions perform area and local security tasks in their support and consolidation areas to assure freedom of maneuver and continuity of operations. They secure lines of communications and mobility corridors between these areas and the MBA. Corps and division commanders assign responsibility for the support and consolidation areas. This is usually a division or a BCT for the consolidation area and a MEB for the support area. These units control the performance of area security tasks within their AOs based on guidance from the commander.

6-138. Units performing security tasks in these two areas display the same initiative, mobility, versatility, and synchronization required for operations in the close and deep areas. The four components of area security in the support and consolidation areas are information collection, base and base cluster self-defense, response operations, and combined-arms TCF operations. (See FM 3-90-2 for a discussion of security tasks.)

Chapter 6

Main Battle Area Forces in an Area Defense

6-139. Commanders normally position their echelon's main body, including the bulk of their combat power, within the MBA where the commander wants to conduct the decisive operation. The commander organizes the main body to halt, defeat, and ultimately destroy attacking enemy forces. Most of the main body deploys into prepared defensive positions within the MBA. However, mobile elements of the force are ready to deploy where and when needed.

6-140. Defending corps and division commanders defeat the enemy's major attacking forces in a number of ways. In most cases, however, they fight a series of engagements within the MBA to—

- Dislocate enemy forces by placing defending friendly forces in unexpected locations that put the enemy in a position of disadvantage in multiple domains.
- Disintegrate the enemy's command and control system by employing lethal and nonlethal effects to target enemy command posts, communications systems, and networks.
- Isolate selected portions of the enemy for destruction in detail or to deny their ability to mutually support each other.
- Destroy critical elements of enemy formations to deny them the ability to mass effects against friendly forces in a coordinated manner and eventually render the overall enemy force ineffective.

6-141. Defending commanders designate their decisive operation or main effort and concentrate combat power in support of it. Corps and division commanders shift their main effort as other threats to the integrity of the defense develop. They continue to do so for as long as is necessary until the enemy culminates. Maneuver units defend, delay, attack, and perform security tasks as required by the situation.

6-142. Defending corps and divisions fight the decisive battle in accordance with their commander's concept of operations. That battle occurs either at the FEBA or within the MBA. Commanders position forces in the MBA to control or to repel enemy penetrations. The division commander assigns each BCT an AO within the MBA on the basis of the defending BCT's capability, the terrain within that AO, and the mission. Assigned BCT AOs usually coincide with a major regimental- or brigade-size avenues of approach. Commanders should not split responsibilities for avenues of approach. The force responsible for the most dangerous AO in the MBA normally receives priority in the initial allocation of artillery, engineer, EW, and CAS capabilities. It is the initial main effort.

6-143. Division commanders strengthen the effort at the most dangerous avenue of approach by narrowing the AO of the BCT assigned to that avenue of approach. The corps or division commander may use cavalry squadrons, Stryker battalions, or other maneuver forces as an economy of force measure in close terrain or to secure one of the division's flanks. (Economy of force requires a sound estimate of what is sufficient for the security mission to allow the massing of combat power at the decisive time and place. It is not the application of as little force as possible.) This allows the division commander to concentrate BCTs on the most dangerous approaches. Corps and division defensive plans must be flexible enough to allow changes in the main effort during the course of the battle.

Reserve in the Area Defense

> *The primary mission of the reserve is to enter the action offensively at the proper place and moment to clinch the victory or exploit success.*
>
> FM 100-5, *Field Service Regulations: Operations*, 1949

6-144. The primary purpose of the reserve in the defense is to strike a decisive blow against the enemy and seize the initiative. The reserve also serves as a hedge against risk when the attacker has the initiative, since no defense can be strong everywhere. Commanders must reconstitute a reserve after committing the existing one.

6-145. Commanders use reserves to counterattack, to exploit enemy weaknesses such as exposed flanks or units vulnerable to defeat in detail, or deny the enemy control of critical terrain. The reserve can also reinforce forward defensive positions, contain enemy penetrations, or react to threats in the support or consolidation areas. Commanders decide on the size, composition, and mission of the reserve as early as possible after determining what risk is acceptable. Commanders down to the BCT level normally retain one third of their maneuver strength in reserve; the higher the degree of uncertainty regarding the enemy, the larger the reserve should be.

6-146. Timing is critical to counterattacks, but a commander may have little latitude to determine with certainty when to commit the reserve. Commanders must anticipate the lead times required to move the reserve to different areas to ensure that when committed it arrives in the right place, in the right combat formation, at the right time. If committed prematurely, the reserve may not be available for a more dangerous contingency and be needlessly exposed to enemy action. If committed too late, it may not arrive in time to matter. Commanders reconstitute their reserve from uncommitted forces or from forces in less threatened locations after committing their original reserve. Situations that demand commitment of the reserve often require an additional reserve to deal with a new problem.

6-147. Commanders must accurately estimate the location, time, and distance for follow-on enemy echelons to close when planning counterattacks. Then they must determine which units can attack, estimate their location and likely status after the counterattack, and what other capabilities in other domains can be used to disrupt and isolate the targeted enemy force. Counterattacking units seek to avoid enemy strength. The most effective counterattacks target the enemy's exposed flanks and rear. Typically, enemy networks, communications, and sustainment capabilities are exposed as they move or displace.

6-148. Reserves are optimally a mixture of mechanized or Stryker units because of their mobility. While artillery and Army aviation assets are not kept in reserve, they should be allocated to support the reserve when it is committed.

6-149. Reserve air assault forces can respond rapidly, if conditions allow assault aviation to fly. In suitable terrain they may reinforce positions to the front or on a flank of an enemy attacking force. In a threatened area they may be positioned in depth. Air assault forces with escorting attack helicopters and armed UASs are suitable against enemy airborne or air-landing units in the corps or division support and consolidation areas. Once committed, however, inserted dismounted infantry have limited ground tactical mobility.

6-150. The mobility and firepower of attack aviation formations often make them the quickest and most effective means for stopping penetrations of BCT MBA defensive positions, particularly against enemy mechanized or motorized formations. Based on the mission variables, attack reconnaissance helicopters employed at the company or battalion level provide lethality, protection, shock effect, and flexibility. However, when facing attacking enemy maneuver forces, employing attack aviation in piecemeal teams in support of multiple ground maneuver forces dilutes combat power and will likely not produce decisive results.

6-151. Corps and division commanders should shift committed BCT forces in the MBA laterally from prepared positions only as a last resort because of three significant risks. First, an attacking enemy can inhibit or prevent such friendly lateral movements with air or artillery interdiction and with EW or OCO. An enemy may even employ chemical munitions. Second, when a force is engaged, even by small probing or reconnaissance actions, it is neither physically nor psychologically prepared to make lateral movements. Third, vacating an MBA position, even temporarily, invites penetration and exploitation by uncommitted enemy forces.

6-152. When deciding where to place the reserve, commanders decide whether to orient the reserve on its most likely mission or its most important mission. Commanders and staffs ensure the commander can effectively use the reserve when needed. Commanders ideally locate the reserve within the AO where it can employ the road network to rapidly displace in response to the greatest number of opportunities or contingencies. Commanders consider terrain, routes, enemy avenues of approach, and probable enemy penetrations when positioning the reserve. Commanders may initially position the reserve in a forward location to deceive the enemy and obscure subordinate unit boundaries, especially those of dissimilar units such as armored and dismounted infantry.

Chapter 6

Control Measures for an Area Defense

6-153. Commanders organize an area defense by designating the MBA and assigning AOs or battle positions to subordinate units located within the MBA. Commanders designate battle positions so that they are astride likely enemy avenues of approach. Commanders create a security area in front of the MBA or around a base of operations. When possible, the boundaries of the subordinate elements of the security force coincide with those of the major defending units in the MBA. Commanders also designate a support area. Commanders have the option of designating a consolidation area.

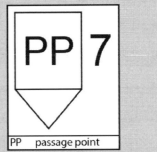

Figure 6-18. Passage point

A *passage point* is a specifically designated place where the passing units pass through the stationary unit (FM 3-90-2). The location of this point is where the commander wants subordinate units to physically execute a passage of lines.

6-154. Area defense maneuver graphic control measures may include phase lines (PLs), EAs, the FEBA, the battle handover line (BHL), strong points, target reference points (TRPs), NAIs, TAIs, decision points, and various other measures, such as final protective fires (FPFs) and obstacle control measures. Tactical mission tasks assigned as part of the mission can also be control measures. (See FM 3-90-1 for more information on these control measures.)

6-155. The *battle handover line* is a designated phase line on the ground where responsibility transitions from the stationary force to the moving force and vice versa (ADRP 3-90). The common higher echelon commander of the two forces establishes the BHL after consulting both commanders. The stationary commander determines the location of the line. The BHL is forward of the FEBA in the defense or the FLOT in the offense. The commander draws it where elements of the passing unit can be effectively supported by the direct fires of the forward combat elements of the stationary unit until passage of lines is complete. The area between the BHL and the stationary force belongs to the stationary force commander. The stationary force commander may employ security forces, obstacles, and fires in the area. Figure 6-21 depicts a BHL used in conjunction with other control measures, including passage points and passage lanes, for a rearward passage of lines. (See figures 6-18 and 6-19.) For clarity the date-time groups associated with contact points in this figure are omitted as are passage lane designations and unit designations on AO boundaries. (See

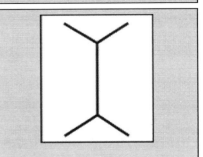

Figure 6-19. Passage lane

A *passage lane* is a lane through an enemy or friendly obstacle that provides safe passage for a passing force (FM 3-90-2). That friendly obstacle may be another unit. That route should allow the passing unit to move rapidly through the stationary unit's area.

FM 3-90-2 for more information on the use of passage points and tactics associated with the conduct of a passage of lines.)

Large-Scale Defensive Operations

6-156. FPFs are the highest type of priority targets, and they take precedence over all other fire requests. (See figure 6-20.) FPFs differ from standard priority targets in that they are fired at the maximum rate of fire until the firing unit is ordered to stop or until all ammunition is expended. They are designed to create a final barrier of fires to prevent the closure of enemy forces on friendly positions. The risk estimate distance for a given delivery system is a factor in how close the FPFs can be placed in front of friendly front lines. Closer FPFs are easier to integrate into direct fire final protective lines. *Danger close* in close air support, artillery, mortar, and naval gunfire support fires, the term included in the method of engagement segment of a call for fire that indicates that friendly forces are within close proximity of the target. (JP 3-09.3) The close proximity distance is determined by the weapon and munition fired.

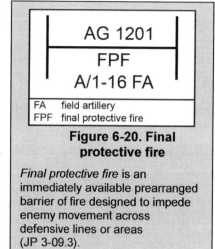

Figure 6-20. Final protective fire

Final protective fire is an immediately available prearranged barrier of fire designed to impede enemy movement across defensive lines or areas (JP 3-09.3).

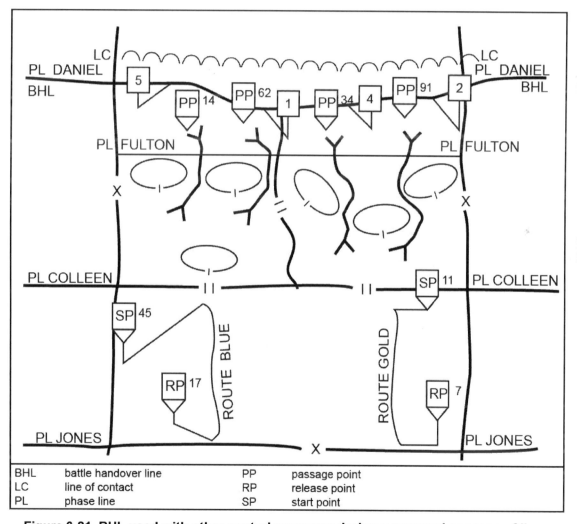

Figure 6-21. BHL used with other control measures during a rearward passage of lines

Executing an Area Defense

6-157. A defending unit within the MBA uses a variety of tactics, techniques, and procedures to accomplish the mission. At one end of the area defensive continuum is a totally static defense oriented on terrain retention. This defense depends on the use of firepower from prepared, protected positions to concentrate combat power against attempted enemy breakthroughs and flanking movements to deny the enemy terrain. At the other end of the area defensive continuum is a dynamic defense focused on the enemy that routinely displaces defending forces from position to position. The defending commander uses mobile forces to cover gaps between defensive positions, reinforces those positions as necessary, and counterattacks to seal penetrations or block enemy attempts at flanking movements. Throughout the area defense continuum, the defending force repeatedly lures or drives the enemy into EAs for destruction in detail. The performance of most area defense falls somewhere between the two extremes.

6-158. Conducting shaping operations in an area defense is similar to shaping operations in the offense. The commander's concept of the operation and the enemy determine how closely the commander synchronizes shaping operations or the supporting effort with the decisive operation or main defensive effort. Commanders conduct shaping operations or supporting efforts to regain the initiative by limiting the attacker's options and disrupting the attacker's plan. Commanders prevent enemy forces from massing and create windows of opportunity for the conduct of spoiling attacks and counterattacks, allowing the defending force to defeat the attacking enemy in detail. Commanders also employ shaping operations and supporting efforts to disrupt enemy operations by attacking enemy CPs at critical stages in the battle or by striking and eliminating key resources, such as wet-gap crossing equipment and supplies in a region that contains numerous unfordable rivers. The performance of reconnaissance and security tasks by subordinate units are normally components of the commander's shaping operations. (See FM 3-90-1 for more information on the area defense task.)

Conducting a Spoiling Attack in the Area Defense

6-159. A *spoiling attack* is a tactical maneuver employed to seriously impair a hostile attack while the enemy is in the process of forming or assembling for an attack (FM 3-90-1). The objective of a spoiling attack is to disrupt the enemy's offensive capabilities and timelines while destroying targeted enemy personnel and equipment, not to seize terrain and other physical objectives. A commander conducts a spoiling attack to—

- Disrupt the enemy's offensive preparations.
- Destroy key assets that the enemy requires to attack, such as fire support systems, fuel and ammunition stocks, and bridging equipment.
- Gain additional time for the defending force to prepare its positions.
- Reduce the enemy's current advantage in the correlation of forces.

6-160. A spoiling attack preempts or seriously impairs the enemy's ability to launch an attack by forcing the enemy to react, disrupting timelines, causing losses, and placing the enemy's plan of attack at risk. Commanders synchronize the conduct of the spoiling attack with other defensive actions. (Figure 6-22 illustrates the concept of a spoiling attack.)

6-161. Commanders conduct a spoiling attack, whenever possible, against an enemy force while it is in assembly areas or attack positions preparing for its own offensive tasks, or when it is temporarily stopped. The forces conducting a spoiling attack require enough combat power to develop the situation. They must be able to defend themselves against those enemy forces that they expect to encounter. A spoiling attack usually employs mechanized forces, attack helicopters, mobile artillery, joint fires, EW, OCO, and MILDEC.

6-162. Commanders can employ reserve forces in a spoiling attack. Commanders assume the risk of temporarily not having a reserve, or they designate another force as the reserve. The following considerations affect the spoiling attack—

- Commanders may want to limit the size of the force used in executing the spoiling attack.
- Spoiling attacks are not conducted if the loss or destruction of the friendly attacking force would jeopardize the commander's ability to accomplish the defensive mission.

Large-Scale Defensive Operations

- The mobility of the force available for the spoiling attack should be equal to or greater than that of the targeted enemy force.
- Operations by artillery or aviation systems to prevent enemy elements not in contact from interfering with the spoiling attack are necessary to ensure the success of the operation.

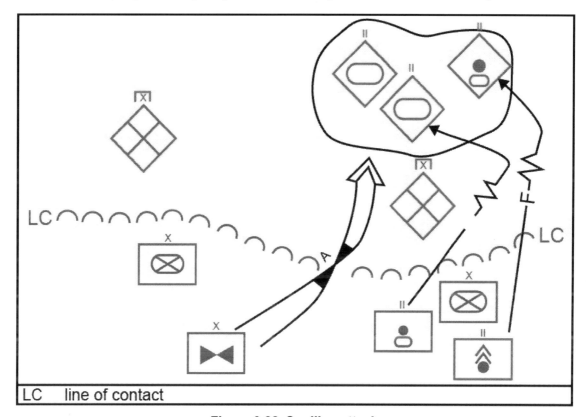

Figure 6-22. Spoiling attack

6-163. There are two conditions that must be met to conduct a successful spoiling attack at an acceptable cost. These conditions are—
- The spoiling attack's objective must be obtainable before the enemy is able to respond to the attack in a synchronized and coordinated manner.
- The commander prevents the force conducting the spoiling attack from becoming overextended.

6-164. If the spoiling attack fails to meet both conditions, it will likely fail. The organization of forces, control measures, and planning, preparation, and execution considerations for a spoiling attack are the same as for any other attack, with the addition of those considerations mentioned in paragraph 6-163.

Conducting a Counterattack During an Area Defense

6-165. A *counterattack* is an attack by part or all of a defending force against an enemy attacking force, for such specific purposes as regaining ground lost or cutting off or destroying enemy advance units, and with the general objective of denying to the enemy the attainment of the enemy's purpose in attacking. In sustained defensive operations, it is undertaken to restore the battle position and is directed at limited objectives (FM 3-90-1). Corps and division commanders direct counterattacks to—
- Defeat or destroy enemy forces.
- Exploit an enemy weakness, such as an exposed flank.
- Regain control of terrain and facilities after an enemy success.
- Regain the initiative from the enemy through offensive action.

Counterattacks are normally conducted from a defensive posture.

Chapter 6

6-166. Counterattack plans include assumptions regarding the size and shape of the anticipated penetration or enemy formation, the strength and composition of the enemy force, and the status of the reserve and forces in the MBA. Commanders plan to conduct counterattacks when and where the enemy is most vulnerable. This is normally when the enemy is fixed while attempting to overcome friendly defensive positions. Other factors that affect a counterattack include the capability to contain an enemy, shaping operations to support the attack, and the strength and responsiveness of the reserve.

6-167. The two levels of counterattacks are major and local counterattacks. In both cases, waiting for the enemy to act first may reveal the enemy's main effort and create an assailable flank to exploit. A defending unit conducts a major counterattack to seize the overall initiative from the enemy through offensive action. Commanders also conduct major counterattacks to defeat or block an enemy penetration that endangers the integrity of the entire defense or to attrit the enemy by the defeat or destruction of an isolated portion of the attacking enemy. Figure 6-23 illustrates the concept of a major counterattack.

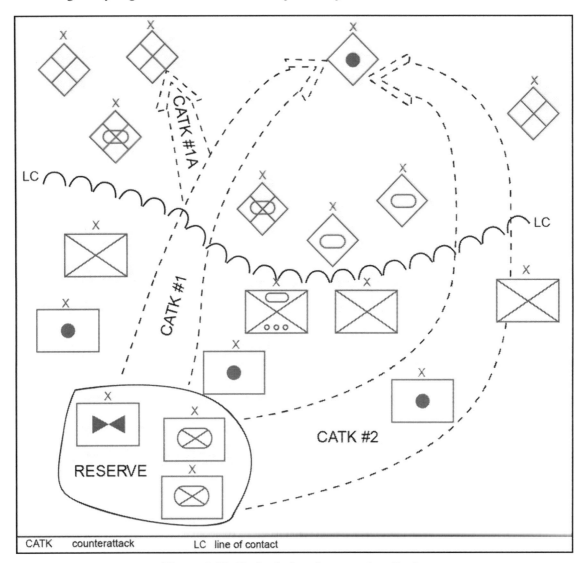

Figure 6-23. Projected major counterattacks

6-168. The counterattacking force maneuvers to isolate and destroy designated enemy forces. It can attack by fire into an EA to defeat or destroy selected enemy forces, restore the original position, or block an enemy penetration. Once launched, a counterattack usually becomes the commander's decisive operation.

6-169. Normally, commanders attempt to retain a reserve or striking force to conduct a decisive counterattack once the enemy main force commits to the attack. Commanders assign objectives to counterattacking forces when they are intended to assault the enemy. Commanders normally assign attack by fire positions when counterattacking using primarily direct and indirect fires. (See FM 3-90-1 for a discussion of the organization of forces, associated control measures, and planning, preparation, and execution considerations associated with performing the counterattack task.)

Consolidation of Gains in the Area Defense

6-170. The unit's defensive plans must address how preparations for, and the conduct of, the area defense impact the civilian population of the AO. This includes the conduct of noncombatant evacuation operations for U.S. civilians and other authorized groups. The commander's legal obligations to that civilian population must be met as long as meeting those obligations does not deprive the defense of necessary resources. Ideally, the host-nation government will have the capability to provide area security for its population and conduct the primary stability tasks. To the extent that a host-nation government is unable to conduct the immediately necessary stability tasks, the defending unit will perform stability tasks within its capability and request further support.

MOBILE DEFENSE

> *Mobile defense is that method of defense in which the forward defensive area is occupied by the minimum forces necessary to warn of impending attack, canalize the attacking forces into less favorable terrain, and block or impede the attacking forces, while the bulk of the defending force is employed in offensive action to destroy the enemy at the time and place most favorable to the defender.*
>
> FM 100-5, *Field Service Regulations: Operations*, 1954

6-171. The *mobile defense* is a defensive task that concentrates on the destruction or defeat of the enemy through a decisive attack by a striking force (ADRP 3-90). The mobile defense focuses on defeating or destroying an enemy by allowing enemy forces to advance to a point where they are exposed to a decisive counterattack by a striking force. The *striking force* is a dedicated counterattack force in a mobile defense constituted with the bulk of available combat power (ADRP 3-90). A fixing force supplements the striking force. Commanders use a fixing force to hold attacking enemy forces in position, to help channel attacking enemy forces into ambush areas, and to retain areas from which to launch the striking force.

6-172. A successful mobile defense requires the orchestration and synchronization of available capabilities across multiple domains to maximize the combat power of the defending units, particularly the striking force. Commanders orchestrate the delivery of their combined effects in order to regain the initiative and seize positions of relative advantage for later exploitation. Corps plans address the performance of information collection, area defensive tasks, and offensive tasks down to the BCT level. Division plans address those same tasks down to the battalion level.

6-173. Plans in a mobile defense address how the conduct of the mobile defense impacts the civilian population of the AO. This is more important during a mobile defense than it is during an area defense because the scope of maneuver tends to be much larger. Civilian attempts to avoid advancing enemy formations and locations where combat occurs will impede the ground maneuver of defending units unless steps are taken to account for their presence and provide alternative routes for these dislocated civilians to use. Commanders communicate these routes to members of the civilian population to ensure they understand. Ideally, host-nation civilian or third-party civilian organizations provide civilian traffic regulation and immediate essential services along those civilian evacuation routes. However, if the host nation cannot perform these tasks, defending units perform them to the extent that their performance does not divert combat power from the defense. Commanders take action to preclude enemy agents from using these routes to infiltrate friendly defensive positions. Simultaneously, commanders meet legal obligations to civilian populations.

6-174. A mobile defense requires an AO with considerable depth. A corps or division commander must be able to shape a battlefield, causing an enemy force to overextend its lines of communications, expose its flanks, and dissipate its combat power. Likewise, commanders must be able to move friendly forces around and behind an enemy force targeted for isolation and destruction.

6-175. A mobile defense is enemy force oriented. A corps or division commander chooses to conduct a mobile defense in two instances: when defending a large AO against a mobile enemy force or when defending against an enemy force with greater combat power but less mobility. A mobile defense incurs great risk, but offers a strong chance of inflicting a decisive defeat and destroying the enemy force.

6-176. When a commander directs a mobile defense, subordinate units conduct either an area defense or a delay to shape the penetration of the enemy attack as part of the fixing force. Operational-level commanders do not assign the mission of a mobile defense to subordinate units smaller than a division because of their lack of resources. Dynamic movement from battle position to battle position characterizes a mobile defense. A commander may direct a BCT to conduct a delay as part of the division mobile defense.

6-177. Commanders may plan to shape the battlefield by defending in one area to deny terrain to the enemy, while they delay in another area to create the illusion of success. Subordinate units conducting a delay can cause the enemy to perceive success, creating an opportunity for the striking force to attack. Commanders may also plan to entice the advancing enemy into an EA by appearing to uncover or weakly defend an area into which the enemy desires to move.

6-178. Commanders may plan one or more spoiling attacks to break up the enemy's momentum, disrupt the enemy's timetable, cause the enemy to shift forces, or allow time for friendly forces to complete their preparations. Planning should ensure the striking force comprises the maximum combat power available to their commander at the projected time of their attack. At a minimum, the striking force's elements have equal or greater combat power than the force they are designed to defeat or destroy. Fire support assets can offset maneuver force shortfalls.

6-179. The plan for the mobile defense incorporates obstacle-restricted zones that allow subordinate units flexibility in positioning. The headquarters' geospatial information support team provides needed terrain products to support the planning process. Specific terrain analysis products help maneuver planning and aid in designing obstacle systems to complement maneuver plans.

Organization of Forces for a Mobile Defense

6-180. Divisions and larger echelon formations normally execute mobile defenses. BCTs and maneuver battalions participate in a mobile defense as part of a fixing force or a striking force because units smaller than a division do not have the capability to fight multiple engagements throughout the width, depth, and height of their AOs while simultaneously resourcing striking, fixing, and reserve forces. BCTs can constitute a portion of the reserve.

6-181. Commanders organize the main body into two principal groups: the fixing force and the striking force. In the mobile defense, reconnaissance and security, reserve, consolidation, and sustaining forces accomplish the same tasks as in an area defense. Commanders complete any required adjustments in task organization before committing subordinate units to combat.

Fixing Force

6-182. The fixing force turns, blocks, and delays the attacking enemy force. The commander organizes it with the minimum combat power needed to accomplish its mission. It tries to shape the enemy penetration or contain the enemy's advance. Typically, it has most of the countermobility assets of the defending unit. The fixing force may conduct defensive actions over considerable depth within the MBA. However, it must be prepared to stop and hold terrain on short notice to assist the striking force on its commitment. The operations of the fixing force establish the conditions for a decisive attack by the striking force at a favorable tactical location. The fixing force executes its portion of the battle essentially as a combination of an area defense and a delaying action. The actions of the fixing force are shaping operations.

Striking Force

6-183. The striking force decisively engages the enemy as attacking enemy forces become exposed in their attempts to overcome the fixing force. The term "striking force" is used rather than reserve because the term

Large-Scale Defensive Operations

"reserve" indicates an uncommitted force. The striking force is a committed force and has the resources to conduct a decisive counterattack as part of the mobile defense. It is the commander's decisive operation.

6-184. The striking force contains the maximum combat power available to a commander at the time of the unit's counterattack. Typically, the striking force in a mobile defense may consist of one-half to two-thirds of the defender's combat power. The striking force is a combined arms force that has greater combat power and mobility than the force it seeks to defeat or destroy. Commanders consider the effects of surprise when determining the relative combat power of the striking force and its targeted enemy unit. The striking force is normally fully task-organized with all functional and multifunctional support and sustainment assets before its actual commitment. Commanders integrate engineer mobility-enhancing assets into the lead elements of the striking force.

6-185. The commander responsible for the overall mobile defense should retain control of the striking force unless communication difficulties make this impossible. Normally this is the corps or division commander. The defending commander's most critical decisions are when, where, and under what conditions to commit the striking force. The defending commander normally accompanies the striking force upon its commitment.

Reserve

6-186. Resourcing a reserve in a mobile defense is difficult and requires commanders to assume risk. Commanders generally use the reserve to support the fixing force. However, if the reserve is available to the striking force, it exploits the success of the striking force.

Control Measures for a Mobile Defense

6-187. A commander conducting a mobile defense uses control measures to synchronize the operation. These control measures include designating the AOs of the fixing and striking forces with their associated boundaries, battle positions, and PLs. The defending commander designates a line of departure or a line of contact as part of the graphic control measures for the striking force. The commander may designate an axis of advance for the striking force. The commander can designate attack by fire or support by fire positions. The commander uses EAs, TRPs, TAIs, and FPFs as necessary. The commander designates NAIs to focus the efforts of information collection assets. This allows the commander to determine the enemy's chosen course of action. The commander designates checkpoints, contact points, passage points, and passage lanes for use by reconnaissance and surveillance assets, security units, and the striking force. (See figure 6-24.)

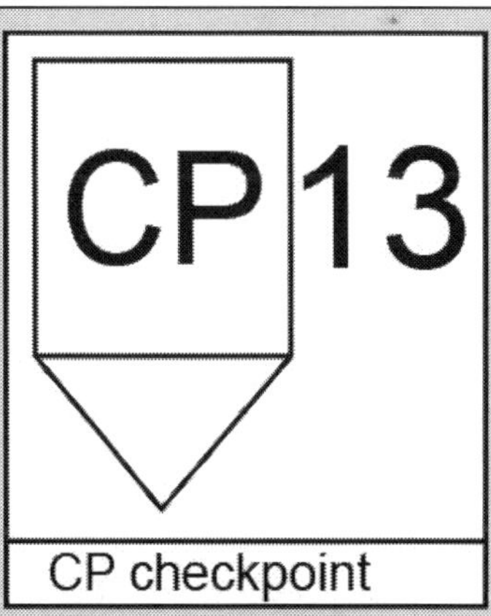

6-188. Commanders and staffs consider not only current operations but also future operations when planning obstacles. Effective obstacles block, turn, or force an enemy to attempt to breach them. Commanders integrate obstacles with fires for greater effectiveness. Commanders design the emplacement of obstacles supporting current operations so that they do not hinder future operations. Any commander authorized to employ obstacles can designate certain obstacles to shape the battlefield as high-priority reserve obstacles. Commanders assign responsibility for preparation to a subordinate unit but retain authority for ordering their completion. One example of a reserve obstacle is a highway bridge over a major river designated for demolition. Such obstacles receive the highest priority in preparation and, if ordered, execution by the designated subordinate unit.

Figure 6-24. Checkpoint

A *checkpoint* is a predetermined point on the ground used to control movement, tactical maneuver, and orientation (ADRP 1-02).

Executing a Mobile Defense

6-189. A commander executing a mobile defense must have the flexibility to yield terrain and shape the enemy penetration. A commander may entice an enemy by appearing to uncover an objective of value to the enemy. The striking force maneuvers to conduct the decisive operation—the

Chapter 6

counterattack—once the results of the actions of the fixing force shape the situation to meet the commander's intent.

6-190. Fixing the enemy establishes the conditions necessary for decisive operations by the striking force. Typically, the commander of the defending force allows the enemy force to penetrate the MBA before the striking force attacks. (Figure 6-25 illustrates this concept.) The fixing force may employ a combination of area defense, delay, and strong point techniques to shape the enemy penetration. The intent of the fixing force is not necessarily to defeat the enemy but to shape the penetration to facilitate a decisive counterattack by the striking force. Commanders ensure that the missions and task organization of subordinate units within the fixing force are consistent with the concept for shaping the enemy penetration. Defensive positions within the fixing force may not be contiguous, since the fixing force contains only the minimum-essential combat power to accomplish its mission.

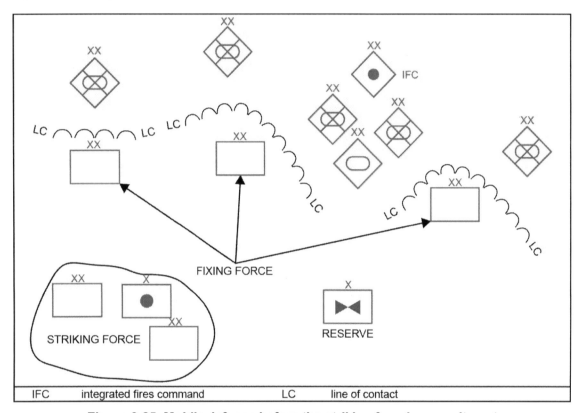

Figure 6-25. Mobile defense before the striking force's commitment

6-191. The defending commander launches the striking force in a counterattack when its offensive power, relative to that of the targeted enemy element, is the greatest. Piecemeal commitment of the striking force in support of local objectives jeopardizes the success of the overall operation. The striking force must execute the counterattack rapidly and violently, employing all combat power necessary to ensure success. The striking force may be committed at a different time than anticipated and in an entirely different area than planned. Thus, it must be able to respond to unexpected developments rapidly and decisively. Figure 6-26 illustrates the mobile defense after commitment of the striking force. (See FM 3-90-1 for additional information on the execution of mobile defense operations.)

RETROGRADE

6-192. The *retrograde* is a defensive task that involves organized movement away from the enemy (ADRP 3-90). An enemy may force these operations, or a commander may execute them voluntarily. In either case, the higher echelon commander of a force executing a retrograde must approve the retrograde operation before its initiation. A retrograde is a transitional operation. It is not conducted in isolation. It is part of a larger scheme of maneuver designed to regain the initiative and defeat the enemy.

Large-Scale Defensive Operations

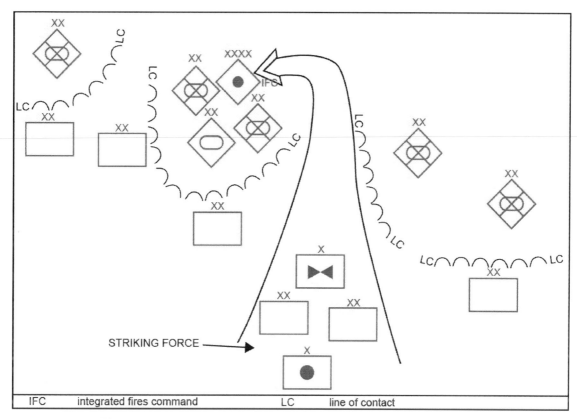

Figure 6-26. Mobile defense after commitment of striking force

6-193. The three forms of the retrograde are delay, withdrawal, and retirement. A *delaying operation* is an operation in which a force under pressure trades space for time by slowing down the enemy's momentum and inflicting maximum damage on the enemy without, in principle, becoming decisively engaged (JP 3-04). In delays, units yield ground to gain time while retaining flexibility and freedom of action to inflict the maximum damage on an enemy. A *withdrawal operation* is a planned retrograde operation in which a force in contact disengages from an enemy force and moves in a direction away from the enemy (JP 3-17). Withdrawing units, whether all or part of a committed force, voluntarily disengage from an enemy to preserve the force or release it for a new mission. A withdrawal is the riskiest of all operations. Deception operations can reduce this operational risk. A *retirement* is a form of retrograde in which a force out of contact moves away from the enemy (ADRP 3-90). In each form of the retrograde, a force not in contact with an enemy moves to another location, normally by a tactical road march. In all retrograde operations, firm control of friendly maneuver elements is a prerequisite for success.

6-194. Friendly forces conducting retrograde operations make maximum use of available combat multipliers to reduce risk. These combat multipliers are not necessarily organic to the retrograding force. Examples of non-organic combat multipliers include the use of EW to obscure portions of the EMS that enemy technical systems need to detect the rearward movement of friendly forces. Non-organic combat multipliers could also include MILDEC activities by other units.

Delay

6-195. The corps or division plan specifies certain parameters to the subordinate units conducting the delay. First, it directs one of two alternatives: delay within their AOs or delay forward of a specified trigger line or terrain feature for a specified time. (See figure 6-27 on page 6-44.) The second parameter is that the order specifies the acceptable risk. The commander prescribes the delaying force's mission, composition, and initial location. The delaying force accomplishes its mission by delaying on successive positions, by delaying on alternate positions, or by a combination of the two. It also attacks, defends, feints, or demonstrates. (See

Chapter 6

chapter 9 of FM 3-90-1 for additional planning, preparation, and execution considerations for conducting a delay.)

6-196. The delay is one of the most demanding ground combat operations. A delay wears down the enemy so that friendly forces can regain the initiative through offensive action, buy time to establish an effective defense, or determine enemy intentions as part of the performance of echelon security tasks. Normally in a delay, inflicting casualties on the enemy is secondary to gaining time. For example, a flank security force conducts a delay operation to provide time for the protected force to establish a viable defense along its threatened flank. Except when directed to prevent enemy penetration of a PL for a specific duration, a force conducting a delay normally does not become decisively engaged. The fixing force in a mobile defense might also conduct a delay to maneuver the attacking enemy force into a location where the striking force can make a decisive attack on enemy force elements to defeat that enemy force in detail.

> A *disengagement line* is a phase line located on identifiable terrain that, when crossed by the enemy, signals to defending elements that it is time to displace to their next position (ADRP 3-90). The commander uses this line in the delay and the defense when the commander does not want the defending unit to become decisively engaged. The commander establishes criteria for the disengagement, such as number of enemy vehicles by type, friendly losses, or enemy movement to flanking locations. Commanders may designate multiple disengagement lines, one for each system in the defense.

6-197. A delay operation can occur when a commander does not have enough friendly forces to attack or defend. It may also occur, based on a unit's mission, in conjunction with a higher echelon commander's intent. The decision to conduct a delay may not be based on the unit's combat power, but because of other reasons. For example, during the performance of security tasks, a commander may conduct a delay as a shaping operation to draw an enemy into an area where the attacking enemy force is vulnerable to a counterattack. Another example is when a commander directs a delay as an economy of force measure to allow the commander to mass combat power and conduct offensive actions elsewhere.

6-198. The ability of a force to trade space for time requires depth within the AO assigned to the delaying force. The amount of depth required depends on several factors, including the—

- Amount of time to be gained.
- Relative combat power of friendly and enemy forces.
- Relative mobility of the forces.
- Nature of the terrain.
- Ability to shape the AO with obstacles and fires.
- Degree of acceptable risk.

Ordinarily, the greater the available depth within an AO, the lower the risk involved to the delaying force and the greater the chance for success.

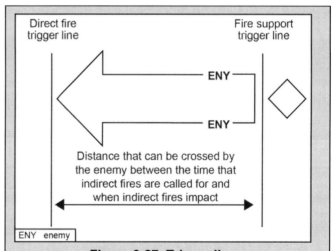

Figure 6-27. Trigger line

A *trigger line* is a phase line located on identifiable terrain that crosses the engagement area that is used to initiate and mass fires into an engagement area at a predetermined range for all or like weapon systems (ADRP 1-02). The commander can designate one trigger line for all weapon systems or separate trigger lines for each weapon or type of weapon system.

Organization of Forces for a Delay

6-199. A commander normally organizes the delaying force into a main body, a security force, and a reserve. The security force usually conducts a screen forward of the initial delay positions. The security force may be a reinforced armored brigade combat team (ABCT) or Stryker brigade combat team (SBCT) for a corps or division conducting a delay. The corps security force could be a division-size element. For a BCT conducting a delay, the security force may consist of the BCT cavalry squadron or a battalion task force.

Control Measures for a Delay

6-200. A delay consists of a series of mutually supporting independent small actions occurring simultaneously across the FLOT. In a delay, subordinates require freedom of action. Common graphics used in a delay include AOs, PLs, battle positions, contact points, checkpoints, EAs, trigger lines, TRPs, and disengagement lines. The delaying commander designates contact points in front of, between, and behind units to assist coordination, ensure continuity of the delay, and draw attention to enemy avenues of approach into unit flanks.

6-201. The unit commander designates additional PLs beyond those established by the higher echelon commander as necessary to control the unit's movement during the delay. A *delay line* is a phase line where the date and time before which the enemy is not allowed to cross the phase line is depicted as part of the graphic control measure (FM 3-90-1). Designating delay lines is a command decision that imposes a high degree of risk on the delaying unit. The delaying unit must do everything in its power, including accepting decisive engagement, to prevent the enemy from crossing that line before the time indicated. A delay line may also be event driven. For example, a commander can order a delaying unit to prevent penetration of the delay line until supporting engineers complete construction of a rearward obstacle belt. (See FM 3-90-1 for additional information on the conduct of delay operations.)

Withdrawal

6-202. Withdrawal planning begins with preparation of the plan for the next mission. Once the new plan is drawn up, the corps, division, and subordinate staffs make plans for the withdrawal, including—
- The location, composition, and mission of security forces.
- The organization of the withdrawing force for combat.
- Control measures, including routes, traffic control posts, and PLs.
- Fire support plans.
- Sustainment priorities.
- MILDEC operations to preserve the force.

6-203. Normally, the corps or division withdrawal plan employs a covering force to preserve the command's integrity. Therefore, when planning to conduct a withdrawal under pressure, a commander resources a strong covering force. The capabilities of ABCTs and SBCTs make them most suitable for use as a covering force during the conduct of a withdrawal. An IBCT needs reinforcement by additional fires, anti-armor, engineer, aviation, and transportation assets before it can act as a covering force.

6-204. Withdrawals are inherently dangerous because they involve moving units to the rear and away from what is usually a stronger enemy force. The heavier the previous fighting and the closer the contact with the enemy, the more difficult the withdrawal. OPSEC is extremely important. A unit usually confines its rearward movement to times and conditions when the advancing enemy force cannot observe the activity so that the enemy cannot easily detect the operation. To help preserve secrecy and freedom of action, for example, commanders must consider visibility conditions and times when enemy reconnaissance satellites can observe friendly movements. OPSEC is especially critical during the initial stages of a delay when most of the functional and multifunctional support and sustainment elements displace.

6-205. A unit withdraws to an assembly area or a new defensive position. Alternatively, it can withdraw indirectly to either area through one or more intermediate positions. When preparing a new position, the commander balances the need for security with the need to get an early start on the defensive effort.

Organization of Forces for a Withdrawal

6-206. Commanders typically organize a withdrawing unit into a security force, a main body, and a reserve. Commanders also organize a detachment left in contact and stay-behind forces if the scheme of maneuver requires them. Commanders avoid changing task organization unless circumstances dictate rapid task

Chapter 6

organization changes immediately before the withdrawal, such as when the unit must conduct an immediate withdrawal to prevent encirclement.

Control Measures for a Withdrawal

6-207. Withdrawing requires careful coordination among all forces involved. Throughout the operation, commanders must tightly control rearward movement and maintain the ability to concentrate decisive combat power at key times and places. The control measures used in a withdrawal are the same as those in a delay or defense. The routes used by each unit in a withdrawal and the block movement times are also withdrawal control measures. (See FM 3-90-1 for additional information on the conduct of withdrawal operations.)

Retirement

> *A retreating force covers its retirement by the detail of a rear guard. When troops have been in action, the rear guard is constituted from the least tried available units.*
>
> Field Service Regulations United States Army, 1923

6-208. A retirement is an operation in which a force out of contact moves away from the enemy. Retirement operations are administrative in nature. However, commanders and staffs consider security throughout the planning process. As in all tactical movements, all-round security of the main body is necessary using advance, flank, and rear security forces. Branches to the plan address enemy capabilities to employ Level I, II, and III threats against retiring units during their movement. Level I threat considerations address terrorist attacks along movement routes.

6-209. Commanders usually conduct retirement operations to reposition forces for future operations or to accommodate the current concept of operations. A retiring unit organizes for combat, but it does not anticipate interference from enemy ground forces. Typically, another unit's security force provides security during the movement of the unit conducting a retirement. However, mobile enemy forces, unconventional forces, air strikes, air assault operations, or long-range fires may attempt to interdict the retiring unit. Commanders must plan for enemy actions and organize the unit for self-defense.

6-210. When a withdrawal from action precedes a retirement, the actual retirement begins after the unit breaks contact and organizes into its march formation. While a force withdrawing without enemy pressure can also use march columns, the difference between the two situations is the probability of enemy interference. Units conduct retirements as tactical road marches where security and speed are the most important considerations.

Organization of Forces for a Retirement

6-211. Commanders normally designate forces to provide advance, rear, and flank security to the main body in a retirement. The formation and number of columns employed during a retirement depend on the number of available routes and the potential for enemy interference. Commanders typically move major elements to the rear simultaneously. However, a limited road net or a flank threat may require staggering a movement in terms of time and ground locations.

Control Measures for a Retirement

6-212. The control measures used in a retirement are the same as those in a delay and a withdrawal. As in a withdrawal, thorough planning and strict adherence to routes and movement times facilitate an orderly retirement. Typically, commanders control movement using movement times, routes, and checkpoints. (See FM 3-90-2 for more information on these movement control measures. See FM 3-90-1 for additional information on the conduct of retirement operations.)

DEFENDING ENCIRCLED

6-213. Defending forces can become encircled at any time during large-scale combat operations. This is especially true during noncontiguous operations. An encircled force can continue to defend encircled, conduct a breakout, exfiltrate toward other friendly forces, or attack deeper into enemy-controlled territory

(see paragraphs 6-223 through 6-224). A commander's form of maneuver once becoming encircled depends on the senior commander's intent and the mission variables, including the—
- Availability of defensible terrain.
- Relative combat power of friendly and enemy forces.
- Sustainment status of the encircled force and its ability to be resupplied, including the ability to treat and evacuate wounded Soldiers.
- Morale and fighting capacity of the Soldiers.

6-214. Encirclement of a friendly force is most likely to occur during highly mobile and fluid operations, or when operating in restrictive terrain. A unit may find itself encircled as a result of its offensive actions, as a detachment left in contact, when defending a strong point, when occupying a combat outpost, or when defending an isolated defensive position. Commanders anticipate becoming encircled when assigned a stay-behind force mission, when occupying either a strong point or a combat outpost, or at the outset of forcible entry operations before a linkup is completed. The principles of defending encircled also apply to base and base cluster defense in support and consolidation areas.

6-215. The senior commander (except Army Medical Department officers) in an encirclement assumes command over all encircled forces and takes immediate action to protect them. When that commander determines the unit is about to be encircled, the commander must decide quickly what assets stay and what assets leave. The commander immediately informs higher headquarters of the situation. Simultaneously, the commander directs the accomplishment of the following tasks:
- Establish security.
- Reestablish a chain of command.
- Establish a viable defense.
- Maintain morale.

6-216. Commanders position security elements as far forward as possible to reestablish contact with the enemy and provide early warning. Vigorous patrolling begins immediately. Each unit clears its position to ensure that there are no enemy forces in the perimeter. Technical assets, such as Joint Surveillance Target Attack Radar System and EW systems, augment local security elements and locate those areas along the perimeter where the enemy is deploying additional forces.

6-217. The encircled commander reestablishes unity of command. The commander reorganizes any fragmented units and places Soldiers separated from their parent units under the control of other units. The commander establishes a clear chain of command throughout the encircled force, reestablishes communications with units outside the encirclement, and adjusts command and support relationships to reflect the new organization.

ORGANIZATION OF FORCES WHEN DEFENDING ENCIRCLED

6-218. To establish a viable defense, the commander of an encircled force establishes a perimeter defense. The commander knows the specific capabilities and limitations of the different friendly units isolated in the encirclement. Therefore, the commander designs the defense to maximize the capabilities of available forces. Forward units establish mutually supporting positions around the perimeter and in depth along principal avenues of approach. Units occupy the best available defensive terrain. Commanders may attack to seize key or decisive terrain so that it is incorporated into the perimeter defense. Once the commander assigns defensive AOs and battle positions, the preparations are the same as in the area defense. Encircled units make their defensive positions as strong as possible, given time and resource constraints. The commander anticipates that the enemy will attempt to split the defenses of the encircled force and defeat it in detail. (See chapter 6 of FM 3-90-1 for more information on the conduct of a perimeter defense.)

6-219. The encircled force commander establishes a reserve that is mobile enough to react quickly to events anywhere along the perimeter. Given the availability of sufficient fuel, the commander constitutes a reserve using armored combat vehicles from ABCT and SBCT elements trapped within the encirclement. The encircled commander centrally positions this mobile reserve to take advantage of interior lines, which exist if the encircled force commander can maneuver the reserve or reinforce threatened positions on the perimeter faster than the enemy can shift location or reinforce. The encircled commander can achieve interior lines through a central position (with operations diverging from a central point), from superior lateral lines of communications, or superior tactical mobility. If only dismounted infantry forces are available, the encircled commander establishes small local reserves positioned at key points around the perimeter to react to potential threats. The encircled commander organizes a mobile anti-armor element from the best available anti-armor systems. If possible, subordinate units also retain their own reserves.

CONTROL MEASURES WHEN DEFENDING ENCIRCLED

6-220. The commander in a perimeter defense designates the trace of the perimeter, battle positions, contact points, and lateral and forward boundaries. The commander designates checkpoints, contact points, passage points, and passage routes for use by local reconnaissance, surveillance, and security elements operating outside the boundary of the perimeter. Commanders can use EAs, TRPs, and FPFs as fire control measures. (See figure 6-28 for a depiction of a perimeter defense employing EAs, TRPs, and FPFs.)

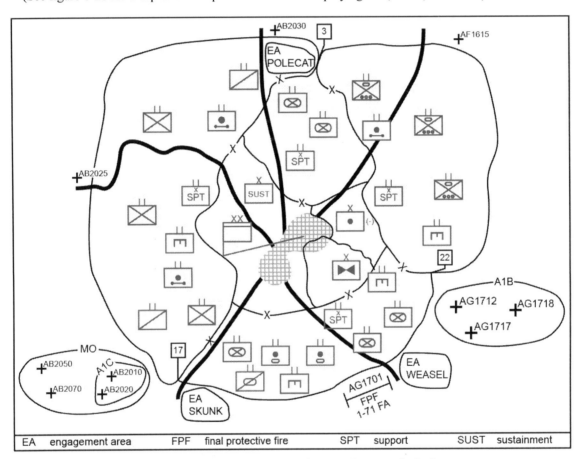

Figure 6-28. Perimeter defense when encircled

BREAKOUT FROM AN ENCIRCLEMENT

6-221. A *breakout* is an operation conducted by an encircled force to regain freedom of movement or contact with friendly units. It differs from other attack only in that simultaneous defense in other areas of the perimeter must be maintained (ADRP 3-90). A breakout can occur in both the offense and the defense. An encircled force normally attempts to conduct breakout operations when one of the following four conditions exist:

- The senior commander directs the breakout or the breakout falls within the intent of a higher echelon commander.
- The encircled force does not have sufficient relative combat power to defend itself against enemy forces attempting to reduce the encirclement.

- The encircled force does not have adequate terrain available to conduct its defense.
- The encircled force cannot sustain itself long enough to be relieved by forces outside the encirclement.

EXFILTRATION

6-222. If the success of a breakout attack appears questionable, or if it fails and a relief operation is not planned, one way to preserve a portion of the force is through organized exfiltration. (See FM 3-90-2 for a detailed description of exfiltration.)

ATTACKING DEEPER INTO ENEMY TERRITORY

6-223. A course of action that an enemy is not likely to expect from an encircled force is to attack deeper into enemy territory to seize key terrain. This type of attack involves great risk, but it may offer the only feasible course of action under some circumstances. Attacking may allow an encircled unit to move to a location where it can be extracted by other ground, naval, or air forces. Attacking deeper is only feasible if a unit can sustain itself while isolated, although some of that sustainment can come from aerial resupply and enemy supply stocks.

6-224. When an enemy is attacking, an encircled friendly force that attacks deeper into the enemy rear area may disrupt the enemy's offense and provide an opportunity for linkup from another direction. If an enemy is defending and the attacking force finds itself isolated through the performance of its own offensive tasks, it may continue the attack toward its assigned objective or a new objective located on more favorable defensive terrain.

LINKUP

6-225. A *linkup* is a meeting of friendly ground forces, which occurs in a variety of circumstances (ADRP 3-90). It happens when an advancing force reaches an objective area previously seized by an airborne attack or air assault. It occurs when an encircled element breaks out to rejoin friendly forces or a force comes to the relief of an encircled force. It also occurs when converging maneuver forces meet. Both forces may be moving toward each other, or one may be stationary. Whenever possible, joining forces exchange as much information as possible before starting an operation. The headquarters ordering the linkup establishes—
- A COP using available mission command systems.
- Command relationship and responsibilities of each force before, during, and after linkup.
- Coordination of fire support before, during, and after linkup, including control measures.
- Linkup method.
- Recognition signals and communication procedures, including pyrotechnics, armbands, vehicle markings, gun tube orientation, panels, colored smoke, lights, and challenge and passwords.
- Operations to conduct following linkup.

CONSOLIDATION OF GAINS IN THE DEFENSE

6-226. Commanders have few resources to devote to the consolidation of gains during tactical operations focused on the defense. There is no new ground to consolidate gains on unless the defense follows an offensive phase. The primary consolidation of gains activities during operations focused on the conduct of defensive tasks ensure that the future conduct of consolidation of gains tasks are adequately addressed in sequel and branch plans to the current operations order. (See chapter 8 for more information on consolidation of gains.)

This page intentionally left blank.

Chapter 7
Large-Scale Offensive Operations

On no account should we overlook the moral effect of a rapid, running assault. It hardens the advancing soldier against danger, while the stationary soldier loses his presence of mind.

Carl von Clausewitz

This chapter begins with a general discussion of the offense, followed by a discussion of how an enemy defense may be arrayed. It continues with a section on how corps and divisions headquarters plan for the offense. This chapter then provides a discussion of forms of maneuver and the four offensive tasks. The chapter concludes with a discussion on the subordinate forms of attack.

OVERVIEW OF LARGE-SCALE OFFENSIVE OPERATIONS

7-1. An *offensive task* is a task conducted to defeat and destroy enemy forces and seize terrain, resources, and population centers (ADRP 3-0). Offensive tasks impose the commander's will on the enemy. Against a capable, adaptive enemy, the offense is the most direct and sure means of seizing, retaining, and exploiting the initiative to gain physical, temporal, and cognitive advantages and achieve definitive results. In the offense, the decisive operation is a sudden, shattering action against an enemy weakness that capitalizes on speed, surprise, and shock. If that operation does not destroy the enemy, operations continue until enemy forces disintegrate or retreat to where they no longer pose a threat. Executing offensive tasks compels the enemy to react, creating or revealing additional weaknesses that the attacking force can exploit.

7-2. There are four primary offensive tasks: movement to contact, attack, exploitation, and pursuit. (Paragraphs 7-120 to 7-227 discuss these tasks.) Units perform these tasks singularly or in combination to impose the commander's will on the enemy. A commander may also conduct offensive tasks to deprive the enemy of resources, seize decisive terrain, deceive or divert the enemy, develop intelligence, or hold an enemy in position. This chapter discusses the offensive basics at the corps and division echelons.

7-3. Commanders conducting offensive tasks employ the four defeat mechanisms—destroy, dislocate, disintegrate, and isolate—in various combinations to accomplish their mission against enemy opposition. (Chapter 1 discusses these four mechanisms.) Commanders seize, retain, and exploit the initiative when performing offensive tasks. Specific offensive tasks may orient on a specific enemy force or terrain feature as a means of affecting the enemy. Even when performing primarily defensive tasks, wresting the initiative from the enemy requires the performance of offensive tasks. The offense can also be used to support friendly operations in other domains, including air, maritime, space, cyberspace, and the information environment by employing the defeat mechanisms against key points or assets that support an enemy's operations.

7-4. Characteristics of the offense include audacity, concentration, surprise, and tempo. The performance of effective offensive tasks capitalizes on accurate and timely intelligence and other relevant information regarding enemy forces, weather, and terrain. Commanders maneuver forces to positions of relative advantage before contact. Contact with enemy forces before the decisive operation is deliberate and designed to shape the optimum situation for the decisive operation. The decisive operation that conclusively determines the outcome of the major operation, battle, or engagement capitalizes on subordinate initiative and shared understanding. Without hesitation, commanders violently execute both movement and fires—within the higher commander's intent—to break the enemy's will or destroy the enemy. (See ADRP 3-90 for a discussion of these four characteristics.)

Chapter 7

7-5. Army corps and divisions conduct offensive tasks within a broad operational scheme. At the operational level, offensive operations directly or indirectly attack the enemy center of gravity. Land component commanders conduct large-scale combat operations to achieve theater-level effects based on tactical actions. Army special operations forces (ARSOF) conduct precision strike operations against key nodes to achieve operational and strategic effects. ARSOF employ indigenous elements and conduct irregular warfare activities in support of joint force objectives. Commanders do this by attacking enemy decisive points, either simultaneously or sequentially. To attain unity of effort, operational commanders clearly describe their intent, identify objectives, and reinforce the relationships among subordinates.

> **Intermittent Communications, Continuous Mission Command: Grant and Sherman Deep Operational Maneuver 1864–1865**
>
> As the 1864 campaign season opened, Union forces attempted to achieve decisive military results after years of operational stalemate in the East by advancing into the heart of the Confederacy along multiple axes to destroy its ability to resist. Though General Ulysses S. Grant in the East and General William T. Sherman in the West would make the main efforts, supporting columns would also tie down defending Confederate forces. As President Lincoln put it, "Those not skinning can hold a leg!"
>
> With the bulk of operationally effective southern forces facing them in Virginia and Georgia, both Grant and Sherman embarked on deep penetration campaigns into the heart of the Confederacy. Grant's drive overland in Virginia drew the strongest opposition in the form of Robert E. Lee's Army of Northern Virginia as it fought its way into the Virginia Tidewater, maneuvered past Richmond, and fixed Lee in defensive positions at Petersburg. Meanwhile, Sherman maneuvered Joe Johnston out of a series of strong defensive positions in the north Georgia mountains before capturing the vital rail center of Atlanta. A supporting column under General Phil Sheridan in Virginia's Shenandoah Valley devastated that grain-producing region, denying vital supplies to Lee's army. After destroying Atlanta, Sherman's command embarked on the "March to the Sea" in late 1864, splitting the eastern Confederacy in two and setting conditions for a subsequent devastating campaign through the Carolinas.
>
> By April 1865, when Grant finally maneuvered Lee out of Petersburg, Sherman's command had marched over 700 miles, destroying agricultural and industrial infrastructure critical to sustaining the Confederate economy and eroding its will to resist. When Lee surrendered, Sherman was only a few days' march away and effectively blocked Lee's escape. By maneuvering through the depths of the Confederacy along multiple independent axes united by a single operational purpose, Union forces created unsolvable military and political dilemmas for the Confederacy and successfully brought the American Civil War to a close.

ENEMY DEFENSE

7-6. During defensive tasks, the enemy typically attempts to slow and disrupt friendly forces with a combination of obstacles, prepared positions, and favorable terrain so that they can be destroyed with massed fires. The enemy is likely to defend in depth, and when provided time, will continuously improve positions in ways that better protect enemy defending units, make attacks against them more costly, and allow the enemy to commit the minimum amount of ground combat power forward. Forward positioned enemy forces are heavily focused on providing observed fires for long range systems and slowing friendly forces long enough to be engaged effectively by those systems. The enemy is likely to conduct a mobile defense whenever capable, using a series of subsequent battle positions to achieve depth. The enemy commander seeks to use fires and obstacles to prevent decisive engagement of the defending ground forces as they reposition, while causing friendly forces to move methodically under continuous fire without ever fixing the

Large-Scale Offensive Operations

enemy's own forces. The enemy can be expected to employ significant electronic warfare (EW), intelligence, surveillance, reconnaissance, and information-related capabilities as part of this defensive effort. Several potential enemies can employ chemical weapons, and some could employ tactical nuclear weapons.

7-7. The enemy main defense zone is organized in a succession of integrated kill zones, obstacles, and battle positons. Figure 7-1, figure 7-2 on page 7-4, and figure 7-3 on page 7-5 show a typical peer enemy defensive scheme.

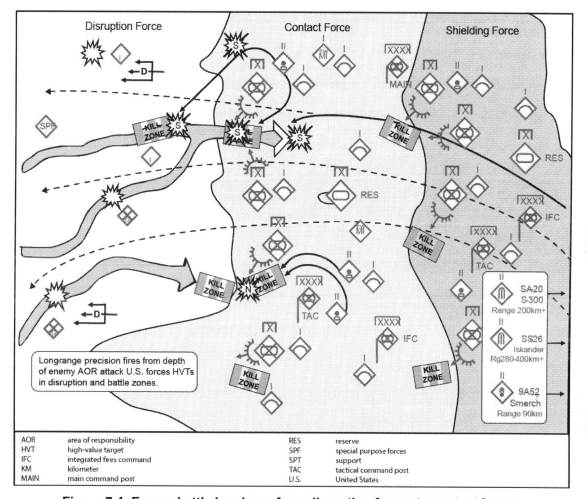

Figure 7-1. Enemy battle handover from disruption forces to contact forces

Chapter 7

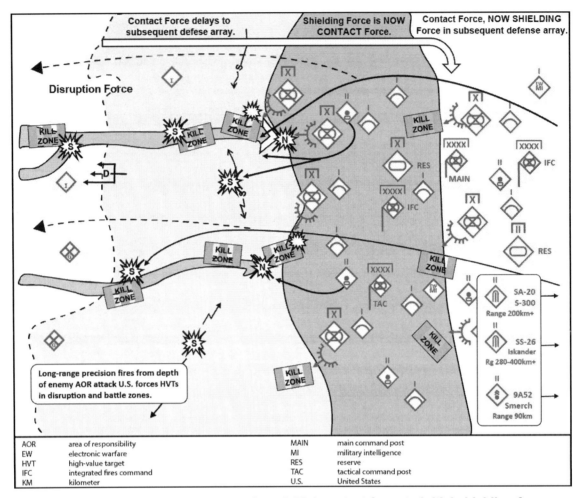

Figure 7-2. Enemy battle handover from initial contact force to initial shielding force

Large-Scale Offensive Operations

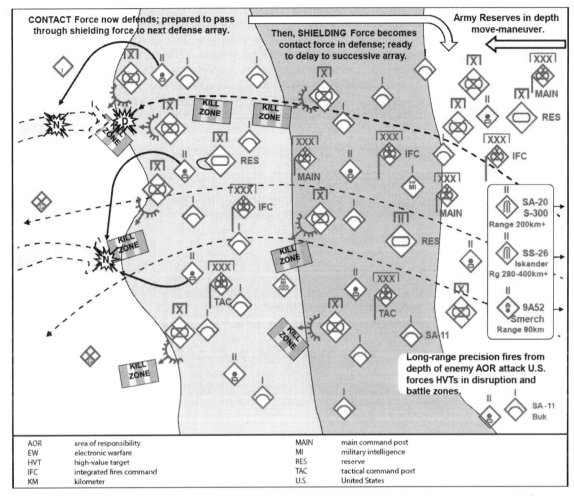

Figure 7-3. Enemy continuous maneuver defense in depth within enemy main defense zone

PLANNING FOR THE CORPS AND DIVISION IN THE OFFENSE

In case of doubt, attack.

General George Patton

7-8. The Army's framework for exercising mission command is the operations process. The major mission command activities performed during the conduct of offensive tasks are planning, preparing, executing and continuously assessing the operation. The offense keeps the enemy off balance, exploits positions of relative advantage, and continuously reduces the enemy commander's options across all domains and the information environment. Planning results in a plan and orders that synchronize the action of forces in time, space, and purpose to achieve objectives and accomplish missions. (See ADRP 5-0 for a discussion of planning.)

7-9. Corps and divisions plan their battles within areas of operations (AOs) assigned by their higher headquarters. The corps or division commander structures an AO using the operational framework to assign objectives, boundaries, phase lines (PLs), and other control measures and authorities within that AO to ensure the successful synchronization and timely convergence of subordinate and supporting formations and capabilities against the enemy. Commanders base many control measures on probable lines of contact in different domains.

7-10. Corps and division headquarters ensure that their offensive tasks and the actions of joint and multinational forces assigned to or supporting their organizations are synchronized to maximize the application of combat power. Normally, corps and division commanders assign specific AOs for performing offensive tasks to define responsibilities to subordinate units. Based on the nature of the enemy and the mission, these subordinate AOs may be contiguous or noncontiguous.

7-11. Corps and division commanders structure the AO according to their vision. In some cases, the corps or division conducts deep operations to isolate the enemy in the deep area, thus setting the conditions necessary for subordinate brigade combat teams (BCTs) to be successful in the close area. In other instances, the corps or division conducts a penetration of enemy defenses in their close area only as a means to maneuver forces deep into the enemy's defenses to destroy enemy forces and capabilities arrayed in depth.

7-12. Commanders consider the mission variables when designating objectives. Special considerations should be given to the enemy's dispositions in depth and the combat effectiveness of subordinate units. Commanders can determine the scheme of maneuver, allocation of available forces, scheme of fires, maneuver support, and scheme of sustainment only after selecting objectives. Commanders also consider the shaping of the AO when the mission transitions from a focus on the offense to the consolidation of gains. For example, targeting an electrical power station for destruction in the offense should be weighed against the need for power when conducting stability tasks during the consolidation of gains.

7-13. After designating objectives, corps and division commanders and staffs consider six complementary elements when planning offensive tasks:
- Scheme of maneuver.
- Continuous deep operations directed against key portions of the enemy defense within the AO, electromagnetic spectrum (EMS), cyberspace, or population.
- Reconnaissance and security tasks conducted forward and to the flanks and rear of the unit's main and supporting attacks.
- Decisive operations and main attacks with shaping operations and supporting actions, such as economy of force activities and the conduct of various reconnaissance and security, movement, mobility, and countermobility tasks as required.
- Reserve operations in support of the offense.
- Sustainment and consolidation of gains operations necessary to maintain offensive momentum.

7-14. The six offensive task planning elements serve as a useful guide in formulating corps and division concepts of operations and in facilitating the orchestration and internal synchronization of activities in the close, deep, support, and consolidation areas. Commanders should consider where friendly force capabilities can converge to achieve the greatest possible relative advantage that can be rapidly exploited. Attempting to engage enemy capabilities symmetrically (for example, using long range fires against enemy long range fires systems) may not generate enough effects to facilitate seizing the initiative.

7-15. *Fire superiority* is that degree of dominance in the fires of one force over another that permits that force to conduct maneuver at a given time and place without prohibitive interference by the enemy (FM 3-90-1). Fire superiority applies to more than the land domain. It applies to the cyberspace domain, the information environment, and the EMS as well. Army and joint forces plan and execute (with proper authorities) effects on, in, and through cyberspace and the EMS. Corps and division plans focus the effects of friendly systems, including those supporting EW, offensive cyberspace operations (OCO), and information-related capability assets assigned to higher echelon headquarters, to achieve fire superiority and allow friendly maneuver forces to breach enemy defensive networks. Corps and divisions must gain and maintain fire superiority at critical points and across multiple domains during their performance of offensive tasks. Having fire superiority allows commanders to maneuver forces without prohibitive losses. Commanders gain fire superiority by using counterfires, precision fires, suppression, and the destruction of key assets, such as radars and enemy collection platforms. Achieving fire superiority requires the commander to take advantage of temporary positions of relative advantage using—

Large-Scale Offensive Operations

- The range, precision, and lethality of available weapon systems.
- A blend of friendly information management, knowledge management, information collection, and EW, OCO, and information operations. (Only forces with proper legal and command authority can create offensive effects in cyberspace or the EMS. Echelons that do not have organic capabilities or authorities for cyberspace or EW operations may integrate supporting effects from forces with those capabilities to support their operations.)
- Maneuver to place the enemy in positions of disadvantage across multiple domains and the information environment where enemy weapon systems and logical entities can be destroyed or disrupted, one or more at a time, with reduced risk to friendly systems.

SCHEME OF MANEUVER

7-16. A scheme of maneuver expands the commander's intent and describes a method of achieving the desired outcome. It describes this in terms of where, when, how, and with what offensive activity needs to occur to achieve the desired end state. Commanders select a scheme of maneuver that seeks to gain a position of advantage over the enemy to rapidly close with the enemy and destroy the enemy's will to resist. Friendly forces following a scheme of maneuver may strike against the enemy's front, flank, or rear and may come from the ground, from the air, from space, from cyberspace, from the EMS, from the information environment, or from a combination of these approaches. Commanders determine the specific form of maneuver or the combination of forms after considering the mission variables (see paragraphs 7-95 through 7-119 for a discussion of forms of maneuver). Commanders select a scheme of maneuver designed to seize the objective most rapidly while facilitating subsequent operations. Surprise and indirect approaches are important considerations when selecting the scheme of maneuver.

7-17. The scheme of maneuver identifies where the decisive operation or main effort occurs. All of a force's resources must operate in concert to assure the success of the decisive operation or main effort. Commanders avoid becoming so committed to the initial main effort that they neglect opportunities. Commanders must be prepared to abandon failed attacks and to exploit unanticipated successes or enemy errors. In some cases the situation may be so uncertain that commanders will not initially designate a main attack.

7-18. Commanders determine available forces allocation concurrently with their scheme of maneuver. They concentrate attacking forces against enemy weaknesses. They weight their decisive operations or main efforts by providing them additional maneuver forces, positioning reserves, by assigning narrower AOs to the forces conducting the decisive operation or main effort, or by assigning priority of fires, support, and other effects to their decisive operation or main effort. The decisive operation or main effort in the offense normally gets priority from close air support (CAS), attack helicopters, armed unmanned aircraft systems (UASs), combat engineers, EW, OCO, information-related capabilities, sustainment, and protection assets. Their deep operations plans also support this decisive operation or main effort. Commanders influence the action by shifting these assets as necessary.

7-19. Small forces attacking at high speed supported by precision munitions capable of area effects, EW, OCO, and military information support operations (MISO) may achieve the same success as larger forces supported with conventional fires. EW, OCO, and MISO may so dissipate the enemy's strength and disrupt the enemy's combined arms team that deep, multiple, and equally weighted attacks may be favorably enhanced or not required.

7-20. The scheme of maneuver addresses exploitation so that units do not pause or lose momentum after seizing an objective. Each unit must plan to maneuver through the enemy's resistance and exploit successes relentlessly. The scheme of maneuver should facilitate rapid dispersion of concentrated units or the introduction of fresh forces to exploit success.

PLANNING OPERATIONS IN DEPTH

Plans for the deep battle must be realistic, complete, and firmly linked to the commander's central concept for an operation.

FM 100-5, *Operations*, 1982

7-21. The purpose of corps and division deep operations might extend beyond being shaping operations that establish favorable conditions for the conduct of close area operations. Deep operations might even be the decisive operation against enemy forces. As such, the corps or division headquarters might only direct subordinate offensive tasks in the close area to facilitate deep operations. Those deep operations may involve a mix of lethal and nonlethal effects.

7-22. Fires combining both lethal and nonlethal effects in the cyberspace, EMS, and information environment are the primary tools of corps and division deep operations. Corps (and divisions to a lesser extent) ensure the full integration of joint assets, particularly the effects of joint fires and joint target acquisition systems, when conducting deep operations. This includes the contributions of cyberspace electromagnetic activities and information-related capabilities (IRCs), which are not generally bound by physical range constraints. The full integration of joint assets may include ARSOF which can significantly enhance a commander's options in the deep fight. These operations may be conducted unilaterally or may include indigenous personnel native to the AO or other unified action partners. ARSOF can also enhance the commanders' appreciation of the deep area by conducting reconnaissance, surveillance, and target acquisition of high-payoff targets (HPTs) or targets of opportunity within the deep area, either unilaterally or with and through indigenous assets or other unified action partners.

7-23. The corps or division conducts deep operations in two main ways. It may isolate subordinate enemy elements in the close area or it may conduct operations to defeat or destroy the enemy's cohesion, nullify the enemy's firepower, disrupt enemy command and control, destroy enemy supplies, or break the morale of enemy commanders and soldiers in the deep area. In the first way, corps and division deep operations isolate their respective close areas by disrupting or severing enemy support systems and enemy command and control when the intent of the corps or division attack is to conduct a decisive close operation. Normally the corps headquarters establishes specific responsibilities for each subordinate division's deep operations. The corps headquarters may provide guidance to divisions on operations within the deep area.

7-24. In the second way, the array of enemy units in the second tactical echelon, reserves, command posts (CPs) and logical entities impact friendly operations. In this case, corps and divisions headquarters mass effects across multiple domains in the close area to enable maneuver forces to conduct a penetration. After the initial penetration, friendly forces exploit their success in sufficient depth to create dilemmas for the enemy commander across the entire depth of the enemy defense. In this case, defending enemy forces move to respond to the friendly force's attack. As elements of the enemy integrated fires command (IFC) and integrated air defense system (IADS) reposition, they become less effective and more vulnerable to attack by joint fires.

Enemy Second Tactical Echelon

7-25. Corps augment division counterfire operations. The corps headquarters can designate its supporting field artillery brigade headquarters as the force field artillery headquarters and have it orchestrate the attack of enemy counterfire targets to attrit, disrupt, and eventually destroy the enemy's IFC and IADS. The corps employs the Army Tactical Missile System, the extended range multiple launch rocket system, armed UASs, and attack helicopters in addition to joint fires to destroy enemy capabilities. This counterfire effort also employs nonlethal effects from IRCs, such as EW, OCO, and MISO. The corps establishes fire support coordination measures (FSCMs) and airspace coordinating measures to ensure expeditious and simultaneous fires and aviation maneuver and to ease and aid the shifting of responsibilities for deep operations between the corps and its subordinate attacking divisions. Those coordination measures are the authorities granted for specific actions in the case of EW, OCO, and MISO.

7-26. In addition to enemy maneuver elements, corps and divisions attack enemy support facilities or key infrastructure, such as cyberspace nodes and logistics units, as part of their deep operations. The detection and targeting of enemy nuclear and chemical-capable delivery systems and their associated special munition supply points are important deep operations goals. These systems are often beyond the range of Army information collection and fires systems, so they require joint coordination for effective targeting.

Mobile Enemy Reserves

7-27. The enemy's reserve and counterattack forces are usually HPTs for corps and division deep operations efforts. Corps and divisions headquarters identify enemy formations possessing sufficient firepower and mobility to influence the offensive task outcomes if they are not countered. The corps or division headquarters must disrupt or preclude such forces from interfering with the penetration or envelopment of static enemy positions.

7-28. Enemy attack helicopter units also pose a formidable threat because of their ability to mass and maneuver in support of the enemy's main defensive positions. Ideally, the corps headquarters identifies and considers such units for destruction while they are still in their assembly areas. This usually involves contributions from echelon above corps intelligence organizations and the employment of joint fires. The corps headquarters also allocates any available short-range air defense assets to the main effort division to provide coverage at the point of penetration where friendly forces are likely to be most concentrated. Effective means of mitigating enemy rotary-wing threats during the offense include a combination of short-range air defense systems, combined arms for air defense, deep strikes on enemy aviation assets, battlefield obscuration, limited visibility conditions, EW, and rapid maneuver.

Enemy Command and Control Nodes and Logical Entities

7-29. The corps headquarters uses joint assets and Army EW systems to jam enemy command and control (C2) networks and disrupt enemy intelligence-collection efforts to preclude and disrupt the flow of information within and between enemy headquarters. Corps and division headquarters typically target identified enemy CPs to disrupt the enemy's capability to provide direction and control. These actions degrade the cohesion of enemy defenses and limit the enemy commander's flexibility to alter the enemy defensive scheme. Commanders employ EW and OCO attacks to disrupt enemy logical entities. Such actions should be a corps priority effort, unless an effective enemy C2 system is necessary to support a friendly military deception (MILDEC) or it is more advantageous to continue to exploit information being obtained from a particular enemy command node.

7-30. Figure 7-4 on page 7-10 illustrates the second way of planning operations in depth. It shows the establishment of two purple kill boxes that expedite the attack of the repositioned IFC and the enemy combined arms army main CP. (See ATP 3-94.2 for more information on deep operations.)

PLANNING RECONNAISSANCE AND SECURITY OPERATIONS

> *Agitate him and ascertain the pattern of his movement. Determine his dispositions and so ascertain the field of battle.*
>
> Sun Tzu

7-31. A corps or division normally conducts reconnaissance forward of its subordinate echelons, or it provides security before making contact with the enemy main body. However, once the corps or division commits its main body forces, that attacking subordinate echelon, a division or BCT, provides its own security. The introduction of another corps- or division-controlled element between the deep area and attacking forces in the close area only tends to increase the coordination necessary between units and complicates the control and execution of close operations.

7-32. In some situations, the attacking subordinate echelon may directly control the higher echelon's security force. For an attacking corps, this would be the division conducting the main effort that is controlling the corps forward security force while it is performing either a forward guard or a cover. Divisions normally designate an advance guard and conduct a flank screen of their most vulnerable flank. However, a corps headquarters may provide a separate security force to perform either a guard or screen on a particularly sensitive flank, such as along one of the corps' boundaries.

Chapter 7

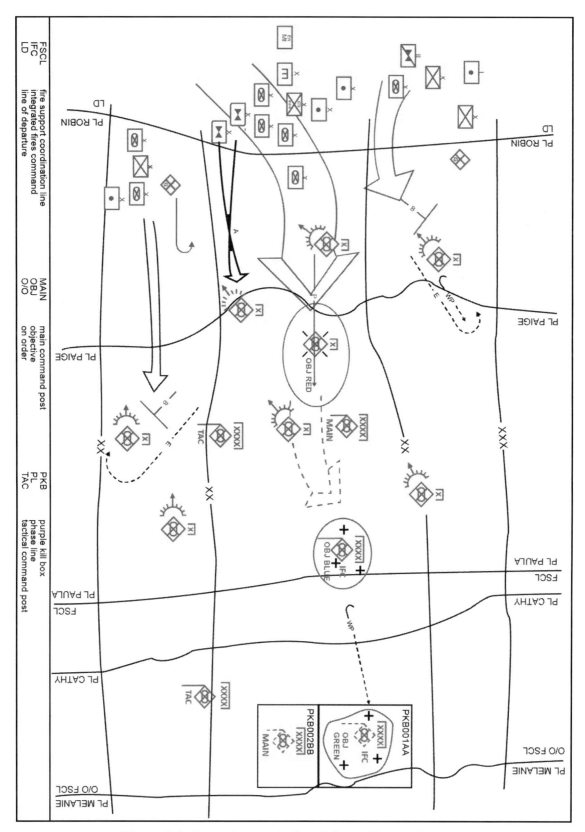

Figure 7-4. Example penetration followed by exploitation

Planning Decisive and Shaping Operations or Main and Supporting Efforts

> *Orders should not attempt to arrange matters too far in advance, for counter measures of the enemy and unexpected contingencies will often make it necessary to recall the original orders and substitute others.*
>
> Field Service Regulations, 1905

7-33. The corps or division commander's visualization allows for the issue of the commander's initial guidance to the staff as to how the commander wants to conduct offensive tasks. One key factor is the AO assigned to the corps or division by a higher headquarters. The corps or division commander structures this AO by assigning objectives, boundaries, phase PLs, and other control measures to ensure the successful synchronization of subordinate unit operations.

7-34. During planning, the commander begins the process by developing planning guidance. The planning guidance states the commander's intent, refines the staff-developed suggestions for the commander's critical information requirements, and directs the staff planning effort. The staff prepares orders for subordinate echelons and establishes the control measures necessary to conduct the corps' or division's decisive operations.

7-35. The commander and staff consider five complementary elements when planning offensive tasks. These elements include—

- How will the corps or division conduct reconnaissance and security operations forward and to the flanks and rear of the corps or division's decisive and shaping operations?
- How will the corps or division conduct shaping operations directed against vital enemy elements regardless of their locations in the AO?
- The initiation of the corps or division's decisive operations requires the reaching of what conditions?
- What is the acceptable degree of risk in regards to the corps or division reserve's composition, size, and location?
- What activities by the other elements of combat power are necessary to maintain offensive momentum?

Commanders and staffs should consider these elements when formulating the concept of operations, scheme of maneuver, and scheme of fires and while synchronizing decisive, shaping, and sustaining operations or whatever other operational framework commanders choose to use. The enemy forces and support systems targeted as part of the offense depend on the mission variables.

7-36. Corps and division commanders normally attempt to perform an envelopment when attacking an enemy in prepared defensive positions. (Paragraphs 7-96 through 7-98 discuss the envelopment.) The decisive operation or main effort either finds an exposed flank or conducts a penetration of the enemy defense to expose a flank. (Paragraphs 7-115 and 7-116 discuss the penetration.) Finding an exposed flank is easier to achieve when the enemy force occupies noncontiguous defensive positions. Without an exposed enemy flank, the corps or division initial main effort focuses on penetrating the weakest point in the main enemy defensive position. Supporting attacks fix any adjacent enemy forces or mobile counterattack forces that could react to the penetration.

7-37. Typically, the main effort would then shift to a trailing friendly unit (usually a division for a corps or a BCT for a division) moving through the area where the penetration occurred once that unit assumes control of the AO and begins its attack against enemy defenses in depth. This shift does not normally occur until at least a BCT of the trailing division or a maneuver battalion of the trailing BCT has passed through the area of penetration and is able to deploy into an attack formation. Priority of the units attacking enemy defenses in depth generally goes to the unit tasked with defeating a counterattack by large mobile enemy formations.

Planning Reserve Operations

7-38. The *reserve* is that portion of a body of troops which is withheld from action at the beginning of an engagement, in order to be available for a decisive movement (ADRP 3-90). It is a commander's principal means of influencing the action decisively once an operation is under way. A reserve is not committed to a particular course of action. It does not have a planned, subsequent mission. Its commitment solely depends on the flow of the battle. The commander establishes its planning priorities for likely contingencies. A reserve reinforces or maintains the attack's momentum by—

- Exploiting success when the opportunity arises.
- Countering tactical reverses, such as an enemy counterattack against committed units.
- Sustaining the attack of a committed unit.
- Countering threats to the corps or division support areas and consolidation areas (if designated) that exceed the capabilities of local forces to defeat.

7-39. The commander should make prior provisions for designating another reserve upon commitment of the command's initial reserve force. This is especially important if that commitment occurs during the early stages of the battle and the original reserve becomes decisively engaged.

7-40. In addition to the actual force designated as the corps or division reserve, there are trailing units within each echelon that, although initially unengaged, are committed to a subsequent course of action. These trailing units can be diverted from their original tasks, if the situation requires.

7-41. Contingencies for the use of the reserve should be part of the corps or division plan. The operations order assigns "be prepared" tasks to the reserve to aid in its planning and execution. Planners must consider the aspects of time and space when recommending the positioning of reserve forces to their commander. Commanders must remember that subordinate commanders require planning and preparation time for each "be prepared" mission.

7-42. Normally, corps planning focuses on probable large-scale enemy counterattacks against committed divisions. Division planning focuses on probable enemy counterattacks against committed BCTs. If it lacks the necessary combat power to defeat an enemy counterattacking force, the reserve force conducts a hasty area defense to block an enemy counterattack. For instance, corps and division planners should anticipate the possibility of a counterattack into their rear by ground maneuver forces from adjacent enemy units not within their respective AOs. This is likely if the corps or division AO overlaps the lateral boundaries of two or more defending enemy units. Given this situation, a brigade-size reserve may only be able to contain or block such a force's attack. It may then be necessary to defeat or destroy the force by diverting assets from committed units.

7-43. Command of the corps or division reserve—and its associated control once committed—is an important consideration. The corps or division commander may opt to retain control of the reserve if it makes contact with the counterattacking enemy force before that enemy force can negatively influence committed divisions or BCTs. If not, the corps or division commander may place the reserve in a command relationship—attached, operational control (OPCON), or tactical command (TACON)—to a committed division or BCT. The key consideration is which headquarters can best control the possible convergence of friendly units and coordinate their fires and movement.

7-44. Commanders may also commit the reserve to sustain the momentum of the attack of a committed unit. There are three basic options for the use of the reserve in this case. First, the reserve continues the attack as a separate force under the control of the corps or division headquarters, and it is given responsibility for the remainder of the committed unit's AO. The second option is to place the reserve in a command relationship—attached, OPCON, or TACON—to the committed unit. This reinforces the committed division or BCT whose commander is probably most knowledgeable of local conditions without needing to change control measures. The third option is for the reserve to assume control of the AO from the committed unit by means of a passage of lines, a relief in place, or a follow on and assume tactical mission task. This allows the rapid introduction of fresh forces. It also allows the higher commander to reconstitute a reserve from the formerly committed unit after that unit has had enough time to resupply and reorganize itself.

7-45. To exploit success, the commander can commit the reserve at any stage of the battle. However, in most instances it is committed during the later stages of the battle once the corps or division's committed forces achieve most of their objectives. The reserve will then probably be given a separate AO and objective that enables the attack to transition into an exploitation and possibly a pursuit.

7-46. The entire reserve or a battalion task force from the reserve may be tasked to perform "be-prepared" missions in the corps or division support and consolidation areas in response to Level III threats as the tactical combat force (TCF). Support and consolidation security tasks are discussed in paragraphs 7-47 through 7-53.

PLANNING SUPPORT AND CONSOLIDATION AREA SECURITY

7-47. The fluidity and rapid tempo of operations pose challenges when planning for the area security of support and consolidation areas. If the corps or division is to maintain the initiative and combat power necessary for the successful performance of offensive tasks, the continued forward movement of units and sustainment support is critical. Maintaining the initiative in the close area often results in significant numbers of bypassed enemy forces and remnants of defeated units as friendly forces maneuver deep into enemy areas by avoiding enemy units in well prepared positions.

7-48. Enemy commanders look for opportunities to counter or at least hinder the performance of corps and division offensive tasks. They attempt to strike deeply into friendly support and consolidation areas using multiple combinations of lethal and nonlethal effects from multiple domains. Corps and division headquarters must keep CPs operating, keep their sustainment capabilities functional, keep their respective lines of communications open on at least a periodic basis, and keep an acceptable stock of supplies. The enemy will almost always employ special purpose forces, irregular forces, EW, long-range artillery, rockets, and missiles, IRCs, and cyberspace electromagnetic activities (CEMA) to disrupt friendly activities. Enemy conventional units and other elements bypassed during the advance of friendly forces must be fixed and destroyed by follow-on forces to prevent interference with friendly activities in support and consolidation areas. The conduct of noncontiguous operations will increase the difficulty of these tasks, as will the lack of friendly host-nation security forces.

7-49. Corps and division commanders assign responsibility for the support area to a maneuver enhancement brigade (MEB) headquarters. The MEB commander designates bases and base clusters for all units operating within the support area. Within that support area, the MEB commander has TACON of all friendly forces for terrain management, movement control, clearance of fires, and security. Division commanders can also choose to establish a support area command post (SACP) with one of the assistant division commanders to provide better coordination and mission command of the support area units and missions. All units located in support and consolidation areas are responsible for defending themselves against enemy Level I threats. All units in support and consolidation areas also perform information collection within their organic capabilities and report that information to the appropriate headquarters.

7-50. Military police (MP) units from the MEB or a functional MP brigade conduct area security within corps and division support areas. They perform route reconnaissance and security along selected ground lines of communications and main supply routes within and external to the support area. (The assistant chief of staff, operations [G-3] coordinates the performance of these tasks outside the support area with the commander of the AO through which the line of communication or main supply routes traverses.) MP units may be assisted in these efforts by any maneuver forces under the command of the MEB commander and any available host-nation security forces. This same group of forces provides reaction forces to help friendly bases and base clusters to defend themselves against enemy Level I and II threats to those bases, base clusters, and lines of communications.

7-51. Corps and division commanders assign subordinate commanders responsibility for AOs constituting the corps or division consolidation areas, respectively. Those subordinate commanders clear their AOs of stay behind forces and bypassed enemy units to ensure friendly freedom of action in those areas as their parent corps or division continues to advance. Initially these division and BCT commanders primarily perform attacks and area security tasks as they clear their AOs of enemy ground forces. These units should be combined arms organizations specifically task-organized for the consolidation of gains requirements in their AOs. These units begin performing selective stability tasks once they establish area security within the consolidation area. Consolidation areas reduce the area security and mission command burden in the support area because they allow the division or corps commanders to decrease the size of the MEB AO.

7-52. The most likely Level III threats to a corps or division consolidation area during the offense is a single large, counterattacking enemy force or the remnants of bypassed enemy units. These units will be intent on—

- Severing or disrupting friendly corps and division CPs, communication sites, and subordinate brigade headquarters.
- Attacking sustainment facilities, supply stocks, and convoys supporting committed units.
- Interdicting lines of communications.
- Destroying airfields, aviation assembly areas, and arming and refueling points.
- Attacking echelons above brigade artillery or engineer assets.
- Disrupting the commitment of the corps or division reserve.

7-53. The corps would employ a TCF of at least brigade size to contend with such a threat. Limited reaction times and extended distances require corps and division TCFs to be extremely mobile and capable of moving by employing a combination of air and ground assets. A TCF is typically a combined arms formation with both air and ground mounted maneuver forces. TCFs also typically contain dedicated engineer (route clearance and mobility) and field artillery assets. A TCF must be able to destroy enemy armored vehicles and dismounted infantry as well as being able to suppress enemy air defense artillery (ADA) systems to allow the employment of friendly attack helicopters and UASs against the Level III threat. The division assigned to the corps consolidation area would be the probable source of a corps TCF. Likewise, the BCT assigned to the division consolidation area would be the probable source of a division TCF.

PREPARING CORPS AND DIVISIONS FOR THE OFFENSE

7-54. Time must be intensively managed when units prepare to perform offensive tasks. Senior commanders should use no more than one-third of the available preparation time for planning and issuing orders. Subordinates must have time to conduct necessary information collection activities, coordination, and troop leading procedures. These preparation considerations apply to all four offensive tasks.

PREPARING FOR MISSION COMMAND IN THE OFFENSE

To be at the head of a strong column of troops, in the execution of some task that requires the brain, is the highest pleasure of war

Lieutenant General William T. Sherman

7-55. Corps and division commanders continually visualize the current situation and formulate plans to get their forces to the intended end state. Commanders go where they can best influence the operation. This is where their personal presence can be felt and where their will for victory can best be expressed, understood, and acted on. Commanders exercise command from wherever they are on the battlefield, and that position should be where they can best direct, lead, and assess decisive operations.

7-56. Corps and division orders for the offense include the minimum control measures required to coordinate the operation. Control measures describe and illustrate the plan, maintain separation of forces, concentrate the effort, provide subordinates freedom of action, assist the commander in the mission command of forces, and add flexibility to the maneuver plan. At a minimum, control measures prescribe a line of departure (LD), a time of attack, and the objective. In addition, commanders could assign AOs and an axis of advance, routes, PLs, checkpoints, and FSCMs. They optimally allow subordinates the maximum freedom of action consistent with necessary synchronization.

7-57. Whenever possible, commanders issue orders face to face to avoid misunderstanding. Subordinate commanders must know what the command as a whole is expected to do, what is expected of them, and what adjacent and supporting commanders are expected to do. Subordinate commanders and staffs must have pertinent information and as much time as possible to prepare their plans. Warning orders that contain as much information as practical are vital in subordinate parallel attack preparations.

7-58. Coordination begins immediately upon receipt of a mission and continues throughout an operation. When time allows, commanders and staffs review subordinate plans, giving additional guidance as needed to ensure a coordinated effort.

7-59. In the end, the success of an operation depends on proper understanding of the commander's intent and aggressive execution at the battalion echelon and below. Soldiers will attack with the required confidence, aggressiveness, and resolve when they understand their own tasks within the scheme of the overall plan and commander's intent, and when they know what their fellow Soldiers on the ground, in the air, in space, in cyberspace, and in the information environment will do.

7-60. Battlefield success requires effective communications. Redundant systems are necessary to allow for communications throughout corps and division AOs in contested environments. Both corps and division commanders must be able to communicate with adjacent units, supporting joint forces, and host-nation and multinational forces in addition to their subordinates. The use of primary, alternate, contingency and emergency (PACE) forms of communication as well as organic unit liaison teams and digital liaison detachments provide commanders options to ensure effective coordination and interoperability. Continuous mission command requires adequate, but not continuous, connectivity.

7-61. An attacking corps commander usually moves with the division conducting the corps decisive operation or main effort and commands the corps from this forward location from a combat vehicle or the tactical CP. Likewise, an attacking division commander may move with the BCT conducting the division main attack. Main and tactical CPs displace forward only when necessary as the attack progresses or the situation permits. To ensure survivability, commanders must reduce the visual and electromagnetic signature of their headquarters. This includes measures such as dispersion of command posts, survivability positions, use of fiber-optic cable, camouflage, cover and concealment, masking antennas with terrain, and remotely locating the headquarters' associated antennas and satellite dishes.

7-62. Sustainment units require sufficient communications to maintain control over their assets operating from dispersed locations—such as supply points, maintenance collection points, ambulance exchange points, and in-transit long-haul convoys—throughout corps and division AOs. Sustainment requirements and the enemy situation will change rapidly in both contiguous and noncontiguous environments. These small, dispersed sustainment assets are extremely vulnerable to enemy physical, cyberspace, EMS denial, and information activities attacks.

PREPARING FOR MOVEMENT AND MANEUVER IN THE OFFENSE

7-63. Corps and division subordinate units preparing to perform offensive tasks remain dispersed until immediately before commencement of the offense, once they occupy their respective AOs. At the prescribed time or event, the corps or division concentrates its subordinate units sufficiently to mass their effects at a specific point in the enemy defense. This achieves decisive results by weighting the main effort and providing only the absolute minimum combat power to shaping operations or supporting efforts. The corps or division or their subordinate units conduct an approach march (over multiple routes when available) to close with the enemy if those units must move a considerable distance to gain contact or attack the enemy. (See FM 3-90-2 for a discussion of the approach march.)

7-64. Corps and division commanders normally conduct deliberate operations when their staffs can develop thorough and timely intelligence on the enemy. Corps and division plans retain flexibility to facilitate transitions to branches and sequels of the base plan, although subordinate unit deliberate attack plans are detailed. The higher the echelon, the further out in terms of time and space that echelon's planners should plan. Corps look deeper in space and time while considering second and third order effects of current and planned operations than divisions, and divisions look deeper than their subordinate brigades.

7-65. Corps and division units vary their tempo of operations, concentrate rapidly to strike the enemy, then disperse and maneuver to subsequent objectives. These actions help keep the enemy off balance and preclude the enemy's effective employment of massed fires, including weapons of mass destruction (WMD). Corps and divisions may conduct a feint, demonstration, or reconnaissance in force to deceive the enemy and test enemy dispositions before the initiation of the decisive operation.

Chapter 7

> ### Movement and Maneuver: The 1973 Arab-Israeli War
>
> After the destruction of its air force and defeat of its ground forces by Israel in the 1967 Six Day War, Egypt reorganized its armed forces in preparation for what it saw as the inevitable next conflict. With help from the Soviet Union, it acquired modern IADSs and anti-tank guided missile systems to counter the Israeli air and armor forces that had been a large factor in its defeats in three previous wars. The new capabilities gave the Egyptian army the opportunity to surprise and disrupt the potent Israeli Defense Forces air-ground team, which it exploited in 1973.
>
> In a well-planned and rehearsed operation, Egyptian forces rapidly assaulted across the Suez Canal, penetrated Israeli Defense Force defensive positions, and established a deliberate defense after a short advance. The Israeli counterattack with aircraft and an armored brigade was defeated with heavy losses from surface-to-air missiles and anti-tank guided missile systems, respectively. The losses were a shock to the previously undefeated Israeli air and armor forces, which needed to quickly reconstitute and adapt to a very different operational environment than 1967. The Israeli Army recognized that a mounted, tank only approach was not going to defeat well trained infantry with modern long range anti-tank missiles; only a combined arms approach to close combat would be effective. Thus, infantry units were task-organized into the armored brigades.
>
> Once the task-organized brigades cleared the Chinese Farm and penetrated Egyptian first echelon defenses, the Israeli Army was able to disrupt and ultimately collapse the IADS umbrella by destroying vulnerable launchers and radars with direct fire. By the time the IADS was largely destroyed on the ground, the Israeli Air Force was again ready for offensive operations. It overwhelmed the Egyptian Air Force and contributed to the complete tactical defeat of the Egyptian army. Ultimately, the Israeli Defense Force counterattacks halted just short of Cairo, due more to world political pressure than military factors.
>
> Israeli over dependence on tank-only formations was a function of casualty aversion and impatient commanders more than doctrine. Mobilization of reservists, even in the Israeli army, takes time and tank units made up the preponderance of ready forces when the war started. The need for speed was influenced by the demands of a two front war that was stressing a small force in terms of threat scope and scale that had not been stressed before.

7-66. Corps and divisions employ a mix of armored brigade combat teams (ABCTs), infantry brigade combat teams (IBCTs), and Stryker brigade combat teams (SBCTs) in addition to their other brigades—such as field artillery, combat aviation, and maneuver enhancement—when preparing to conduct offensive tasks. To capitalize on each type of BCT's unique capabilities while minimizing their limitations, commanders must determine—

- The tasks they are to accomplish.
- The appropriate level where task organization should occur.
- The appropriate command or support relationships.
- The required amount and type of augmentation and support they are to provide to the force.
- The sustainment concept.

7-67. Augmentation normally is required if a commander attaches combined arms and Stryker battalions to IBCTs. Command and support relationship considerations and capabilities are essential in ensuring the feasibility of task-organizing between different types of BCTs. Dismounted infantry within SBCTs and IBCTs can—

Large-Scale Offensive Operations

- Infiltrate enemy positions and establish support by fire positions to facilitate a subsequent attack by combined arms (tank and infantry fighting vehicle) and mounted Stryker battalions.
- Conduct operations in urban and complex terrain.
- Rapidly respond to threats directed against echelon support and consolidation areas when supported by assault aviation.
- Conduct air assaults to seize key terrain or destroy enemy formations,
- Perform economy of force missions.

7-68. Dismounted infantry forces may require augmentation to increase their anti-armor capability and mobility. Dismounted infantry battalions and companies can be attached to ABCTs and SBCTs and vice versa. However, dismounted infantry units should not be so overburdened with augmentation that they lose their unique capabilities. The scheme of sustainment must ensure that organizations retain the ability to support task organization changes, particularly with regard to maintenance, fueling, rearming, and recovery operations.

7-69. Friendly dismounted infantry forces can deny enemy heavy forces easy access through close terrain. This forces the enemy to fight dismounted to protect its tanks and other armored vehicles and to engage friendly dismounted infantry. The corps or division commander can then use ABCTs, SBCTs, and attack helicopters and armed UASs to strike decisively at chosen times and places.

7-70. ABCTs can lose their ability to maneuver when confronted by enemy forces dominating key locations in close terrain along friendly routes of advance. However, SBCTs and IBCTs can conduct dismounted attacks over close terrain to close with and destroy enemy forces and seize these key locations. They can also use air assaults to secure decisive or key terrain to aid the movement of mounted forces. Staffs must consider several factors when planning this type of operation. Dismounted infantry forces will need—

- Additional transportation assets to rapidly move dismounted infantry combat elements into forward assembly areas or attack positions.
- Additional artillery to enhance their organic field artillery battalion's capability to engage the enemy with indirect fires and to execute counter fires.
- Additional intelligence assets.
- Support from attack reconnaissance helicopter and UAS assets.
- Allocation of CAS sorties.
- Additional anti-armor assets to protect against an armored threat.
- Sustainment support of dismounted infantry forces once they are inserted or infiltrated.

7-71. Corps and divisions allocate additional engineer assets to the decisive operation or main effort to give it greater mobility. The engineer focus in the offense is mobility, then countermobility. Commanders place engineer units forward to augment the divisions or BCTs conducting the decisive operation or main effort. Engineering tasks performed by these reinforcing engineers include—

- Conducting combined arms breaching.
- Conducting area and route clearance.
- Conducting gap crossing.
- Constructing and maintaining combat roads and trails.
- Providing general engineering for follow-on forces and logistic units.
- Augmenting reconnaissance forces in terrain analysis, especially in bridge classification and mobility analysis for routes of advance.
- Emplacing obstacles on unit flanks directed against likely enemy avenues of approach into and throughout the AO.
- Maintaining key facilities, such as airfields and landing strips.
- Constructing protective positions to protect key assets, such as aviation assembly areas; forward arming and refueling points; petroleum, oils, and lubricants (POL) supplies; and ammunition points.
- Hardening other designated critical facilities by their priority.

7-72. Keys to effective counter obstacle operations are the effective suppression, obscuration, security, and reduction of an obstacle to enable the assault. This requires detailed contingency planning, well-rehearsed breaching operations, and trained engineers familiar with unit standard operating procedures. These engineers are integrated into the attack formation. Complex obstacles require detailed engineer estimates and appropriate engineer assets.

7-73. Corps and division planners must anticipate breaching requirements in time to adequately provide breaching units with additional engineer assets, such as plows, rakes, and supplementary artillery munitions for obscuration and counterfire. They should request large-scale obscuration, EMS and EW obscuration effects when the capabilities are available. When possible, all units conduct breaching operations in-stride to allow the force to maintain the attack's momentum.

7-74. Engineer units operating in the corps or division support and consolidation areas primarily conduct general engineering, survivability, and route and area clearance tasks. These engineer units may also be assigned responsibility for base camp defense, area defense, or terrain management. However, commanders must weigh the risks associated with assigning these missions. In addition to removing them from their engineer-specific mission, these engineer units require time to assemble because they are normally dispersed when conducting engineer missions on an area basis. They also require fire support and anti-armor augmentation.

PREPARING FOR INTELLIGENCE IN THE OFFENSE

7-75. The intelligence effort helps commanders decide when and where to concentrate combat power. Collection assets answer the corps or division commander's priority intelligence requirements and other information requirements, which flow from the intelligence preparation of the battlefield (IPB) and the war-gaming process. Information required may include—

- Enemy centers of gravity or decisive points.
- Location, orientation, and strength of enemy defenses.
- Location of enemy reserves, fire support, and other attack assets in support of defensive positions.
- Enemy air avenues of approach, and likely enemy engagement areas (EAs).
- Key terrain, avenues of approach, and obstacles.

The assistant chief of staff, intelligence (G-2) also identifies threats to the corps and division support and consolidation areas, such as enemy special purpose forces and irregular activities, which may interfere with corps or division control of the attack.

7-76. The G-2 assists the G-3 in synchronizing the capabilities of joint, multinational, and national assets into the collection effort. The G-2 recommends specific reconnaissance tasks for corps or division controlled reconnaissance forces, realizing the commander may task these forces to conduct a security or attack mission. A focused approach in allocating collection assets maximizes the capability of the limited number of assets available to the corps or division.

7-77. The G-3 synchronizes intelligence operations with combat operations to ensure all corps and division information collection means provide timely information in support of operations. The G-3 tasks information collection assets to support the targeting process of decide, detect, deliver, and assess in keeping with the corps or division commander's priority intelligence requirements. Collection assets locate and track HPTs and pass targeting data to fire support and other elements. These other elements could be organizations supporting the corps or division that has the authority to conduct OCO or that can conduct EW.

PREPARING FOR FIRES IN THE OFFENSE

> *The bulk of the artillery is employed in support of the main attack. However, provision is made for the mutual support of the artillery with both the main and holding attacks unless these attacks are beyond mutual supporting distance.*
>
> FM 6-20, *Field Artillery Field Manual*, 1940

7-78. Allocating and synchronizing all fires, especially joint fires and nonlethal effects, complements and weights the corps or division's decisive operation or main effort. This synchronization effort also helps the corps or division commander control the tempo of the attack. Attacking commanders employ fires to—
- Conduct intense and concentrated preparatory fires before and during the initial attack stages.
- Conduct suppressive fires to isolate the objective of the decisive operation or main effort and to help fix enemy forces during shaping operations and supporting attacks.
- Provide continuous suppression to allow attacking formations to close with the enemy.
- Conduct suppression or destruction of enemy air defense.
- Supplement subordinate unit counterfires to diminish or stop an enemy's ability to effectively employ artillery and rockets.
- Execute deep operations in concert with other corps and division assets.
- Deny, through electronic attack and OCO, enemy use of critical C2, fire support, and intelligence systems and networks.
- Obscure friendly forces and key terrain.

7-79. Corps and division CPs must always know the locations of their friendly units. Commanders must ensure that organic fire support agencies and supporting joint assets clearly understand the scheme of maneuver so that they can maximize their capabilities in the scheme of fires for the greatest effect.

Fire Support in the Offense

7-80. Fire support preparation creates conditions that improve the friendly forces' chances for success. It facilitates and sustains transitions, including those to branches and sequels. Fire support in the offense is generally more decentralized than in the defense in order to sustain momentum. Preparation requires action by fire support personnel at every echelon. Mission success depends as much on fire support preparation as on fire support planning. Fire support rehearsals help staffs, units, and individual fire support personnel to better understand their specific role in upcoming operations, synchronize execution of the fire support plan, practice complicated tasks before execution, and ensure equipment and weapons are properly functioning.

7-81. Commanders are responsible for the clearance of fires. *Clearance of fires* is the process by which the supported commander ensures that fires or their effects will have no unintended consequences on friendly units or the scheme of maneuver (FM 3-09). The center of this process is in the current operations integrating cell. Clearance of fires may be assisted through staff processes, control measures, embedded in automation control systems, or through active or passive recognition systems.

7-82. Timely execution of joint fires is critical when conducting offensive tasks. All fire support providers must understand FSCM and procedures for controlling fires. The fire support coordination line (FSCL) gives sister Services greater freedom of action in the area beyond the FSCL and facilitates operations in depth. Placing the FSCL too far forward limits joint fires assets and slows response times to deliver effects due to the required clearance of fires procedures. The FSCL is normally located along the corps-controlled forward security force or the division's forward boundary, although the FSCL's exact location depends on several factors. These factors include—
- The location of enemy forces.
- The anticipated rate of the friendly advance.
- Artillery and rocket capabilities of friendly advancing forces.
- The scheme of maneuver, including the maneuver of Army aviation units.
- The desired tempo of operations.

Air and Missile Defense in the Offense

7-83. Unless augmented with air and missile defense (AMD) capabilities, corps and divisions primarily conduct passive air defense measures and employ combined arms for air defense against enemy fixed- and rotary-wing aircraft and UASs. The theater Army air and missile defense command (AAMDC) or an ADA brigade may provide counter-rocket, artillery, and mortar systems or short-range air defense systems to provide protection to critical sites. Man-portable air defense systems should be distributed in a weapons hold status to provide defense against attack or surveillance by enemy aerial platforms. Effective use of these systems requires extensive training and effective control to enable friendly aerial freedom of maneuver and reduced fratricide risk.

7-84. AMD elements within the corps and division headquarters coordinate with the AAMDC or ADA brigade to ensure as much integration of the area air defense commander's (AADC's) air defense plan for coverage of corps and division key assets as possible. These key assets include CPs, aviation assembly areas, field artillery locations, sustainment bases, and reserve assembly areas. The AMD element synchronizes priority AMD defense coverage of the decisive operation, particularly in areas where the attack is vulnerable to enemy air action, such as a wet-gap sites and canalizing terrain. The AAMDC or ADA brigade work for the AADC. The AADC is normally also the joint force air component commander (JFACC). The supported maneuver commander positions available short-range air defense to protect attack positions from air attack and units moving forward.

PREPARING FOR SUSTAINMENT IN THE OFFENSE

7-85. Commanders integrate sustainment and tactical plans and ensure sustainment operators maintain situational awareness during the offense. Supported units keep their sustainment elements informed of the tactical situation, their current sustainment status, and projected future operations. This may require tactical units to physically deliver tactical updates and reports to their supporting sustainment organizations when use of the EMS and cyberspace is not possible or is degraded. Successful corps and division operations depend in large part on the initiative and innovation of sustainment planners and operators.

7-86. In the offense, commanders generally position sustainment units and facilities as close as possible to the tactical units they support in order to reduce travel time and increase responsiveness. Consumption of Class III supplies is typically higher in the offense, while the consumption of Class IV and V supplies is typically less. Sustainment elements providing essential support (including ammunition and POL) should move forward during the period just prior to the beginning of the offense. If possible, such movements should occur at night or when friendly actions hinder enemy detection capabilities.

7-87. Sustainment planners coordinate assistance from stationary units when the corps or division plan calls for units to conduct a forward passage of lines during the offense. Stationary units may be able to assist those units moving forward through their positions with POL, medical aid, or other support. The requirement for POL from sustainment units during the attack decreases when sustainment operators can keep their own bulk POL transporters full by having other units top off combat vehicles prior to an attack.

7-88. Anticipating sustainment requirements effectively requires an understanding of the type of battle the corps and division commanders expect to fight. For example, fighting peers with air forces requires large quantities of AMD munitions. Consumption may be relatively light in attacks against weakly defended locations. Units fighting enemy infantry in restricted and urban terrain use large quantities of small arms and artillery ammunition and water. Breaching and wet-gap operations consume large amounts of obscurants. Units attacking enemy armored forces require large quantities of anti-armor munitions and fuel. Ensuring that maneuver and sustainment commanders share a common visualization of how a battle should unfold enables more responsive sustainment. Sustainers operationalize their support based on the mission variables unique to each case.

7-89. Sustainment planners need to proactively prepare to support maneuver units during the offense. One technique they can use is to prepare push packages of ammunition, POL, and repair parts. If communications are degraded, sustainers automatically dispatch these push packages to supported units.

7-90. Lengthening lines of communications requires frequent forward movement of stocks and sustainment units and the establishment of support areas within BCT AOs. Displacing sustainment units normally cannot provide support. Sustainment planners must ensure continuous support when planning the displacement of sustainment units and capabilities. This is particularly necessary with regard to Army health support.

7-91. Corps and division subordinate units can use captured supplies and materiel or locally obtained supplies to supplement their supply stocks and to increase their operating and safety levels. Commanders can make the location of enemy and local supplies and materiel as information requirements for their information collection assets. Explosive ordnance disposal (EOD) personnel provide subject matter expertise on the disposition of captured and recovered enemy and foreign ordnance stores, or large cache recoveries. This expertise includes support to weapons technical intelligence that enhances force protection and supports intelligence based operations. In a similar fashion, a veterinarian will need to clear captured foodstuffs and POL specialists need to clear captured fuel for friendly use. Civil affairs (CA) area studies and open source materials can help pinpoint potential locations of host-nation local supplies and materiel.

7-92. Water can be as critical a commodity as ammunition. Combat takes a tremendous amount of energy and substantially increases the human body's requirement for water. Vehicles operating under constant mechanical strain also use large amounts of water. Planners consider water sources along the route of advance. Planners consider the impact of the accidental or intentional release of toxic industrial materials and use of chemical munitions on the available water supply. Operations in desert areas can be limited by the sources of water and the means to transport it.

7-93. Planners select and use supply routes carefully. When possible, truck convoys moving supplies forward use secondary routes to avoid detection, or they should be integrated into the movement of maneuver units. Sustainment activities also avoid setting a pattern which could disclose the locations of maneuver and other units being supported. Maneuver and other units—such as artillery and engineers—carry as much materiel as possible into battle because supply system disruptions will occur. Sustainment planners and commanders review the authorized stockage list and the prescribed load list of each unit to ensure units only transport combat essential items. Commanders authorize their maintenance teams to make maximum use of controlled substitutions to return maintenance failures and battle losses to action. Maintenance leaders train their mechanics on battle damage assessment and repair and recovery operations.

PREPARING FOR PROTECTION IN THE OFFENSE

7-94. Unit movement into attack positions is thoroughly coordinated and planned in detail to preserve surprise. Force concentrations take place quickly and make maximum use of operations security (OPSEC). Units use cover and concealment, signal security, and military deception (MILDEC) actions. Indicators that will alert the enemy to the coming attack are avoided, since deception is vital to success. The attacking force organizes to cope with the environment. This may include attacking across obstacles and rivers, during snow or rain, at night, or on battlefields containing nuclear and chemical hazards. Engineer reconnaissance units as well as chemical, biological, radiological, and nuclear (CBRN) reconnaissance units support maneuver units throughout their attack. CBRN reconnaissance units orient on command objectives identified during the IPB process. Commanders position available decontamination assets to support their scheme of maneuver and array available obscuration means to counter enemy sensors.

FORMS OF MANEUVER

7-95. *Forms of maneuver* are distinct tactical combinations of fire and movement with a unique set of doctrinal characteristics that differ primarily in the relationship between the maneuvering force and the enemy (ADRP 3-90). The Army has six forms of maneuver: envelopment, flank attack, frontal attack, infiltration, penetration, and turning movement. Maneuver units accomplish their missions by synchronizing the contributions of all warfighting functions to execute these forms of maneuver. Commanders generally choose one form on which to build a course of action. A higher echelon commander rarely specifies the specific form of maneuver. However, that higher commander's guidance and intent, along with the mission and any implied tasks, may impose constraints—such as time, security, and direction of attack—that narrow the forms of maneuver to one alternative. Additionally, the AO's characteristics and the enemy's dispositions also help commanders determine the form of maneuver. A single operation may contain several forms of maneuver, such as a frontal attack to clear an enemy's disruption zone followed by a penetration to create a gap in enemy defenses. Then, the commander might use a flank attack to expand that gap and destroy the enemy's first line of defense.

Forms of maneuver
- Envelopment
- Flank attack
- Frontal attack
- Infiltration
- Penetration
- Turning movement

Chapter 7

ENVELOPMENT

The deep envelopment based on surprise, which severs the enemy's supply lines, is and always has been the most decisive maneuver of war. A short envelopment which fails to envelop and leaves the enemy's supply system intact merely divides your own forces and can lead to heavy loss and even jeopardy.

General Douglas MacArthur

7-96. *Envelopment* is a form of maneuver in which an attacking force seeks to avoid the principal enemy defenses by seizing objectives behind those defenses that allow the targeted enemy force to be destroyed in their current positions (FM 3-90-1). Envelopments are common when employing a noncontiguous operational framework. At the tactical level, envelopments focus on seizing terrain, destroying specific enemy forces, and interdicting enemy withdrawal routes. The commander's decisive operation focuses on maneuvering around an assailable flank. It avoids the enemy's strength—the enemy's front—where the effects of enemy fires and obstacles are greatest. Generally, a commander prefers to conduct an envelopment instead of a penetration or a frontal attack because the attacking force tends to suffer fewer casualties while having the most opportunities to destroy the enemy. Envelopment also produces great psychological shock to the enemy. If no assailable flank is available, the attacking force creates one through the conduct of a penetration.

7-97. The four varieties of envelopment are the single envelopment, double envelopment, encirclement, and vertical envelopment. (Figure 7-5 illustrates a single envelopment.) A *single envelopment* is a form of maneuver that results from maneuvering around one assailable flank of a designated enemy force (FM 3-90-1). The VII and XVIII Corps executed a single envelopment of the Iraqi Army during Operation Desert Storm in 1991. A *double envelopment* results from simultaneous maneuvering around both flanks of a designated enemy force (FM 3-90-1). Encirclement operations are where one force loses its freedom of maneuver to an opposing force that is able to isolate it by controlling all ground lines of communications and reinforcement. *Vertical envelopments* are tactical maneuvers in which troops, either air-dropped or airlanded, attack the rear and flanks of a force, in effect cutting off or encircling the force (JP 3-18). Figure 7-6 shows an example of control measures used when conducting a single envelopment.

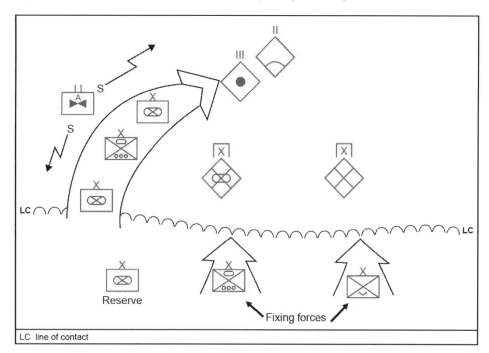

Figure 7-5. Example of a single envelopment

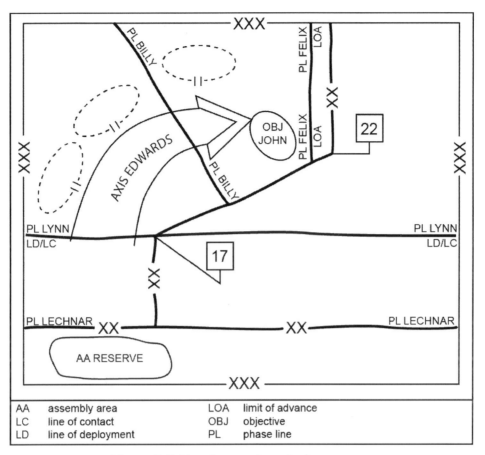

Figure 7-6. Envelopment control measures

7-98. The commander, at a minimum, designates AOs for each unit participating in the envelopment by using boundaries. The commander also designates PLs, support by fire positions, attack by fire positions, contact points, and appropriate FSCMs, such as a restrictive fire line or boundary between converging forces, and any other control measures necessary to control the envelopment. (See FM 3-90-1 for a discussion of the organization of forces and planning, preparation, and execution considerations associated with the performance of the envelopment task. See FM 3-90-2 for a discussion of encirclement operations.)

FLANK ATTACK

Everyone knows the moral effects of an ambush or an attack in flank or rear.

Carl von Clausewitz

7-99. A *flank attack* is a form of offensive maneuver directed at the flank of an enemy (FM 3-90-1). (See figure 7-7 on page 7-24.) Flank attacks occur when employing both contiguous and noncontiguous operational frameworks. A flank is the right or left side of a combat formation, and it is not oriented toward the enemy. It is usually not as strong in terms of forces or fires as the front of a combat formation. A flank may be created through the use of fires or by a successful penetration.

7-100. A flank attack is similar to an envelopment, but it is generally conducted on a shallower axis. It is designed to defeat an enemy force while minimizing the effect of the enemy's frontally-oriented combat power. Flank attacks are normally conducted with the main effort directed at the flank of the enemy. Usually, a supporting effort engages the enemy's front by fire and maneuver, while the main effort maneuvers to attack the enemy's flank. This supporting effort diverts the enemy's attention from the threatened flank. A flank attack is often used for a hasty attack or meeting engagement where speed and simplicity are crucial to maintaining battle tempo and the initiative.

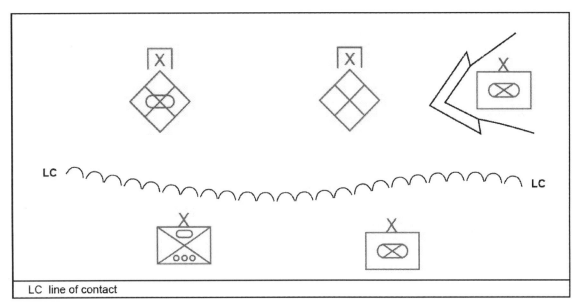

Figure 7-7. Flank attack

7-101. The primary difference between a flank attack and an envelopment is one of depth. A flank attack is an envelopment delivered squarely on the enemy's flank. An envelopment is an attack delivered beyond the enemy's flank and into the enemy's support areas, but short of the depth associated with a turning movement.

7-102. Just as there is a relationship between unit size and the ability of a friendly force to execute a turning movement instead of an envelopment, this relationship extends downward between an envelopment and a flank attack. Corps and divisions are the most likely echelons to conduct turning movements. Divisions and BCTs are the echelons most likely to conduct envelopments—single or double. Smaller-sized tactical units, such as maneuver battalions, companies, and platoons, are more likely to conduct flank attacks than larger tactical units. This is largely a result of troop-to-space ratios and sustainment and mission command constraints. The Confederate attack by the four divisions of Stonewall Jackson's II Corps of the Army of Northern Virginia on the Union XI Corps of the Army of the Potomac at Chancellorsville on 2 May 1863 is a historical example of a flank attack.

7-103. The control measures associated with a flank attack are similar to those addressed in the envelopment discussion in paragraph 7-98. The primary difference between these forms of maneuver is which portion of the enemy position is attacked.

FRONTAL ATTACK

Are there not other alternatives than sending our armies to chew barbed wire in Flanders?
Winston S. Churchill

7-104. A *frontal attack* is a form of maneuver in which an attacking force seeks to destroy a weaker enemy force or fix a larger enemy force in place over a broad front (FM 3-90-1). Frontal attacks occur most often when the commander employs a contiguous operational framework. Figure 7-8 illustrates a frontal attack.

Large-Scale Offensive Operations

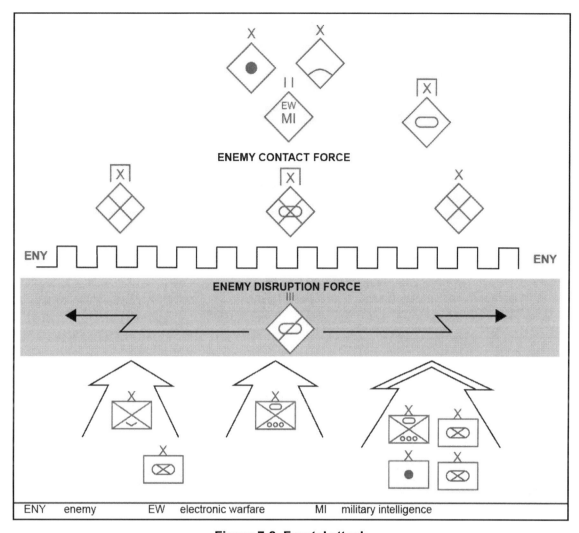

Figure 7-8. Frontal attack

7-105. At the tactical level, an attacking force can use a frontal attack to rapidly overrun a weak enemy force. A commander commonly uses a frontal attack as a shaping operation in conjunction with other forms of maneuver. A commander normally employs a frontal attack to—
- Clear enemy security forces.
- Overwhelm a shattered enemy during an exploitation or pursuit.
- Fix enemy forces in place as part of a shaping operation.
- Conduct a reconnaissance in force.

7-106. Frontal attacks are also necessary when assailable flanks do not exist. Where a penetration is a sharp attack designed to rupture the enemy position, the commander designs a frontal attack to maintain continuous pressure along the entire front until either a breach occurs or the attacking forces succeed in pushing the enemy back. Frontal attacks conducted without overwhelming combat power are seldom decisive. Consequently, the commander's choice to conduct a frontal attack in situations where the commander does not have overwhelming combat power is rarely justified unless the time gained is vital to the operation's success. The Vth and VIIth Corps amphibious landings on Omaha and Utah Beaches on 6 June 1944 are examples of a frontal attack. (See FM 3-90-1 for a discussion of the organization of forces, control measures, and planning, preparation, and execution considerations associated with the performance of the frontal attack task.)

Chapter 7

7-107. A commander conducting a frontal attack may not require any additional control measures beyond those established to control the overall mission. This includes an AO, defined by unit boundaries, and an objective, at a minimum. The commander can also use any other control measure necessary to control the attack, including—

- Attack positions.
- Line of departure (LD).
- PLs.
- Assault positions.
- Limit of advance (LOA).
- Direction of attack or axis of advance for every maneuver unit.

An *attack position* is the last position an attacking force occupies or passes through before crossing the line of departure (ADRP 3-90). An attack position facilitates the deployment and last-minute coordination of the attacking force before it crosses the LD. An *assault position* is a covered and concealed position short of the objective from which final preparations are made to assault the objective (ADRP 3-90). These final preparations can involve tactical considerations, such as a short halt to coordinate the final assault, reorganize to adjust to combat losses, or make necessary adjustments in the attacking force's dispositions.

7-108. Figure 7-9 shows the control measures listed in paragraph 7-107. It does not show a direction of attack since this control measure is used primarily at the battalion level and below. A unit conducting a frontal attack normally has a wider AO than a unit conducting a penetration.

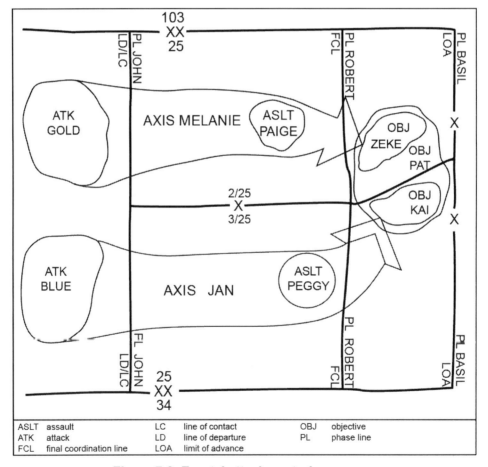

Figure 7-9. Frontal attack control measures

INFILTRATION

7-109. An *infiltration* is a form of maneuver in which an attacking force conducts undetected movement through or into an area occupied by enemy forces to occupy a position of advantage behind those enemy positions while exposing only small elements to enemy defensive fires (FM 3-90-1). Infiltrations can occur when using a contiguous or noncontiguous operational framework. Historically, the scope of the mission for an infiltrating force has been limited.

> *Note.* Infiltration is also a march technique used within friendly territory to move forces in small groups at extended or irregular intervals. (See FM 3-90-2 for a discussion of infiltration as a march technique.)

7-110. Infiltration occurs by land, water, air, or a combination of means. Moving and assembling forces covertly through enemy positions takes a considerable amount of time. To successfully infiltrate, the force must avoid detection and engagement. Since this requirement limits the size and strength of the infiltrating force—and infiltrated forces alone can rarely defeat an enemy force—infiltration is normally used in conjunction with and in support of the other forms of offensive maneuver.

7-111. Commanders order an infiltration to move all or part of a unit through gaps in enemy defenses to—
- Reconnoiter known or templated enemy positions and conduct surveillance of named areas of interest (NAIs) and targeted areas of interest (TAIs).
- Attack enemy-held positions from an unexpected direction.
- Occupy a support by fire position to support the decisive operation.
- Secure key terrain.
- Conduct ambushes and raids to disrupt the enemy's defensive structure by attacking reserves, fire support and air defense systems, communication nodes, and sustainment.
- Conduct a covert breach of an obstacle or obstacle complex.

7-112. Special operations forces (SOF) and light infantry units up to brigade size are best suited to conduct an infiltration. In some circumstances, armored and Stryker-equipped forces operating in small units can conduct an infiltration. However, as the proliferation of technology leads to increased situational understanding, the ability of these forces to avoid enemy contact and move undetected through enemy positions is reduced. To increase the likelihood of a successful infiltration of larger forces, a commander may conduct an infiltration in coordination with precision fires, deception operations, and electronic and cyber-attacks as a prelude to an infiltration. (See FM 3-90-1 for a discussion of the organization of forces and the planning, preparation, and execution considerations associated with the performance of the infiltration task.)

7-113. Since infiltrating forces are dispersed, they require a significant number of control measures. Control measures for an infiltration include, as a minimum—
- An AO for the infiltrating unit.
- One or more infiltration lanes.
- An LD or point of departure.
- Movement routes with their associated start points and release points, or a direction or axis of attack.
- Linkup or rally points, including objective rally points.
- Assault positions.
- One or more objectives.
- An LOA.

7-114. Commanders can impose other measures to control the infiltration, including checkpoints, PLs, and assault positions on the flank or rear of enemy positions. If it is not necessary for the entire infiltrating unit to reassemble to accomplish its mission, the objective may be broken into smaller objectives. Each infiltrating element then moves directly to its objective to conduct operations.

Chapter 7

PENETRATION

7-115. A *penetration* is a form of maneuver in which an attacking force seeks to rupture enemy defenses on a narrow front to disrupt the defensive system (FM 3-90-1). Penetrations most often occur when a commander employs a contiguous operational framework. They also occur during the reduction of enemy encircled pockets. Destroying the continuity of that defense allows the enemy's subsequent isolation and defeat in detail by exploiting friendly forces. The penetration extends from the enemy's security area through main defensive positions into the enemy support area. Commanders employ a penetration when there is no assailable flank, enemy defenses are overextended, weak spots are detected in the enemy's positions, or time pressures do not permit envelopment. The First U.S. Army's Operation Cobra (the breakout from the Normandy lodgment in July 1944) is a classic example of a penetration. Figure 7-10 illustrates potential correlation of forces or combat power for a penetration. Figure 7-11 shows an example corps penetration. (See FM 3-90-1 for a discussion of the organization of forces, control measures, and planning, preparation, and execution considerations associated with the performance of the penetration task.)

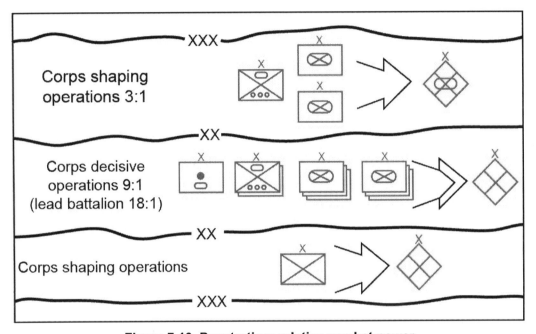

Figure 7-10. Penetration: relative combat power

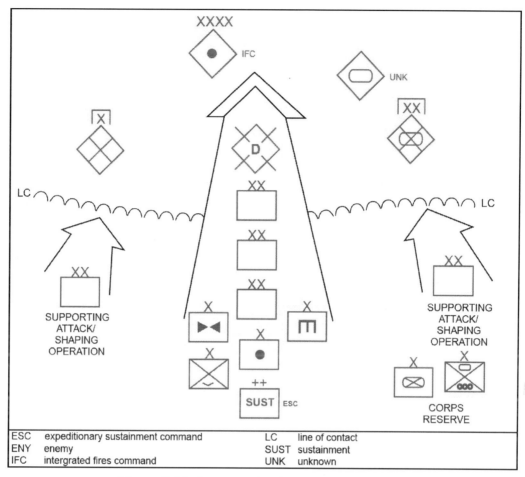

Figure 7-11. An example of a corps penetration

7-116. The control measures associated with a penetration include an AO for every maneuver unit, a LD or line of contact; a time of the attack or time of assault; PLs; an objective; and an LOA to control and synchronize the attack. (A commander can use a battle handover line instead of an LOA if the commander knows where the likely commitment of a follow-and-assume force will occur.) (Figure 7-12 on page 7-30 shows these control measures.) Commanders draw the lateral boundaries of the unit making the decisive operation narrowly. Commanders locate the LOA beyond the enemy's main defensive position to ensure completing the breach. If an operation results in opportunities to exploit success and pursue a beaten enemy, commanders adjust existing boundaries to accommodate the new situation.

Chapter 7

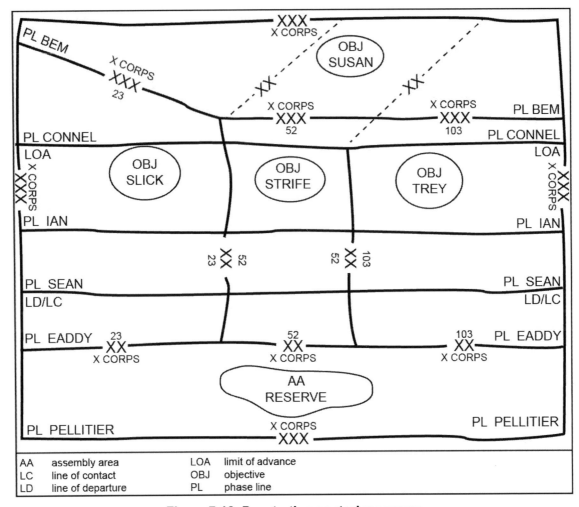

Figure 7-12. Penetration control measures

TURNING MOVEMENT

7-117. A *turning movement* is a form of maneuver in which the attacking force seeks to avoid the enemy's principle defensive positions by seizing objectives behind the enemy's current positions thereby causing the enemy force to move out of their current positions or divert major forces to meet the threat (FM 3-90-1). Turning movements can occur when the commander uses a contiguous or noncontiguous operational framework. However, a commander can employ a vertical envelopment using airborne or air assault forces to effect a turning movement. An *airborne assault* is the use of airborne forces to parachute into an objective area to attack and eliminate armed resistance and secure designated objectives (JP 3-18). An *air assault* is the movement of friendly assault forces by rotary-wing or tiltrotor aircraft to engage and destroy enemy forces or to seize and hold key terrain (JP 3-18). It can also be conducted using waterborne or amphibious means. (See JP 3-02 for a discussion of amphibious operations. See figure 7-13 for a graphic depiction of a turning movement.)

Large-Scale Offensive Operations

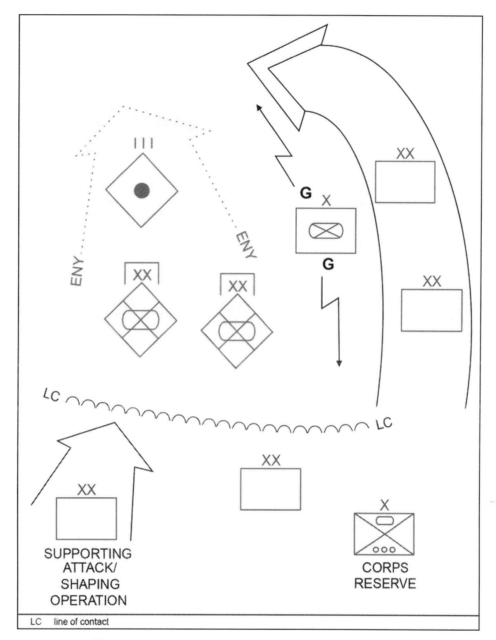

Figure 7-13. Example of a corps turning movement

7-118. The commander designates the AOs for each unit participating in the turning movement by establishing boundaries. The commander also designates additional control measures as necessary to synchronize the subordinate force operations. These additional control measures include PLs, contact points, objectives, LOAs, and appropriate FSCMs. A *contact point* in land warfare is a point on the terrain, easily identifiable, where two or more ground units are required to make contact (JP 3-50). It contains the date-time group when contact is to occur. (See figure 7-14.)

Figure 7-14. Contact point

7-119. A commander uses a turning movement to seize vital areas in the enemy's support area before the main enemy force can withdraw or receive support or reinforcements. Commanders frequently transition this form of maneuver from the attack

Chapter 7

into an exploitation or pursuit. A turning movement differs from an envelopment because the force conducting a turning movement seeks to make the enemy forces displace from their current locations, whereas an enveloping force seeks to fix and destroy the enemy forces in their current locations from an unexpected direction. (See FM 3-90-1 for a discussion of the organization of forces, control measures, and planning, preparation, and execution considerations associated with the performance of the turning movement task.)

OFFENSIVE TASKS

7-120. There are four basic offensive tasks—movement to contact, attack, exploitation, and pursuit. These apply to the tactical and operational levels of warfare. The execution of these four tasks pose significantly different challenges. Although the names of these offensive tasks convey the overall aim of a selected offense, each typically contains elements of the other. Corps and division commanders combine these tasks with the forms of maneuver based on their intent and their higher echelon commander's concept of operations.

7-121. The organization of forces for each offensive task is slightly different. Corps and division commanders organize their subordinate units and assign them to their respective AOs based on their concept of operations for defeating the enemy. Corps and division commanders accept risk to mass the effects of combat power for their decisive operations or main effort. Corps and divisions headquarters reinforce the subordinate units conducting the decisive operation or main effort with additional fires, information collection, engineer, attack aviation, air defense, and sustainment support.

7-122. During the performance of offensive tasks there may be enemy counterattack penetrations into the flanks of advancing divisions and BCTs. The divisions and BCTs employ security forces on exposed flanks to provide early warning and reaction time. If flank security forces cannot defeat or fix enemy counterattacks, Divisions and BCTs affected by these enemy counterattacks quickly transition to a defense or a delay. The corps or division commander determines size, composition, direction of attack, probable objective, and rate of movement of the counterattacking enemy force. This allows the corps or division commander to assess risk and develop courses of action to defeat the counterattacking enemy force. These courses of action will employ a mix of dynamic defensive actions to turn the enemy force into engagement areas and block its further advance. This may require changing task organization. It will require adjusting subordinate unit boundaries, assigned missions, and control measures. The commander directs the execution of situational or reserve obstacles, the shifting of priority of fires, and the assumption of risk. Corps and division commanders employ offensive actions to isolate and destroy a counterattacking enemy force once it is fixed in these engagement areas.

MOVEMENT TO CONTACT

During the advance to contact every agency of intelligence, reconnaissance and security is utilized to ensure that the main forces are engaged under the most favorable conditions.

FM 100-5, *Field Service Regulations: Operations*, 1944

7-123. *Movement to contact* is an offensive task designed to develop the situation and establish or regain contact (ADRP 3-90). Commanders conduct a movement to contact when an enemy situation is vague or not specific enough to conduct an attack. A movement to contact employs purposeful and aggressive reconnaissance and security operations to gain contact with the enemy main body and develop the situation. The movement to contact force defeats enemy forces within its capability. If the movement to contact force meets a superior force that it is unable to defeat, the movement to contact force conducts security or defensive tasks as necessary to further develop the situation. The fundamentals of a movement to contact are—

- Focus all efforts on finding the enemy.
- Make initial contact with the smallest force possible, consistent with protecting the force and avoiding decisive engagement while retaining enough combat power to develop the situation and mitigate the associated risk. (Contact may be initiated in cyberspace or the information environment.)
- Task-organize the force and use movement and combat formations to deploy and attack rapidly in any direction.

- Keep subordinate forces within supporting distances to facilitate a flexible and mutually supporting response.
- Maintain contact regardless of the course of action adopted once contact is gained.

7-124. Close air support, air interdiction, Army aviation reconnaissance, security and attacks, cyberspace electromagnetic activities, IRCs, space operations, and counter air operations are essential to the success of large-scale movements to contact. Local air superiority or, at a minimum, air parity is vital to the operation's success. A successfully executed movement to contact creates favorable conditions for subsequent tactical actions.

7-125. A unit may conduct a movement to contact with its subordinate elements abreast, with each subordinate element in turn conducting a movement to contact in its AO. As an alternative, a unit may lead with one subordinate element and follow with another. Figure 7-15 illustrates a corps with two divisions conducting a movement to contact with one division leading.

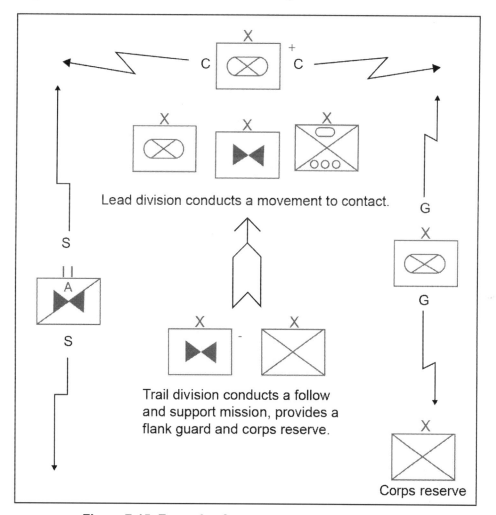

Figure 7-15. Example of a corps movement to contact

7-126. In figure 7-15 the lead division is the decisive operation and the trail division follows and supports. The lead division engages and destroys enemy security forces. The trail division with the follow and support mission eliminates bypassed enemy forces, secures key terrain, screens the flanks of the corps, and conducts minimum essential stability tasks, if required. It also provides one IBCT as the corps reserve. Depending upon the mission variables, the trail division may conduct a forward passage of lines through the lead division and conduct the corps decisive operation. In this example, the corps reserve is an IBCT positioned in a heavy pickup zone in the corps AO. The corps provides aviation support obtained by tasking the combat aviation brigades (CABs) of its subordinate divisions and coordinates with the United States Air Force (USAF) and theater aviation brigade to air move or air assault the reserve upon commitment. A corps normally uses PLs to control an operation as its divisions advance. A corps transitions to a deliberate operation if the lead division encounters a prepared enemy defense. A corps conducts hasty operations and exploits any favorable situations if an enemy is unprepared or defending weakly.

7-127. A movement to contact may result in a meeting engagement. A *meeting engagement* is a combat action that occurs when a moving force, incompletely deployed for battle, engages an enemy at an unexpected time and place (FM 3-90-1). Once contact is made with an enemy force, the commander has five options: attack, defend, bypass, delay, or withdraw. Movement to contact is a primary means to develop the situation through action. (See FM 3-90-1 for additional information on the performance of a movement to contact.)

Organization of Forces for a Movement to Contact

7-128. A movement to contact is organized with a forward security force—either a covering force or an advance guard—and a main body. Based on the mission variables, a commander may increase the unit's security by resourcing an offensive covering force and an advance guard for each column, as well as flank and rear security (normally a screen or guard).

7-129. A movement to contact mission requires the commander not to have contact with the enemy main body. However, the commander may still know the location of at least some enemy reserve and follow-on forces. If the corps or division commander has enough information and intelligence to target enemy uncommitted forces, reserves, or sustaining activities, the commander normally designates forces, such as long-range artillery systems, attack helicopters, extended-range UASs, and fixed-wing aircraft to engage known enemy elements regardless of their locations within the AO. At all times the forward security element and the main body perform information collection. Reconnaissance is largely force oriented and not as detailed as a typical zone reconnaissance that gathers detailed information on both the enemy and terrain. The terrain reconnaissance tasks are minimized to only those necessary to facilitate mobility of the force.

Security Elements

7-130. A commander conducting a movement to contact typically organizes the forward security element into a covering force or an advanced guard to protect the movement of the main body and to develop the situation before committing the main body. This forward security element is normally the unit's initial main effort. A covering force is task-organized to accomplish specific tasks independent of the main body such as conduct mobility and selected countermobility tasks in accordance with the mission variables. This covering force reports directly to the establishing commander. If it is unable to resource a covering force, a force conducting a movement to contact may use an advance guard in the place of a covering force. The advanced guard accomplishes similar tasks as the covering force, but it must remain with supporting range of the main body, resulting in less protection and reaction time for the main body.

7-131. A corps conducting a movement to contact normally task-organizes its offensive covering force around a base provided by an ABCT or a SBCT. The corps forward covering force is usually reinforced with field artillery, attack reconnaissance aviation units, short-range air defense systems, engineers, and sustainment assets.

7-132. A division conducting a movement to contact separate from a corps operation normally task-organizes a division offensive covering force around a base provided by a combined arms battalion or a Stryker battalion. The division covering force is also reinforced with field artillery, rotary-wing attack-reconnaissance aviation, short-range air defense, and engineers.

7-133. Air-ground operations are essential to successful movements to contact. At the division level, the attack-reconnaissance battalion and Gray Eagle UAS company organic to the CAB may remain under division control until contact is made. The heavy attack reconnaissance squadron with its organic AH-64 troops and Shadow UAS platoons from the CAB may be under OPCON or direct support to the covering or advance guard force leading the movement to contact. The heavy attack reconnaissance squadron provides maximum reconnaissance forward, while the division's retention of control of the attack-reconnaissance battalion enables agile and lethal combat power employment to enable the decisive operations once the situation is developed. If a CAB is task-organized to a corps, the corps headquarters may employ CAB components in a similar manner.

7-134. Military intelligence units and systems conduct operations to locate enemy units and systems throughout the performance of the movement to contact task. Signals intelligence (SIGINT) and EW systems usually operate with the covering force and flank guard. The covering force commander directs EW against enemy C2 and fire support nets. The commander may use electronic deception to deceive the enemy as to the location of the main body. Space-based and cyberspace systems and activities also support the security force in locating and determining the presence of enemy disruption forces, contact forces, and shielding forces.

7-135. Each column within either the corps or division main body may designate an advance guard to assist the covering force and to prevent premature deployment of the main body. This advance guard develops the situation along an axis of advance or along designated routes. A *route* is the prescribed course to be traveled from a specific point of origin to a specific destination (FM 3-90-1). A column advance guard is usually a tank heavy or balanced battalion task force moving in front of each BCT. It operates within range of the field artillery moving with the main body. (See figure 7-16.)

7-136. Flank and rear guards may be used to protect the main body from ground observation, surprise, and direct fire. A flank or rear guard usually consists of one or two company teams for each BCT within the main body.

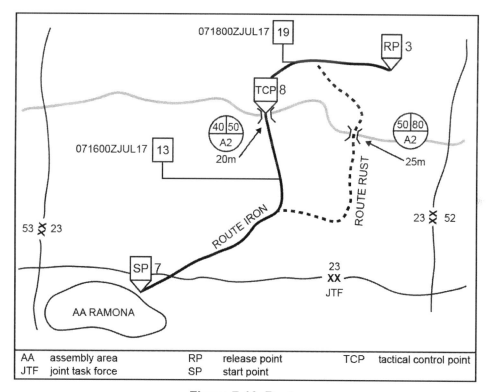

Figure 7-16. Route

Main Body

7-137. The main body of a corps or division conducting a movement to contact consists of corps or division forces not detailed to security duties. It is normally the main body that conducts the decisive operation within the conduct of the movement to contact. The main body may conduct either a tactical road march or an approach march. (See FM 3-90-2 for a discussion of these two tasks.) The combat elements of the main body prepare to respond to enemy contact with the unit's security forces.

7-138. Each BCT within the corps or division main body normally requires the use of at least two routes as it moves forward. A division main body normally moves with at least two BCTs abreast moving along two generally parallel axes of advance. A division commander may prefer for all subordinate BCTs to move abreast when enemy forces are not organized in depth. This assumes the presence of adequate mobility corridors. Likewise, a corps conducting a movement to contact normally moves with a least two of its divisions abreast of each other.

7-139. BCT commanders position one or two of the batteries in each BCT's field artillery battalion well forward in each BCT's column. Engineer mobility assets from each BCT's brigade engineer battalion also move with the BCT's leading companies.

7-140. The corps or division commander can assign a follow and support mission to a subordinate unit within the main body. This allows that subordinate unit to relieve forward security forces from such tasks as observing bypassed enemy forces, handling dislocated civilians, and clearing routes. This prevents main body forward and flank security forces from being diverted from their primary mission.

Reserve

7-141. Commanders designate a portion of the main body for use as the reserve. The size of the reserve is based on the mission variables and the amount of uncertainty concerning the enemy. The more vague the enemy situation, the larger the reserve. The reserve in a movement to contact typically constitutes approximately one-fourth to one-third of the force. On contact with the enemy, the reserve provides flexibility to react to unforeseen circumstances and allows the unit to quickly resume its movement.

Control Measures for a Movement to Contact

7-142. Commanders and staffs use control measures to focus movement, control fires, integrate airspace use, and enable freedom of action when executing a movement to contact. These measures include designation of an AO with left, right, front, and rear boundaries (in contiguous operations) or a separate AO bounded by a continuous boundary (in noncontiguous operations). The commander further divides the AO into subordinate unit AOs to facilitate subordinate unit actions.

7-143. The operation usually starts from an LD at the time specified in the operation order (OPORD). Commanders control a movement to contact by using PLs, contact points, and checkpoints as required. Commanders control the depth of the movement to contact by using a LOA or a forward boundary. The limit of advance or forward boundary is typically placed on suitable terrain for the force to establish a hasty defense. It should not be in locations dominated by enemy-controlled terrain features that would place the force at risk once the LOA is achieved.

> A *line of departure* is a phase line crossed at a prescribed time by troops initiating an offensive task (ADRP 3-90). The purpose of the LD is to coordinate the advance of the attacking force, so that its elements strike the enemy in the order and at the time desired.

7-144. Corps, division, or BCT commanders use boundaries to separate various organizational elements and clearly establish responsibilities between different organizations. Commanders can designate a series of PLs that can successively become the new rear boundary of the forward security element as that force advances. Each rear boundary becomes the forward boundary of the main body and shifts as the security force moves forward. The rear boundary of the main body designates the limit of responsibility of the rear security element. This line shifts as the main body moves forward. (See ADRP 3-90 for a discussion of these control measures.)

7-145. Alternatively, a commander could designate intermediate and primary objectives to limit the extent of the movement to contact and orient the force. However, these are often terrain-oriented and used only to guide movement. Although a movement to contact may result in taking a terrain objective, the primary focus remains on finding the main enemy force. If the location of significant enemy forces is known prior to initiation of the movement to contact, the commander plans the conduct of another offensive task.

Large-Scale Offensive Operations

Specific Planning Considerations for a Movement to Contact

7-146. A corps or division conducts a movement to contact to find the enemy and determine the enemy's dispositions when the intelligence picture is incomplete or stale. This occurs when friendly information collection is incomplete or when the enemy effectively contests use of space, such as in a degraded, denied, and disrupted space operations environment (known as D3SOE), the EMS, and cyberspace. The commander is increasingly dependent upon ground and Army aviation reconnaissance in such circumstances.

7-147. Commanders seek to gain contact by using the smallest elements possible. These elements are normally ground scouts or aerial scouts performing reconnaissance, but they may also be UASs or other reconnaissance and surveillance assets. The commander may task-organize the unit's scouts with additional combat power to allow them to develop the situation. The commander selects the unit's movement and combat formations to make initial contact with the smallest force possible. Movement formations should also provide for efficient movement of the force and effective placement of the reserve. The commander can choose to have all or part of the force conduct an approach march as part of the movement to contact to provide that efficient movement. An approach march can facilitate the commander's decisions by allowing freedom of action and efficient movement of the main body. (See FM 3-90-2 for a discussion of an approach march.)

Information Collection in a Movement to Contact

7-148. The commander conducts information collection activities to determine the enemy's location and intent while conducting security tasks to protect the main body. This includes the use of available fixed-wing aircraft, EW, Army aviation manned and unmanned attack and reconnaissance aircraft, space-based surveillance platforms, and cyberspace assets. This allows the main body to focus its planning and preparation, including rehearsals, on the conduct of hasty attacks, bypass maneuvers, and hasty defenses. Planning addresses not only actions anticipated by the commander based on available information and intelligence, but also includes the conduct of meeting engagements at anticipated times and locations—probable lines of contact—where they might occur in multiple domains. The commander tasks the forward security force with conducting hasty route reconnaissance of routes the main body will traverse.

7-149. Information collection may be more rapid in a movement to contact than in a zone reconnaissance because the emphasis is on making contact with the enemy. However, increasing the speed of the information collection effort may result in increased risk. If the mission requires it, the corps or division commander can task subordinate BCTs and CABs or the cavalry squadrons in those subordinate BCTs to further develop the situation during actions on contact. (See FM 3-90-1 for a discussion of actions on contact.)

7-150. The G-2, assisted by the engineer and air defense staff representatives, must carefully analyze the terrain, including air avenues of approach. The G-2 identifies the enemy's most dangerous course of action in the war gaming portion of the military decision-making process. Because of the force's vulnerability, the G-2 must not underestimate the enemy during a movement to contact. A thorough analysis of the enemy during IPB—by developing the modified combined obstacle overlay (including intervisibility overlays) and other products, such as the event templates—enhances the force's security by indicating danger areas where the force is most likely to make contact with the enemy. It also helps to determine movement times between PLs and other locations. Potential danger areas are likely enemy defensive locations, EAs, observation posts, and obstacles. The fires system targets these areas by designating target reference points, and they become on-order priority targets placed into effect and cancelled as the lead element can confirm or deny enemy presence. (See figure 7-17.) The intelligence annex of the movement to contact order must address coverage of these danger areas. If reconnaissance and surveillance forces cannot clear these areas, more deliberate movement techniques are required.

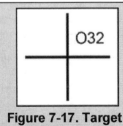

Figure 7-17. Target reference point

A *target reference point* is a predetermined point of reference, normally a permanent structure or terrain feature that can be used when describing a target location (JP 3-09.3).

Chapter 7

Scheme of Maneuver for a Movement to Contact

7-151. Commanders develop decision points to support changes in the force's movement formation or a change from an approach march to a combat formation. Using both human and technical means to validate decision points, commanders must determine the acceptable degree of risk based on the mission. A commander's confidence in the products of the IPB process and the acceptable risk determine the unit's combat formation and scheme of maneuver. In a high-risk environment, it is usually better to increase the distance between forward elements and the main body than to slow the speed of advance.

7-152. The frontage assigned to a unit in a movement to contact must allow it to apply sufficient combat power to maintain the momentum of the operation. Reducing the frontage normally gives the unit adequate combat power to develop the situation on contact while maintaining the required momentum. Covering force and advance guard commanders should have uncommitted forces available to develop the situation without requiring the deployment of the main body.

7-153. Bypass criteria should be clearly stated, and it depends on the mission variables. For example, an ABCT or SBCT commander in an open desert environment may direct that no mounted enemy force larger than a platoon can be bypassed and that all other forces will be cleared from the brigade axis of advance. Any force that bypasses an enemy unit must maintain contact with it until that force can hand that contact off to another friendly element, usually a force assigned a follow and support mission.

7-154. Corps and divisions can execute shaping operations in support of their subordinate BCTs as part of a movement to contact, although, by definition, a force conducts a movement to contact when the enemy situation is vague or totally unknown. Shaping operations to support a movement to contact occur when the information regarding enemy reserves and follow-on forces is available, but information regarding those enemy forces in close proximity to friendly forces is not available. As in any other type of operation, commanders plan to focus operations on finding the enemy and then delaying, disrupting, and destroying enemy forces as much as possible before arriving in direct-fire range.

7-155. In a movement to contact, the commander may not designate a decisive operation until subordinate forces make contact with the enemy. In this case, the commander retains resources under direct control to reinforce the decisive operation when it is designated. In turn, the commander may designate the decisive operation during the initial stages of a movement to contact because of the presence of a key piece of terrain or an avenue of approach. Once subordinate units make contact with the main enemy force and determine enemy dispositions, the corps or division commander maneuvers other forces into the battle to seize key terrain and exploit positions of relative advantage.

Scheme of Fires for a Movement to Contact

7-156. Commanders rely primarily on fire assets and Army aviation assets to weight the lead element's combat power, but commanders also provide it with the additional combat multipliers it needs to accomplish the mission. The fires system helps develop fire superiority when organized correctly to fire immediate suppression missions to help maneuver forces get within direct-fire range of the enemy.

7-157. Corps- and division-controlled EW assets attack enemy tactical communications before maneuver forces make physical contact with enemy forces. Electronic surveillance assets are primarily employed to provide early detection and location of enemy forces in concert with maneuver forces. Once found, the staff assists the commander in identifying enemy units in contact, determining their strength, dispositions, posture, and intentions. Information collection resources must accomplish these tasks early in order for the commander to concentrate EW and OCO nonlethal effects to develop and influence the situation in synchronization with lethal effects.

Executing a Movement to Contact

7-158. Each element of a corps or division performing a movement to contact synchronizes its actions with adjacent and supporting units. Each element maintains contact and coordinates as prescribed in orders and unit standard operating procedures.

Large-Scale Offensive Operations

Execution of a Movement to Contact by Security Forces

7-159. Each column's advance guard maintains contact with the corps or division's offensive covering force. Rear and flank security forces maintain contact with and orient on the movement of the main body. These security forces prevent unnecessary delay of the main body and prevent the deployment of the main body as long as possible. Reconnaissance elements operate to the front and flanks of each column's advance guard and maintain contact with the security force. The commander may instruct follow and support forces from each column to eliminate small pockets of resistance bypassed by the column's advance guard.

7-160. The commander of the corps or division offensive covering force chooses a combat formation to make contact with the smallest possible force while providing flexibility for maneuver. This is typically a wedge or a box formation, depending on whether the offensive covering force has three or four major subordinate maneuver commands. The covering force commander ensures that the route or axis of advance traveled by the main body is free of enemy forces. The offensive covering force may move continuously (using traveling and traveling overwatch) or by bounds (using bounding overwatch). It moves by bounds when contact with the enemy is imminent and the terrain is favorable.

7-161. Advance and flank guards keep enough distance between themselves and the offensive covering force to maintain flexibility for maneuver. This distance varies with the level of command, the terrain, and the availability of information about the enemy.

7-162. SIGINT and EW systems deploy well forward as part of the forward security force. From these positions, they provide support to maneuver units constituting the security force. These assets displace forward using alternate bounds to maintain continuous support. SIGINT systems monitor radios and other emitters associated with enemy reconnaissance forces, contact forces, and fire support systems. EW assets jam enemy command and control nets, especially those between enemy reconnaissance units and their control headquarters. They also jam enemy weapons systems which rely on electronic guidance or control.

7-163. A corps or division performing a movement to contact must attempt to cross any obstacles it encounters without loss of momentum by conducting hasty (in-stride) breaches. Commanders use forward security forces in an attempt to seize intact bridges whenever possible. Lead security elements bypass or breach obstacles as quickly as possible to maintain the momentum of the movement. Commanders direct subsequent elements of the main body to bypass an obstacle site and take the lead, if these lead elements cannot overcome obstacles. Following forces can also reduce obstacles that hinder the unit's sustainment flow. (See FM 3-90-1 for additional information on the performance of the movement to contact task.)

Execution of a Movement to Contact by the Main Body

7-164. Movement should be as rapid as the terrain, the mobility of the force, and the enemy situation permit. Open terrain provides maneuver space on either side of the line of march and facilitates high-speed movement. It also allows for greater dispersal and usually permits more separation between forward security elements and the main body than restricted terrain allows. A corps or division commander should never commit the main body to canalizing terrain before forward security elements have advanced far enough to ensure that the main body will not become fixed within that terrain. The enemy may have also established FSCMs that allow the enemy to employ non-observed harassing and interdiction fires to interdict friendly forces traversing these choke points. As the enemy situation becomes known, the commander may shorten the distance between main body elements to decrease reaction time or deploy the force to prepare for contact.

7-165. The corps or division main body advances over multiple parallel routes with numerous lateral branches to remain flexible and reduce the time needed to initiate maneuver behind these forward security elements. In a movement to contact, the main body's march dispositions must allow maximum flexibility for maneuvering during movement and when reacting to forward or flank security forces making contact with enemy forces.

7-166. The corps or division main body deploys once the enemy's location is determined. The main body may execute a tactical road march or an approach march for all or part of the movement to contact to efficiently use the available road network or reduce the time needed to move from one location to another. CPs and support elements travel along high-mobility routes within the AO and occupy hasty positions as necessary.

Chapter 7

7-167. The commander moves well forward in the movement formation. Once the formation makes contact with the enemy, the commander can move quickly to the area of contact, analyze the situation, and direct elements aggressively. The unit's security elements conduct actions on contact to develop the situation once they find the enemy. Once they make contact with the enemy, a number of actions occur—deploy and report, evaluate and develop the situation, choose a course of action, execute selected course of action, and recommend a course of action to the higher echelon commander. (Units equipped with a full set of digital information systems may be able to combine or skip one or more of the steps in that sequence. Those units will conduct maneuver and remain within supporting distance of each other with a significantly larger AO than units equipped with analog systems.) These actions normally constitute a major portion of the unit's shaping operations.

7-168. The commander's fire support systems tend to focus on suppression missions to disrupt enemy forces as they are encountered and smoke missions to obscure or screen exposed friendly forces when conducting a movement to contact. The commander schedules the movements of fire support systems in synchronization with the movement of the rest of the force. Fire support systems that cannot match the cross-country mobility of ground maneuver units cause them to slow their rate of advance. In this case, if ground maneuver units do not slow down, they run the risk of outrunning their fire support. The commander synchronizes the employment of CAS to prevent the enemy from regaining balance while the commander's ground fire support assets are repositioning. The main body updates its priority target list during a movement to contact.

7-169. The considerations in paragraph 7-168 are similar to the considerations that apply to AMD. A corps or division commander conducting a movement to contact remains aware of the AMD umbrella provided by Sentinel radars and AAMDC Patriot systems, short-range air defense systems, UAS defeat jammers, and the combat air patrol provided by USAF, U.S. Navy, and U.S. Marine Corps fighter aircraft.

7-170. The unit's tempo, momentum, tactical dispersal, use of cover and concealment, and attention to electromagnetic emission control complicate the enemy's ability to detect and target the main body until contact is made. Once the force makes contact and concentrates its effects against detected enemy forces, it becomes vulnerable to strikes by enemy conventional weapons and WMD. It must concentrate its combat effects rapidly in a meeting engagement and disperse again as soon as it overcomes resistance to avoid enemy counteractions, if the movement to contact is to continue. However, the results of that meeting engagement will determine the specific course of action selected.

ATTACK

The enemy say that Americans are good at a long shot but cannot stand the cold iron. I call upon you to give a lie to the slander. Charge!

Brigadier General Winfield Scott

7-171. An *attack* is an offensive task that destroys or defeats enemy forces, seizes and secures terrain, or both (ADRP 3-90). Attacks incorporate coordinated movement supported by fires. They may be part of either decisive or shaping operations. A commander may describe an attack as hasty or deliberate, depending on the time available for assessing the situation, planning, and preparing. A commander may decide to conduct an attack using only fires, including EW, OCO, and other IRCs. An attack differs from a movement to contact because, in an attack, commanders know at least part of an enemy's disposition. This knowledge enables commanders to better orchestrate the warfighting functions and converge the effects of available combat power more effectively.

7-172. Attacks take place along a continuum defined at one end by fragmentary orders that direct the execution of rapidly executed battle drills by forces immediately available. The other end of the continuum includes published, detailed orders with multiple branches and sequels, detailed knowledge of all aspects of enemy dispositions, a force that has been task-organized specifically for the operation, and the conduct of extensive rehearsals. Most attacks fall between the ends of the continuum as opposed to either extreme. (ADRP 3-90 discusses this continuum between hasty and deliberate operations.)

7-173. Typically, a corps employs multiple divisions in the attack. Normally, a corps attacks with two or more divisions abreast and one or two divisions following, either in support (follow and support) or prepared to continue the offense (follow and assume). A corps commander normally retains at least one BCT as the

Large-Scale Offensive Operations

corps reserve. However the size of the reserve reflects the commander's risk assessment—more risk requires a larger reserve.

7-174. If the situation permits, a corps fixes the main enemy defense using feints and limited attacks, while conducting a turning movement with two divisions. Figure 7-18 illustrates a four-division corps attack course of action using two divisions to conduct a turning movement. The corps commander weights the turning movement (decisive operation) attacking the enemy IFC and IADS using a second division with two BCTs to follow and support. The corps' field artillery brigade positions itself to support the turning movement and also attack the enemy IFC and IADS. The divisions in the turning maneuver receive all available corps and joint assets. In this example, the corps commander retained one BCT in reserve.

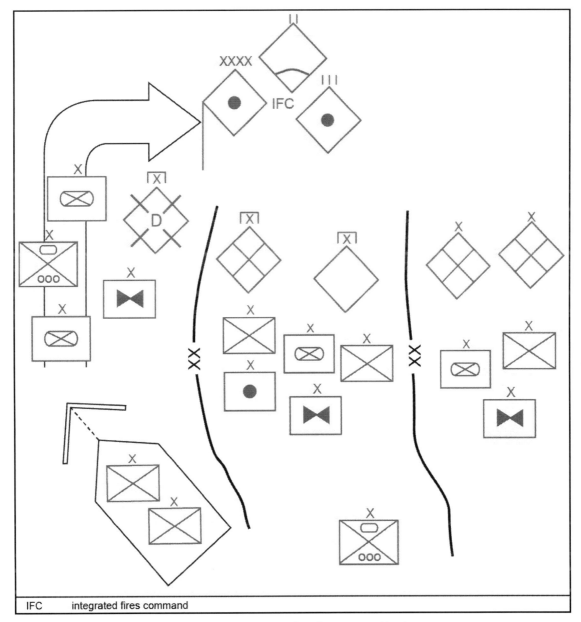

Figure 7-18. Example of a corps attack

7-175. If a corps cannot envelop or turn an enemy's defense, the corps conducts a penetration. Typically, a corps concentrates one division in a narrow AO and places a second, larger division in position to follow and assume. The corps makes the penetration division the main effort and the follow-on division the decisive

6 October 2017   FM 3-0   7-41

Chapter 7

operation as soon as the penetration allows exploitation. The other divisions conduct shaping attacks, military deception, and economy of force measures. Paragraphs 7-115 and 7-116 discuss the penetration as a form of maneuver. (See FM 3-90-1 for additional information on the performance of an attack.)

Organization of Forces for an Attack

7-176. Once a commander determines the scheme of maneuver, the commander task-organizes the force to give each unit enough combat power to accomplish its mission. Commanders normally organize the attacking force into a security force, a main body, and a reserve, all supported by some type of sustainment organization. Commanders should complete any changes in task-organization in time to allow units to conduct rehearsals with their attached and supporting elements. The best place and time for an attacking force to task-organize is when it is in an assembly area.

Security Forces for an Attack

7-177. Commanders resource dedicated security forces during an attack only if the attack will uncover one or more flanks or the rear of the attacking force as it advances. In this case, the commander designates a flank or rear security force and assigns it a guard or screen mission, depending on the mission variables. Normally an attacking unit does not need extensive forward security forces; most attacks are launched from positions in contact with the enemy, which reduces the usefulness of a separate forward security force. An exception occurs when the attacking unit is transitioning from the defense to an attack and had previously established a forward security area as part of the defense.

Main Body in an Attack

7-178. Commanders organize the main body into combined arms formations to conduct the decisive operation and necessary shaping operations. Commanders focus the decisive operation toward the decisive point which can consist of the immediate and decisive destruction of the enemy force, its will to resist, seizure of a terrain objective, or the defeat of the enemy's plan. The scheme of maneuver identifies the focus of the decisive operation. All of the force's available resources operate in concert in multiple domains to assure the success of the decisive operation. The subordinate unit or units designated to conduct the decisive operation can change during the course of the attack. If commanders expect to conduct a breach operation during the attack, commanders designate assault, breach, and support forces.

7-179. If it is impractical to determine initially when or where the decisive operation will be, commanders retain flexibility by arranging forces in depth, holding out strong reserves, and maintaining centralized control of long-range fire support systems, attack reconnaissance helicopters, EW, OCO, and other IRCs. As soon as the tactical situation is clear enough to allow the commander to designate the decisive operation, the commander focuses available resources to support those subordinate forces conducting that decisive operation so that they can achieve their assigned objectives. Enemy actions, minor changes in the situation, or the lack of success by other attacking elements cannot be allowed to divert those forces or supporting effects from the decisive operation.

7-180. Commanders may need to designate a unit or units to conduct shaping operations to create windows of opportunity for executing the decisive operation. Commanders allocate the unit or units assigned to conduct shaping operations the minimum combat power necessary to accomplish their missions. Overwhelming combat power cannot be employed everywhere during large-scale combat operations. Units conducting shaping operations usually have a wider AO than those conducting a decisive operation. If a commander has sufficient forces to conduct shaping operations, the commander can assign the tasks of follow and assume or follow and support to subordinate units. (FM 3-90-1 defines these two tactical mission tasks.)

Reserve in the Attack

7-181. Commanders use their reserves to exploit success, defeat enemy counterattacks, or restore momentum to a stalled attack. Once committed, the reserve's actions normally become or reinforce the decisive operation. Commanders make every effort to reconstitute another reserve from units made available by the revised situation. Often a commander's most difficult and important decision concerns the time, place, and circumstances for committing the reserve. The reserve is not a committed force; it is not used as a follow and support force or a follow and assume force.

7-182. Attacking commanders allocate combat power to the reserve primarily on the level of uncertainty about the enemy, especially the strength of any expected enemy counterattacks. A reserve has mobility equal to or greater than the most dangerous enemy ground threat to be able to counter that threat. The reserve should also possess at least comparable combat power to the most likely enemy counterattacking force, if it is going to attack and defeat that enemy counterattack and allow its parent corps or division to continue its attack.

7-183. In addition, the strength and composition of the reserve vary with the reserve's contemplated missions, the forces available, the form of offensive maneuver selected, the terrain, and the risk accepted. The commander only needs to resource a small reserve to respond to unanticipated enemy reactions when detailed information about the enemy exists. Commanders should not constitute the reserve by weakening the decisive operation.

Control Measures for an Attack

7-184. Units conducting offensive tasks are assigned an AO within which to operate. Within the AO the commander normally designates the following control measures regardless of whether the attack takes place in a contiguous or noncontiguous environment:

- AOs for subordinate units of battalion size or larger.
- A PL as the LD, which may also be the line of contact. (See figure 7-19.)
- The time to initiate the operation.
- The objective.

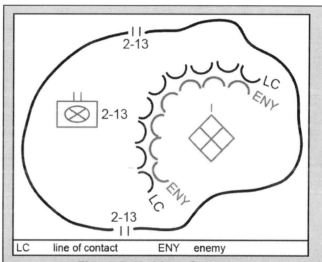

Figure 7-19. Line of contact

A *line of contact* is a general trace delineating the location where friendly and enemy forces are engaged (FM 3-90-1). In the offense, a line of contact is often combined with the LD.

7-185. If necessary, a commander can use either an axis of advance or a direction of attack to further control maneuver forces. Authorities granted to a commander act as control measures for EW, OCO, and the employment of other IRCs.

7-186. A commander can use any other control measures necessary to control the attack. Short of the LD or line of contact, a commander may designate assembly areas and attack positions where the unit prepares for offensive actions or waits for the establishment of the required conditions to initiate the attack. Beyond the LD or line of contact, the commander may designate checkpoints, PLs, probable lines of deployment, assault positions, direct fire control measures, and FSCM. Between the probable line of deployment and the objective a commander can use a final coordination line, assault positions, attack by fire and support by fire positions, and a time of assault to further control the final stage of the attack. (See figure 7-20 on page 7-44.) A commander can impose an LOA beyond the objective if the commander does not want the unit to conduct exploitation or pursuit operations. (Appendix A of FM 3-90-1 discusses these control measures.)

7-187. Commanders need to maintain tight control over the movement of all attacking elements during attacks conducted under limited-visibility conditions to prevent fratricide due to a loss of situational awareness by small units. These conditions require commanders to impose additional control measures beyond those used in daylight attacks. These additional measures may include using a point of departure and a direction of attack.

Executing an Attack

7-188. An attack must be violent and rapid to shock the enemy and to prevent the enemy's recovery until destruction of the integrity of the enemy defense occurs. An attack may require a powerful and violent direct assault of enemy-occupied key or decisive terrain, although in a noncontiguous environment indirect approaches may result in fewer casualties. The attacker must minimize exposure to enemy fires by using maneuver and counterfire, avoiding obstacles, taking advantage of the terrain, maintaining security, and insuring continued orchestration of the warfighting functions across multiple domains. Synchronized movement and fires coupled with the contributions of all warfighting functions across multiple domains are used to achieve superior combat power at the point of assault. Artillery preparation, suppressive fires, isolation of the enemy force, concentration of combat power, and physically overrunning enemy positions all combine to destroy a defending enemy. An attacking force needs to remain organized as a cohesive force for the battle on its objective and the exploitation of battlefield success.

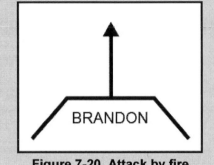

Figure 7-20. Attack by fire position

An *attack by fire* position is the general position from which a unit conducts the tactical task of attack by fire (ADRP-3-90.) The purpose of an attack by fire position is to mass the effects of direct fire systems for one or multiple locations toward the enemy.

7-189. Commanders must be well forward for psychological reasons as well as for effective control of their units. They are best able to make assessments and issue instructions by remaining near the source of combat information. While on the move, forces communicate primarily by combat net radio, but they remain prepared to continue their forward maneuver when the enemy contests use of the EMS. Commanders must guard this and other means of communications and demand discipline in passing only essential information. Brevity reduces the EMS signature; clarity remains essential. Commanders facilitate coordination by monitoring the reports of adjacent units.

7-190. Army corps and divisions attack under a variety of circumstances. In large-scale combat operations they most often attack from defensive positions that form a perimeter around their initial lodgment within a joint operations area (JOA) or theater of operations. That perimeter defense may or may not have withstood an enemy attack. The enemy may or may not have contested that initial lodgment. Corps and divisions will attack after the joint force commander (JFC) builds up sufficient forces within the lodgment. The breakout from the lodgment may take the form of a movement to contact, if the enemy has defeated U.S. information collection means in multiple domains. Corps and divisions conduct a meeting engagement following their movement to contact in this case.

7-191. Corps and division commanders employ their information collection assets, primarily reconnaissance and security, to further develop the situation if the meeting engagement is not fully successful. When conducting more deliberate operations, corps and division commanders ensure that their forces do not become disorganized or vulnerable to an enemy counterattack by conventional forces, EW, cyberspace operations, information activities, and WMD. Commanders make necessary adjustment decisions and direct portions of the force to establish hasty defensive positions when necessary. Other parts of the force may occupy assembly areas. Rapid dispersion is essential, but it must be orderly enough to allow the attack to continue after a brief period of reorganization and resupply with concurrent multi-echelon planning and coordination.

7-192. Commanders concentrate all available firepower on the enemy-occupied positions at the beginning of the assault. These fires shift to targets beyond the assault objective as troops move on to the objective. This requires detailed planning, precise execution, and considerable discipline in the fire support force as well as in the assault force. Dismounted assault forces should move as close behind their own fires as possible. Armored forces should assault under the close cover of overhead artillery fire.

7-193. The attacker overcomes enemy resistance by violent, concentrated firepower and a rapid advance upon reaching the assault objective. Speed during this phase of the attack is absolutely essential to reduce casualties and to avoid being stalled in the enemy's kill zones. The assault must move completely across the enemy-occupied position. Fortified positions on the assault objective are then attacked from the flank or rear after the assaulting force has passed around them.

7-194. Corps and division headquarters orchestrate a coordinated effort to suppress enemy IFC, IADS, and C2 across multiple domains in support of assaulting forces. Those enemy forces in depth posing the greatest threat to the attacking force are nuclear or chemical delivery systems, C2 facilities of the force being attacked, reserve forces, and the IFC. Corps and divisions target these enemy assets for dislocation and isolation until BCT maneuver units can maneuver closely to destroy these enemy assets with direct and indirect fire.

7-195. Attacking BCT maneuver forces must disperse rapidly and initiate an exploitation once they take enemy occupied positions. The enemy will target its positions against the possibility of their loss. Exploitation occurs either by continuing the attack with the same force or by the forward passage of another force. ABCTs and SBCTs normally exploit success experienced by dismounted infantry attacks on fortified defenses. Any pause in the conduct of offensive tasks increases the force's vulnerability to conventional counterattack or counterattack by unconventional means, such as EW, cyberspace, and nuclear or chemical fires. A pause also gives the enemy time to reorganize successive defensive arrays.

EXPLOITATION AND PURSUIT

7-196. Exploitation and pursuit proceed directly from the attack and are initiated from attack dispositions. An exploitation or a pursuit is planned as a sequel to an attack. Exploitation is the bold continuation of an attack to maximize success. Pursuit is the relentless destruction of retreating enemy forces who have lost the capability to effectively resist. Corps and division commanders continually monitor the progress of their attack to determine if they can transition to pursuit or exploitation. They closely monitor the situation to detect the development of any significant enemy threat to their command by reinforcing enemy formations. Corps and division deep operations during the performance of exploitation and pursuit normally focus on—

- Isolating the retreating enemy force and preventing its reinforcement.
- Attacking the enemy forces at critical chokepoints to facilitate their destruction.
- Disrupting, turning, and stopping lead elements of the retreating enemy force.
- Attacking enemy supporting enablers and repositioning reinforcements.

Exploitation

7-197. Exploitation forces drive swiftly for deep objectives, seizing enemy command posts, severing enemy escape routes, and striking at enemy reserves, artillery, and logistics units to prevent the enemy from reorganizing an effective defense. Exploitation forces should be large and reasonably self-sufficient combined arms organizations, such as BCTs. They should be able to change direction of movement on short notice. Exploitation forces receive support from joint fires, Army aviation, and echelons above corps EW and OCO assets. Commanders and staffs coordinate to meet the sustainment requirements of exploitation forces, including aerial resupply to move emergency lifts of POL and ammunition and medical evacuation. During the exploitation, the lead exploiting force usually conducts the main attack, while follow and support units conduct supporting attacks.

7-198. A commander normally employs permissive FSCM during exploitation. A coordinated fire line ensures rapid response. (See figure 7-21 on page 7-46.) Movement of the coordinated fire line is particularly important to provide adequate support as the force continues to advance. Even if the culmination of the exploitation is not anticipated, establishing a forward boundary is important to facilitate operations beyond that boundary by a higher echelon headquarters. Commanders should consider that placing the forward boundary too deep limits the ability for other higher echelon forces to generate effects to enable the exploitation or pursuit forces. Commanders can use additional control measures, such as targets and checkpoints, as required.

Mission Command in Exploitation

7-199. Planning for exploitation begins during the preparation for an attack. Corps and division commanders tentatively identify forces, objectives, and AOs for the exploitation force before the attack begins to avoid losing critical time. Subordinates report indications of the disintegration of enemy defenses. Commanders quickly issue warning orders, group forces, establish control measures, and arrange to meet the exploiting forces' sustainment requirements when the opportunity to exploit occurs. Commanders of committed forces act quickly to exploit local advantages. Corps and divisions should aim for an early and rapid transition from attack to exploitation. Most exploitations are initiated from the front rather than directed from the rear. When possible, the leading forces of the attack retain the lead for the exploitation. Commanders normally designate exploiting forces by issuing fragmentary orders during the course of an attack.

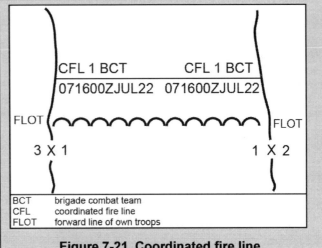

Figure 7-21. Coordinated fire line

The *coordinated fire line* is a line beyond which conventional surface-to-surface direct fire and indirect fire support means may fire at any time within the boundaries of the establishing headquarters without additional coordination (JP 3-09). The purpose of the coordinated fire line is to expedite the surface-to-surface attack of targets beyond the coordinated fire line without coordination with the ground commander in whose area the targets are located.

7-200. Assigned missions for an exploiting force include seizing objectives deep in the enemy rear, cutting enemy lines of communications, isolating and destroying enemy units, and disrupting enemy C2. The exploiting force commander must have the greatest possible freedom of action to accomplish the mission. This commander takes the initiative and moves aggressively and boldly. The objective—a critical enemy communications center, a mountain pass, or key terrain that significantly contributes to destruction of organized enemy resistance—may often be some distance away. At other times, the objective may simply be a point of orientation. Exploitation operations are decentralized.

7-201. Corps or division commanders maintain sufficient control to alter the direction of the exploiting force or to prevent its overextension. Those commanders rely on their subordinate commanders to find the fastest way to their objectives, to deploy as necessary to fight, and to seize all opportunities to damage the enemy or to accelerate the pace of the operation. A corps or division commander uses minimum control measures, but issues clear instructions concerning seizure of key terrain and the size of enemy forces which may be bypassed. Exploiting force commanders must be careful not to dissipate combat power in achieving minor tactical successes or in reducing small enemy forces. Their aim is to reach the objective with the maximum strength as rapidly as possible. Higher echelon commanders move division and BCT rear boundaries forward as appropriate to avoid the size of these echelon AOs becoming significantly larger than their areas of influence. At some point the land component command may also need to move the corps rear boundary forward.

7-202. Control is vital to prevent overextension of either the exploiting force or the logistics required to sustain it. This control is especially important if the enemy is capable of regrouping unexpectedly to attack the exploiting force. Available joint fires target enemy forces in multiple domains and the information environment that cannot be bypassed or contained. Space control activities are used to defeat enemy efforts that interfere with or attack U.S. or allied space systems and negate enemy space capabilities. Rapid advances provide security from enemy counterattacks employing conventional or unconventional—cyberspace and nuclear—means by keeping enemy forces off balance and degrading their intelligence and surveillance capabilities. Exploitation continues day and night as long as the opportunity remains. Commanders must give specific attention to the control of logistic units and convoys to sustain an exploitation and to ensure that supplies and support reach the force safely and on time. Exploiting commanders must call these sustainment assets forward and guide them around bypassed enemy positions and obstacles. Aerial resupply by either Army aviation or air drop may be a viable option to sustain an exploitation.

7-203. Corps and division commanders must exercise aggressive and demanding leadership to keep units advancing during an exploitation because troops are usually tired when the exploitation opportunity occurs. Meeting and overcoming repeated light resistance can actually add to the momentum of well-trained units, even if they are fatigued. Corps and division commanders should continue to exploit with fresh forces when fatigue, disorganization, or attrition weakens the original exploiting force. The introduction of fresh forces is also important when the original exploiting force must hold ground or resupply. Corps and division commanders ensure that the introduction of fresh forces does not congest routes and create lucrative targets as they move forward.

Maneuver in Exploitation

7-204. Exploitation forces normally advance rapidly on a wide front toward their objectives. Leading elements maintain only those local reserves necessary to ensure flexibility of operation, maintain continued momentum in the advance, and provide essential security. ABCTs and SBCTs are best suited for exploitation on the ground. Air assault forces are extremely useful in seizing defiles, crossing obstacles, and otherwise capitalizing on their mobility to attack and cut off disorganized enemy elements. They can also seize key terrain, such as important wet-gap crossing sites or vital enemy communications nodes along an exploiting force's route or in the enemy rear. Attack helicopter units can interdict and harass enemy armored and other forces and delay their retreat.

7-205. Exploitation actions are rapid, bold, and aggressive in information collection, prompt to use firepower, and quick to employ uncommitted units. An exploiting force clears only enough of its AO to permit its advance. It contains, bypasses, or destroys those enemy pockets of resistance too small to jeopardize the mission. It reports bypassed enemy forces to adjacent units, following units, and higher echelon headquarters. It attacks from the march and overruns enemy formations or positions strong enough to pose a threat to the mission. If an enemy is too strong for leading elements of an exploitation force to destroy and cannot be bypassed, succeeding elements of the force mount hasty attacks.

7-206. In exploitation and pursuit operations, follow and support forces—
- Widen or secure the shoulders of the initial penetration.
- Destroy bypassed enemy units.
- Relieve supported units that have halted to contain enemy forces.
- Block the movement of enemy reinforcements.
- Open and secure lines of communications.
- Guard prisoners, key areas, and installations.
- Control dislocated civilians.

A commander can assign all of these tasks, except widen or secure the shoulders of the initial penetration, to a unit assigned to a consolidation area to enable the continued freedom of action for friendly forces.

7-207. A follow and support force in an exploitation is not a reserve. It is a committed force, and it is provided the appropriate artillery, engineer, and sustainment support. BCTs may have missions to follow and support in division operations. In corps exploitations, divisions may follow and support other divisions.

7-208. An exploiting force depends primarily on its speed and enemy disorganization for security. Overextension is a risk inherent in aggressive exploitation. While corps and division commanders must be concerned about overextension, they guard against being overcautious as well. They rely on aggressive reconnaissance by aircraft, UASs, SOF, and joint technical systems, including space-based systems. In addition, echelon above corps EW and OCO assets can seek out enemy counterattack forces and jam enemy C2, fire support, and intelligence nets and disrupt the logical entities necessary for these enemy elements to function.

7-209. Echelons above brigade engineers integrate into exploiting BCT operations to help breach obstacles and keep forces moving forward. These engineers also keep supply routes open and unimpeded for use in sustaining these BCTs.

7-210. During an exploitation and pursuit, the extended distances and changing rates of advance make it more feasible for the leading elements (the exploiting force and encircling force) to control security to the corps' front and forward flanks. The more uncertain the situation, the more the corps or division headquarters must rely on its aviation assets, information collection assets, and joint reconnaissance and surveillance capabilities for the information necessary for situational understanding. Simultaneously, the corps or division headquarters rely on lethal effects delivered by attack helicopters, armed UASs, and fixed wing aircraft for rapid response.

7-211. Flank security in an exploitation is difficult for the divisions to resource and coordinate. Therefore, the corps should provide assets to screen the corps' most vulnerable flank. The distance of a flank may be too extensive for a single unit. Therefore, the corps may elect to cover the flank of its leading echelons and to direct follow and support forces to cover their own flanks.

7-212. During an exploitation (or pursuit), corps and division units operate at greater distances and normally move much more rapidly than during an attack. Extended distances make it difficult for a corps or division reserve to provide responsive support to the leading element of the exploiting force. The presence of attack aviation and armed UAS elements and dismounted infantry with necessary assault aviation elements in the corps or division reserve to conduct air assault operations reduces those difficulties.

7-213. The need to position a follow and support force or a direct-pressure force immediately behind the lead elements makes it difficult to locate a corps or division reserve in a position to support the echelon exploiting force. Consequently, a greater decentralization of assets and control of the battle to these lead elements is usually necessary. The exploiting forces should be sufficiently weighted with enough combat power so that they can resource their own local reserves. These local reserves can then sustain the attacks of the exploiting force and continue to exploit success.

Fires in Exploitation

7-214. Field artillery units should always be available to fire into and beyond retreating enemy columns. In some cases, field artillery battalions from the corps field artillery brigade are placed in a support relationship to the organic field artillery battalion in exploiting BCTs. Precision munitions may be useful for destroying an enemy IFC, IADS, or reserve forces. They can also close routes of escape and engage suitable targets of opportunity.

7-215. Air defense arrangements for the initial attack should remain effective for the exploitation. Man-portable air defense systems distributed before an attack continue to be transported by advancing maneuver forces. Those same short-range systems accompany and protect convoys, trains, and CPs. The AADC may choose to position other AMD systems forward to provide coverage for corps and division forces as the exploitation gains ground. The air defense coverage by the AAMDC becomes less effective as the corps and division AOs extend during the exploitation, and the limited numbers of AMD assets must cover more area. Thus, it is particularly important during an exploitation that the AADC's counterair operations establish local air superiority over corps and division AOs.

Sustainment During Exploitation

7-216. Exploitations and pursuits may be limited more by vehicle failures and the need for fuel than by combat losses and ammunition. Supplies and the transportation assets to carry the supplies necessary to sustain the force become increasingly important as an exploitation progresses. As supply lines lengthen, security of routes will also become a problem. The largest possible stocks of fuel, spare parts, and ammunition should accompany the exploiting force so that momentum does not slow for lack of support. When possible, echelons above brigade sustainment assets should follow an exploiting force for distribution along lines of communications. Organic maintenance teams within the attacking BCTs repair disabled vehicles or send them to collection points along designated main supply routes for evacuation and repair.

7-217. Sustainment support arrangements must be extremely flexible. Some sustainment units and capabilities may be placed in a command relationship to corps and division maneuver forces in deep or diverging exploitations.

Large-Scale Offensive Operations

Pursuit

In pursuit you must always stretch possibilities to the limit. Troops having beaten the enemy will want to rest. They must be given as objectives, not those that you think they will reach, but the farthest that they could possibly reach.

Field-Marshal Viscount Allenby

7-218. Normally, a pursuing commander maintains pressure on the enemy to prevent the reconstruction of an orderly withdrawal while simultaneously dispatching forces to encircle or to cut off the enemy. Pursuit requires great energy and resolution on the part of an attacking commander. Fatigue, dwindling supplies, diversion of friendly units to other tasks, and approaching darkness may all be reasons to discontinue an attack, but commanders must insist on continuous pursuit as long as the enemy is disorganized and friendly forces can continue. Figure 7-22 on page 7-50 illustrates a corps pursuit concept.

7-219. Control measures associated with a pursuit include an AO for each maneuver unit. Commanders use PLs to designate a forward and rearward boundary for the direct-pressure force. Commanders can designate a route, an axis of advance, or an AO to allow the encirclement force to move parallel to and eventually get ahead of the fleeing enemy. Commanders can designate a terrain objective as a guide for encircling forces. The commander establishes a boundary or a restrictive fire line between the force conducting the encirclement and the force exerting direct-pressure before the encircling force reaches its objective. Other FSCMs around the area currently occupied by the force conducting the encirclement are established to relieve the encircling force of unnecessary fire support coordination responsibilities. The overall commander directs security operations beyond the encircling force, allowing it to engage the withdrawing enemy without devoting resources to flank and rear security. The overall commander establishes additional control measures to control the convergence of both elements of the friendly force, such as restrictive fire lines, PLs, and contact points. Figure 7-23 on page 7-51 illustrates these control measures.

Transition to a Pursuit

7-220. As the enemy becomes demoralized, and the enemy defense begins to disintegrate, exploitation may develop into pursuit. Commanders of all exploiting units must anticipate the transition to pursuit and consider new courses of action as enemy resistance breaks down. Successful pursuit requires unrelenting pressure against the enemy to prevent enemy reorganization and preparation of defenses. Pursuit completes the destruction of an enemy force that loses the ability to defend and attempts to break contact with friendly forces. The commander may designate a terrain objective for orientation, but the enemy force is the primary objective. Attacking corps or division commanders may be able to launch a pursuit after an initial assault, if they promptly exploit precision lethal and multi-domain nonlethal effects.

Mission Command in Pursuit

7-221. Commanders locate themselves well forward to ensure continuation of the pursuit's momentum and apply continuous pressure on displacing enemy units. The pursuit allows greater risks than other types of offensive tasks because of the enemy's disorganization. Pursuits are aggressive and decentralized. Troops in a pursuit push themselves and their equipment to the utmost endurance limits during both daylight and darkness. Corps and division commanders focus their attention forward. They delegate greater authority than usual to subordinates so that they can aggressively exploit windows of opportunity as they occur. Commanders use minimal control measures to enable flexibility, but they issue clear instructions concerning seizure of key terrain and the size of enemy forces which may be bypassed.

Chapter 7

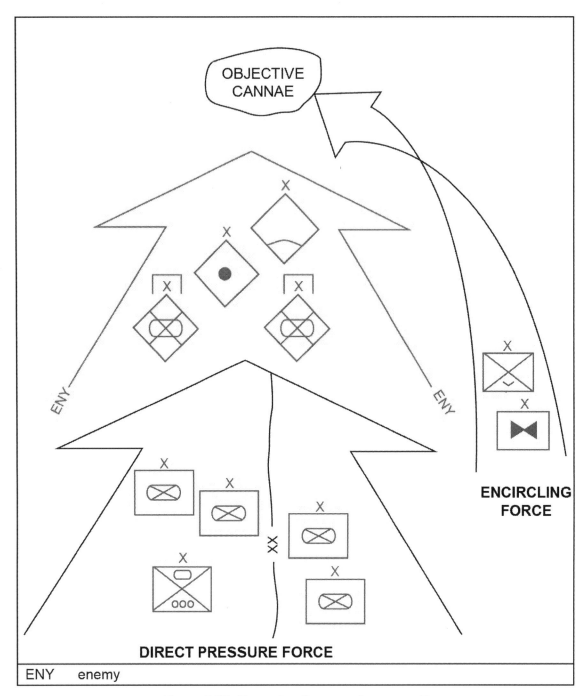

Figure 7-22. Example of a corps in a pursuit

Large-Scale Offensive Operations

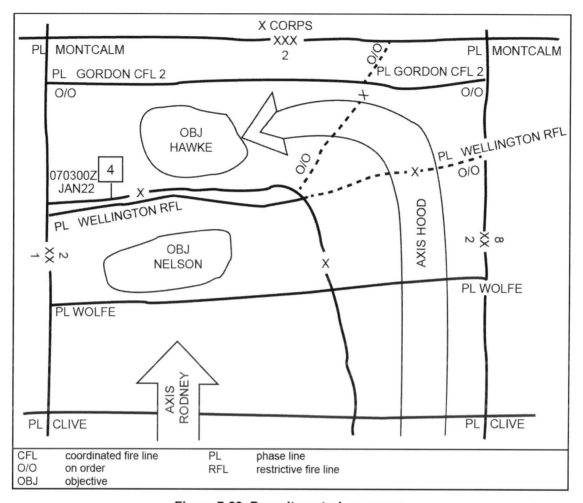

Figure 7-23. Pursuit control measures

Movement and Maneuver in Pursuit

7-222. In the pursuit, the commander relentlessly maintains direct pressure on retreating enemy forces while an encircling force cuts enemy lines of retreat. The pursuing force attempts double envelopments of the retreating main enemy force when conditions permit. It makes maximum use of available air assault and attack aviation capabilities in addition to joint fires. During the pursuit, artillery, engineer, and sustainment units will often be attached to the maneuver units they support. Pursuit operations require—

- A direct-pressure force that keeps enemy units in flight, denying them any chance to rest, to regroup, or to resupply.
- An encircling force to envelop the fleeing force, cut its escape route, and, in conjunction with the direct-pressure force, to attack and to destroy the enemy force.

7-223. The direct-pressure force conducts hasty attacks to maintain enemy contact and its forward momentum until the complete destruction of the retreating enemy force. In the pursuit, the direct pressure force usually conducts the main attack until the enemy force is destroyed or encircled. The direct-pressure force prevents enemy disengagement and subsequent reconstitution of the enemy's defense and inflicts maximum casualties. The direct-pressure force should consist of armor-heavy forces. Its leading elements move rapidly along all available roads and contain or bypass small enemy pockets of resistance. Follow and support units reduce these small contained or bypassed enemy forces. At every opportunity, the direct pressure force envelops, isolates, and destroys enemy detachments, provided such actions do not interfere with its primary mission.

7-224. An enveloping force gets to the enemy's rear area as swiftly as possible by the most advantageous routes to cut off the enemy's retreat and blocks the enemy's escape. This results in the retreating enemy force's encirclement and destruction between the direct-pressure and enveloping forces. The enveloping force advances along or flies over routes that parallel the enemy's line of retreat to reach defiles, communications centers, bridges, and other key terrain or choke points ahead of the enemy's main force. This requires that the encircling force be at least as mobile as the enemy. Air assault forces are ideally suited for this role. The enveloping force organizes a hasty defense once positioned on this favorable terrain. If it cannot outdistance the enemy, the enveloping force attacks the main body of the retreating enemy force on one or both of its flanks. If the attempt to cut the enemy's escape routes fails, the corps or division commander immediately dispatches a new enveloping force.

7-225. An encirclement can completely destroy the enemy. A corps or division commander may see possibilities for cutting off and encircling enemy elements during the exploitation of successful penetrations into the enemy's rear. However, that commander must not initiate an encirclement prematurely or divert from more important missions. To encircle an enemy force, the pursuing force must overtake the retreating enemy and block the enemy's escape. Strong forces should position themselves to block enemy breakthrough attempts. Air assault forces supported by attack helicopters, armed UASs, and joint fires can perform this mission. Once the enemy is encircled, the corps and division main and supporting efforts may become the reduction of the encircled forces.

7-226. Friendly forces must prevent enemy attempts to break out at any point, including through forests, broken terrain, swamps, and across water obstacles. All of these areas should be secured or covered against enemy exfiltration. Fires or barriers, including scatterable mines, may block gaps if sufficient combat troops are not available. Maneuver and fires of all forces involved in the encirclement must be coordinated. Constant pressure, including EW, MISO, and OCO, will weaken the enemy.

7-227. The enemy commander must not have time to organize an all-around defense. If the enemy is able to form a perimeter, the pursuit force must repeatedly split it into smaller elements until the final destruction or capitulation of the encircled enemy force. If time is not critical, the commander can keep the encirclement closed, stop enemy breakout and break-in attempts, and weaken the enemy by fires alone. The collapse of a large and encircled enemy force can be accelerated considerably by the massed employment of precision munitions assisted by EW, OCO, MISO, and other IRCs. (See FM 3-90-2 for additional information on the reduction of encircled enemy forces.)

SUBORDINATE FORMS OF THE ATTACK

7-228. Subordinate forms of the attack have special purposes and include the ambush, counterattack, demonstration, feint, raid, and spoiling attack. The commander's intent and the mission variables determine which of these forms of attack commanders employ. Commanders can conduct each of these forms of attack, except for a raid, as either a hasty or a deliberate operation. Chapter 6 discusses the counterattack and spoiling attack, so those tasks are not discussed in this chapter.

Forms of the attack
- Ambush
- Counterattack
- Demonstration
- Feint
- Raid
- Spoiling attack

AMBUSH

> *Ambushes are prepared by constructing concealed obstacles along the most probable routes of hostile advance and siting fixed weapons to sweep them with fire. Outguards provide for the security of the command, preparing ambushes in advanced positions to break the attack before it reaches the defensive position or to enable capture of hostile patrols.*
>
> FM 7-5, *Infantry Field Manual-Organization and Tactics of Infantry: The Rifle Battalion,* 1940

7-229. An *ambush* is an attack by fire or other destructive means from concealed positions on a moving or temporarily halted enemy (FM 3-90-1). An ambush stops, denies, or destroys enemy forces by maximizing the element of surprise. Ambushes can employ direct fire systems and other destructive means, such as command-detonated mines, indirect fires, and supporting nonlethal effects. They may include an assault to close with and destroy enemy forces. In an ambush, ground objectives do not have to be seized and held. Small tactical echelons execute ambushes.

DEMONSTRATION

> *A demonstration is an operation designed to confuse the enemy, to delay or reduce the effectiveness of his dispositions, and to cause him to commit his reserves against the demonstrating force.*
>
> FM 60-5, *Amphibious Operations: Battalion in Assault Landings*, 1951

7-230. A *demonstration* in military deception is a show of force similar to a feint without actual contact with the adversary, in an area where a decision is not sought that is made to deceive an adversary (JP 3-13.4). The planning, preparing, and executing considerations for a demonstration are the same as for the other forms of attack. For example, the 4th and 5th Marine Expeditionary Brigades remained aboard their amphibious ships in the Persian Gulf during Operation Desert Storm. In this capacity they served as a demonstration to divert Iraqi attention and forces from the actual location of the coalition force's ground attack. Central Command exploited the U.S. Marine Corps' high media profile to distract Iraqi planners until the war's end. The threat posed by their amphibious capabilities caused the Iraqis to keep seven divisions focused on the coast to stop a landing that never came.

FEINT

7-231. A *feint* in military deception is an offensive action involving contact with the adversary conducted for the purpose of deceiving the adversary as to the location and/or time of the actual main offensive action (JP 3-13.4). In a feint the commander assigns the force an objective limited in size or scope. Forces conducting a feint make direct fire contact with the enemy, but they avoid decisive engagement. The planning, preparing, and executing considerations for a feint are the same as for the other forms of attack. For example, the German Army Group B's attack into the Netherlands and northern and central Belgium in May 1940 can be considered a feint. (It can also be considered as a supporting attack.) It convinced the British and French high command that the German main effort was in the north instead of German Army Group A's advance through the Ardennes toward Sedan.

RAID

7-232. A *raid* is an operation to temporarily seize an area in order to secure information, confuse an enemy, capture personnel or equipment, or to destroy a capability culminating in a planned withdrawal (JP 3-0). Raids are usually small, involving battalion-sized or smaller forces. (See FM 3-90-1 for additional information on performing the raid task.)

OTHER TACTICAL CONSIDERATIONS

7-233. Corps and division commanders have other tactical considerations when conducting offensive tasks. They usually address these tactical considerations as part of their shaping operations or supporting efforts.

FORWARD PASSAGE OF LINES

7-234. The offense begins with a forward passage of lines, a task in which one friendly unit moves forward through positions held by another. It must be well planned and coordinated to ensure minimum congestion and confusion. When possible, a passage should be through elements that are not in contact.

7-235. The respective subordinate unit commanders coordinate specific details of the passage. Normally, the overall commander assigns boundaries to designate areas through which subordinate elements will pass. Such boundaries usually correspond to those of the stationary force. The stationary unit occupies contact and passage points, supplies guides, and provides information concerning the enemy, such as the location and composition of minefield and other obstacles.

7-236. The stationary unit may provide the initial logistic support to the forward passing unit to ensure continuous support without increasing battlefield clutter. Once started, the passing unit moves through the

Chapter 7

stationary unit as quickly as possible to minimize the vulnerability of the two forces. The moving force must assume control of the battle as soon as its lead elements have passed through the stationary force. Artillery supporting the stationary force and its direct fires should be integrated into the fire support plan of the passing unit. (See FM 3-90-2 for more information on passage of lines.)

USE OF TERRAIN

7-237. Attacking forces should select avenues of approach that permit rapid advance, allow for maneuver by the attacking force, provide cover and concealment, permit lateral shifting of reserves, allow good communications, resist constriction efforts by enemy engineers, and orient on key terrain.

7-238. Battalion task forces and company teams advance from one covered and concealed position to the next. Corps and divisions move along maneuver corridors and lines of communications that provide for rapid advance of all combined arms and supporting forces. To sustain forward momentum, all elements of combat power—leadership, information, mission command, movement and maneuver, intelligence, fires, sustainment, and protection—move forward in an orchestrated manner as part of the combined arms team.

7-239. Terrain chosen for the decisive operation's main effort should allow rapid movement into the enemy's rear. Corps and division commanders should normally identify and avoid terrain that will hinder a rapid advance. However, if the enemy can be surprised, an initial maneuver over difficult terrain may be desirable. Where possible, commanders at all echelons should personally reconnoiter the terrain, particularly terrain where the decisive operation or main effort is to be conducted.

7-240. Corps and division commanders orchestrate the forward movement of their subordinate combat elements to provide as much mutual support as possible. However, rapidly moving forces should not generally be held back to preserve a uniform rate of advance. Elements may seize intermediate objectives to provide mutual support and protection for adjacent units moving on more exposed routes. Key terrain along the route of advance must be either seized or controlled by fire. When present, decisive terrain becomes the attack's focal point.

FLANK SECURITY

7-241. Attacking corps and division commanders should not ignore the threat to their flanks. That risk increases as the attack progresses. They must assign responsibility for flank security to attacking units or designate other forces to perform the task of cover, guard, or screen. It may sometimes be necessary to reduce flank protection to maintain forward momentum. The speed of attack itself offers a degree of security because it makes enemy defensive reactions less effective.

7-242. Obstacles covered by security forces can improve flank security. Commanders should be mindful of how they affect maneuver options when directing the establishment of obstacle zones and obstacle belts along unit flanks.

WET-GAP CROSSING OPERATIONS

7-243. Commanders must plan to quickly cross whatever rivers or streams are in the path of advance. A wet-gap crossing requires special planning and support. The size of the obstacle and the enemy situation will dictate how to make the crossing. Attackers should strive to cross rivers without loss of momentum regardless of how they get across. Only as a last resort should the attacking force pause to build up forces and equipment. (See ATP 3-90.4 for more information on wet-gap crossing operations.)

BREACHING OPERATIONS IN THE OFFENSE

7-244. Units conducting offensive tasks can execute three types of breaches—deliberate, hasty, and covert. Breaching begins when friendly forces detect an obstacle and begin to apply the breaching fundamentals. Breaching ends when battle handover occurs between follow-on forces and the unit conducting the breach. A breaching activity includes the reduction of minefields, other explosive hazards, and other obstacles. Generally, breaching requires significant combat engineering support to accomplish. (See ATP 3-90.4 for additional information on breaching.)

ENCIRCLEMENT OPERATIONS

7-245. Corps and division commanders conduct offensive encirclements to isolate targeted enemy forces. Typically, encirclements result from penetrations and envelopments, or are extensions of exploitation and pursuit operations. As such, they are not a separate form of the offense but an extension of an ongoing operation. They may be planned sequels or result from exploiting an unforeseen opportunity. They usually result from the linkup of two encircling arms conducting a double envelopment. However, they can occur in situations where the attacking commander uses a major obstacle, such as a shoreline, as a second encircling force. Although a commander may designate terrain objectives in an encirclement, isolating and defeating enemy forces are the primary goals. Ideally, an encirclement results in the rapid surrender of the encircled force. This minimizes friendly force losses and resource expenditures.

Organization of Forces for an Offensive Encirclement

7-246. An encirclement operation usually has at least two phases—the actual encirclement and actions taken against the isolated enemy. Commanders consider adjusting subordinate unit task organizations between phases to maximize unit effectiveness in each phase. The first phase is the actual encirclement that results in the enemy force's isolation. The organization of forces for an encirclement is similar to that of a movement to contact or an envelopment. The commander executing an encirclement operation organizes encircling forces into a direct pressure force and one or more encircling arms. ABCTs, SBCTs, and air assault, and airborne units are especially well suited for use as an encircling arm since they have the tactical mobility to reach positions that cut enemy lines of communications. Bypassed and encircled enemy forces on the flanks and rear of advancing friendly forces require these encircling forces to maintain all-around security, which includes local security measures and security forces.

7-247. One commander should direct the encirclement effort. However, there must also be unity of command for each encircling arm. The encircling force headquarters may name one of its subordinate units as the headquarters for an encircling arm. Alternatively, that force's headquarters may create a temporary command post from organic assets, such as its tactical CP, to control one or more arms of the encirclement. If an encircling arm has subordinate inner and outer arms, each of them also requires separate subordinate commanders. The missions and spatial orientation between the inner and outer encircling arms are sufficiently different; therefore, one force cannot act in both directions at once.

7-248. If there is no possibility of the encircled forces receiving relief from enemy forces outside the encirclement, commanders only organize an inner encircling arm. If there is danger of an enemy relief force reaching the encircled enemy force, commanders organize both inner and outer encircling arms. Commanders assign the outer encircling arm a security mission, an offensive mission to drive away any enemy relief force, or a defensive mission to prevent the enemy relief force from making contact with the encircled enemy force. Once the encirclement is complete, these inner or outer encircling arms form a perimeter.

Offensive Encirclement Control Measures

7-249. Control measures for an encirclement are similar to those of other offensive tasks, especially an envelopment, but with a few additional considerations. If the commander uses both an inner and an outer encircling arm, the commander must establish a boundary between them. The commander places the boundary so that each element has enough space to accomplish the mission. The inner force requires enough space to fight a defensive battle to prevent the encircled force from breaking out. The outer force requires adequate terrain and depth to its AO to defeat any attempt to relieve the encircled force.

Figure 7-24. Free-fire area

A *free-fire area* is a specific area into which any weapon system may fire without additional coordination with the establishing headquarters (JP 3-09).

7-250. The commander who controls both converging forces establishes a restrictive fire line between them. Commanders may also establish a free-fire area, which encloses the area occupied by a bypassed or encircled enemy forces. (See figure 7-24.) Commanders may establish a no-fire area over locations and facilities required for the consolidation of gains. (See figure 7-25. See appendix A of FM 3-90-1 for more information on the use of restrictive fire line, free-fire areas, and other FSCMs.) Commanders may also establish contact points.

Executing an Encirclement

7-251. One result of an envelopment might be an encirclement. Encirclements are most common during the conduct of double envelopments. However, encircled enemy forces can result from a single envelopment or flank attack that pins an enemy force against a major obstacle. Conducting offensive tasks in a noncontiguous environment—such as occur during the conduct of exploitations and pursuit—can also result in the encirclement of bypassed enemy forces.

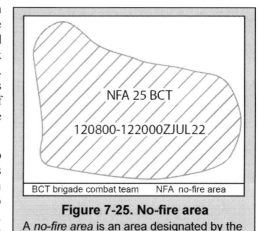

Figure 7-25. No-fire area

A *no-fire area* is an area designated by the appropriate commander into which fires or their effects are prohibited (JP 3-09.3).

7-252. Commanders conduct encirclement operations to deprive enemy forces of their freedom of maneuver. This denies them of the capability to defend or delay in an organized fashion. Encirclement operations also seek to cutoff enemy evacuation and reinforcement routes. Encirclements may be deliberately planned, or they may occur during the performance of all offensive tasks.

7-253. Prior planning is probably the most important consideration of encirclement operations. The encircling commander identifies and sets the conditions of the encirclement before it develops in order to deny the enemy force as many advantages—such as securing advantageous terrain—as possible before surrounding it. Encirclements can occur as a consequence of an operational action, a tactical action, or a combination of both. However, the reduction of an encirclement is strictly a tactical action. An encirclement consists of five actions:

- Penetrating the enemy's defenses.
- Exploiting and attacking on converging axes.
- Linking up forces and establishing the inner encirclement.
- Establishing an outer ring to counter enemy reserves and continue the exploitation.
- Destroying the encircled enemy force.

7-254. The initial state of the encirclement is the penetration of the enemy's defenses in several locations. Armored, Stryker, and air assault infantry elements—if the enemy air defense situation allows the use of assault aviation forward of the forward line of own troops (FLOT)—quickly advance into the depths of the enemy's defenses and secure the inner encirclement ring. Follow-on forces support the inner ring and focus on destroying the enemy pocket. Follow-on forces establish an outer ring, while armored, Stryker, and air assault forces continue the exploitation and develop the offensive on the external front.

7-255. Encirclement of the enemy is complete only after the creation of the interior ring, the outer ring, and the organization of an air, EW, cyberspace, and information blockade (and naval blockade where applicable). The friendly force must prevent the enemy from resupplying or otherwise supporting encircled enemy forces.

7-256. Commanders and staffs must plan for dealing with bypassed enemy forces. This includes the challenges that detainees and dislocated civilians cause. Once a decision is made to reduce an encirclement, the commander normally uses one of two possible methods of reduction—fire alone or fire and movement.

7-257. Reduction by fire alone implies that the encircling friendly commander will use fire support as the primary or sole means of reducing the encirclement. This includes employing artillery, direct fires, CAS, EW, OCO, IRCs, and attack helicopters and armed UASs against the enemy perimeter. Reduction by fire alone also reduces the number of casualties suffered by the friendly encircling force. However, it may increase the number of civilian casualties and amount of collateral damage which may be important later as friendly forces seek to consolidate gains.

7-258. Unfortunately, reduction by fire alone has disadvantages. Reduction by fire requires a significant amount of weapons, ammunition, and time. Friendly forces using lethal and nonlethal effects against the encircled force are not available for other missions. Another disadvantage of reduction by fire is its inability to guarantee results. Fires alone might not be sufficient to force the surrender of the encircled enemy force.

7-259. Reduction by fire and maneuver uses a combination of fire and ground maneuver forces to attack and destroy an encirclement. It is the surest method of reduction because it forces the enemy to surrender, displace, or face annihilation. This method also allows the friendly encircling force commander to retain the initiative. The major drawback of reduction by fire and maneuver is that it reduces the strength of the encircling force through attrition.

7-260. Commanders must determine which reduction technique to use after selecting the reduction method. This is how commanders will employ that reduction method. Reduction by fire alone uses only one technique, the application of overwhelming fire, and it requires decisions on selection of munitions, delivery means, and targets. Reduction by fire and maneuver incorporates at least four techniques: reduction by continuous external pressure, divide-and-conquer, selective reduction, and reduction by infiltration.

7-261. The first technique, reduction by continuous external pressure, is the classic siege technique. The encircling friendly force contains the encircled enemy force. The friendly force uses fires against the enemy pocket, and it conducts ground attacks against the encircled enemy's perimeter in a battle of attrition.

7-262. This technique has some disadvantages. The encircled enemy force usually has the advantage of strong defensive positions and interior lines. This allows the enemy commander to quickly transfer forces within the defensive perimeter. Last, as a result of these two defensive advantages, the attacking friendly force can expect to suffer a greater number of casualties than the defending enemy. In comparison with the other techniques, reduction by continuous external pressure has few, if any, advantages, unless the encircling force has an overwhelming force advantage.

7-263. The divide-and-conquer technique is a more viable and less costly technique. It is also the technique German and Soviet armies used against pockets of resistance during World War II. Once a force surrounds and contains a pocket, the encircling force launches a penetration to divide the pocket in two. Another penetration then divides these pockets into smaller ones. These penetrations and sub-divisions continue until resistance subsides. This technique eliminates the enemy's advantage of interior lines.

7-264. The third technique, selective reduction, attacks the cohesion of an encircled force by focusing on the sequential destruction of specific targets (for example, in a situation where the encircled force is strong in air defense and artillery assets). The friendly encircling force might focus on eliminating the pocket's AMD systems first. It then uses Army aviation, CAS, and ground maneuver forces to eliminate the encircled enemy's artillery. The encircling friendly force could then use armored fighting vehicles to attack enemy logistics assets. Dismounted infantry attacks against enemy armored elements could follow. The objective is the eroding of the total combined arms strength of the pocket by eliminating specific capabilities of the encircled enemy combined arms team. Commanders and staffs can use this technique in combination with the other reduction techniques.

7-265. The fourth technique, reduction by infiltration, moves friendly forces through the perimeter of the encirclement. This isolates and reduces small portions of the pocket without external interference.

7-266. The encircling commander identifies special planning considerations for the entire force and for specific members of the combined arms team in addition to selecting reduction methods and techniques. These planning considerations include—
- The effects of a pause to reorganize.
- Maneuver and fire support control measures.
- Continuous reconnaissance.
- Encirclement isolation.
- MISO.
- EW.
- Enemy use of WMD.
- Creation and employment of a mobile reaction force.
- Sustainment.
- Dealing with an outside enemy force attempting to assist the encircled enemy force.

CONSOLIDATION OF GAINS IN THE OFFENSE

7-267. As corps and divisions continue to advance at some point in time they will transition from one phase of the major operations or campaign plan to another and begin executing a sequel to their previous offensive order. The end of offensive tasks may not be the decisive act. Consolidation of gains may be the decisive operation in the major operation or campaign. The transition to a focus on the conduct of area security and stability tasks from the conduct of offensive tasks cannot be an afterthought. Setting the conditions for the consolidation of gains may have significant impact on the planning and execution of offensive-centric actions. (See chapter 8 for more information on consolidation of gains.)

Chapter 8
Operations to Consolidate Gains

It is no doubt a good thing to conquer on the field of battle, but it needs greater wisdom and greater skill to make use of victory.

Polybius

The chapter begins with an overview of operations to consolidate gains. Next it describes threats and challenges to the consolidation of gains. An expanded description of the operational framework and the consolidation area follows. The chapter concludes with a description of consolidation activities and the roles of theater army, corps, division, and brigade combat teams (BCTs) in operations to consolidate gains.

OVERVIEW OF OPERATIONS TO CONSOLIDATE GAINS

Loss of hope, rather than loss of life, is the factor that really decides wars, battles and even the smallest combats. The all-time experience of warfare shows that when men reach the point where they see, or feel, that further effort and sacrifice can do no more than delay the end, they commonly lose the will to spin it out and bow to the inevitable.

Sir Basil Liddell Hart

8-1. *Consolidate gains* are the activities to make enduring any temporary operational success and set the conditions for a stable environment allowing for a transition of control to legitimate authorities (ADRP 3-0). Commanders continuously consider activities necessary to consolidate gains and achieve the end state. Consolidate gains is integral to winning armed conflict and achieving enduring success. It is essential to retaining the initiative over determined enemies because it ultimately removes both the capability and will for further resistance. It is the final exploitation of tactical success. Army forces, when integrating and reinforcing the efforts of all unified action partners, provide the joint force commander (JFC) significant capability to consolidate gains.

8-2. Consolidation of gains occurs in portions of an area of operations (AO) where large-scale combat operations are no longer occurring. Consolidation of gains activities consist of security and stability tasks and will likely involve combat operations against bypassed enemy forces and remnants of defeated units. Therefore units may initially conduct only minimal essential stability tasks and then transition into a more deliberate execution of stability tasks as the primary mission as security improves. Operations to consolidate gains require combined arms capabilities and the ability to employ fires and manage airspace, but at a smaller scale than large-scale combat operations. Units in the close area involved in close combat do not conduct consolidation of gains activities. Consolidation of gains activities are conducted by a separate maneuver force in the designated corps or division consolidation areas.

8-3. Army forces conduct consolidation of gains throughout the range of military operations. The U.S. Army has always been required to consolidate gains. It did so with varying degrees of success in the Indian Wars, after the Civil War during Reconstruction, after the Spanish American War, during World War II, Korea and Vietnam, and more recently in Haiti, Afghanistan, and Iraq. How successful its efforts to consolidate gains were informed how the outcomes of those wars can be viewed today. Each conflict was unique and involved an Army role in the governance of an area for periods of time that were not predicted beforehand. As such, planning to consolidate gains is essential to any operation.

8-4. Consolidation of gains is not a synonym for stability, counter-insurgency, or nation-building. It describes activities designed to make the achievement of the military objective enduring. As such, it encompasses a broad array of tasks combined in variable ways over time in a specific operational context.

Chapter 8

Offensive, defensive, and stability tasks all contribute to the consolidation of gains. Consolidate gains is a continuous planning consideration not tied to a specific timeline or phase. It is about exploiting tactical success by establishing security and stability in a manner decisive enough to achieve national strategic aims.

8-5. To consolidate gains in the most effective and efficient manner requires a clear understanding of both the purpose of the operation and all potential enemy capabilities to resist. The critical assumption for planning is that enemy forces will use all means at their disposal and seek new means to protract a conflict even as they direct remaining forces in the field. It is the simultaneous exploitation of existing advantage and the rapid pursuit of remaining means of resistance that deny an enemy the ability to prolong conflict after enemy initial forces in the field are defeated. This exploitation and pursuit can involve many axes of attack including—

- Physical seizure of storage areas containing weapons, munitions, and fuel, seizure of server farms, radio and television stations, barracks and police stations, and seizure of key terrain that controls movement like bridges and border crossings.
- Control of enemy security services, to include accountability of those not already captured or casualties; however, trained manpower should be encouraged to continue service to country as part of a better future wherever possible.
- Rapid physical control of population centers and the establishment of public order backed by force, in accordance with the rules of engagement and the law of land warfare.
- Rapid and comprehensive use of information operations to shape public opinion, discredit enemy narratives, and promote friendly narratives.

8-6. During the conduct of large-scale combat operations, the size of consolidation areas generally increase over time, as does the echelon of command responsible for those areas. For example, a division may initially assign a BCT responsibility for the division's consolidation area. However, conditions could require the commitment of additional forces to conduct consolidate gains activities in multiple division consolidation areas. As division boundaries shift to better enable success in the deep and close areas, a corps may establish a corps consolidation area. Corps would then assign a division the responsibility for consolidating gains in the corps' consolidation area. Eventually, most Army units can expect to conduct consolidate gains activities when large-scale combat operations are complete. Operations to consolidate gains represent a significant transition in focus for the echelon of command tasked with the mission.

8-7. Operations to consolidate gains require the dynamic execution of area security and stability tasks based on the desired operational end state that supports the strategic objective of the campaign. Consolidate gains activities include the relocation of displaced civilians, reestablishment of law and order, providence of humanitarian assistance, and restoration of key infrastructure. Concurrently, Army forces must be able to accomplish such activities while sustaining, repositioning, and reorganizing forces for ongoing or future operations. Commanders make a conscious shift in emphasis from defeat of enemy forces in the field to those measures that address the long-term security and stability of a particular nation or area and its population. The goal is to transition control over territory and populations to legitimate authorities in a way that allows U.S. forces to make the strategic position of relative advantage gained during combat operations enduring.

8-8. Commanders must analyze all available sources of capability and capacity to facilitate the consolidation of gains. When properly coordinated, unified action partners, including U.S. and foreign government agencies, international agencies, nongovernmental organizations, and contractors can provide significant resources. The theater army has the most significant role in the planning, coordination, and allocation of resources. It is most likely the entry point for both coordination and introduction of unified action partner capabilities into an AO.

8-9. Army forces must deliberately plan and prepare for consolidating gains to capitalize on operational success. Planning considerations include tactical, operational, and strategic risk, changes to task organization, and new or additional assets required to achieve the desired end state. These assets may include engineers, military police (MP), explosive ordnance disposal (EOD) units, civil affairs (CA), medical support required for stability tasks, or additional maneuver forces to conduct area security tasks. Initially, it is likely that Army forces will be responsible for integrating and orchestrating consolidate gains activities. As the security situation improves, Army forces may transition to support of other organizations, such as the United Nations. To consolidate gains, Army forces take specific actions. These actions include—

- Conduct area security: Forces conduct security tasks to defeat enemy remnants and protect friendly forces, routes, critical infrastructure, populations, and actions within an assigned AO.
- Conduct stability tasks: Forces initially execute minimum-essential stability tasks, and then through executing primary stability tasks they provide essential governmental services, emergency infrastructure reconstruction, and humanitarian relief.
- Influence local and regional audiences: Commanders communicate credible narratives to specific audiences to prevent interference and ultimately generate support for operations.
- Establish security from external threats: Commanders ensure sufficient combat power is employed to prevent physical disruption from threats across the various domains and the information environment.

8-10. Consolidating gains is not the same as unit consolidation. Army forces routinely conduct consolidation upon occupying a new position on the battlefield or achieving military success. *Consolidation* is the organizing and strengthening a newly captured position so that it can be used against the enemy (FM 3-90-1). Consolidation is a security related set of tasks that represent the initial steps in operations to consolidate gains. Consolidation activities include—
- Conducting reconnaissance.
- Establishing security.
- Eliminating enemy pockets of resistance.
- Positioning forces to enable them to conduct a hasty defense by blocking possible enemy counterattacks.
- Adjusting fire planning.
- Preparing for potential additional missions.
- Shaping the information environment.

(See FM 3-90-1 for an additional discussion of consolidation.)

8-11. When consolidating gains, establishing and sustaining security is the unit's first priority. Without security the accomplishment of many stability tasks becomes problematic. Security operations will likely involve combat operations against enemy remnants or irregular forces fighting on or from among the local population and in remote areas, as well as criminal elements taking advantage of the lack of civil control forces in a given area. A critical consideration for Army units executing area security tasks is to avoid activities that create the perception of doing more harm than good. During the transition from large-scale ground combat to consolidation of gains, commanders may implement more restrictive rules of engagement than those used during large-scale combat operations. If commanders do not manage the application of force carefully, the perceptions of friendly forces and the local population about what constitutes acceptable levels of violence may rapidly diverge. Thus, commanders establish and maintain communications with the population to assist them in understanding the overall goal of military actions and how those actions benefit them. Capabilities such as military information support operations (MISO), public affairs, and combat camera can assist in this effort.

8-12. The window of opportunity for setting a geographic area on a desirable path to consolidate gains is potentially narrow. Commanders should expect counterattacks below the threshold of open state-on-state hostilities, conducted by remnants of defeated enemy forces. These remnant enemy forces may be conventional, irregular, or criminal. Enemy actions will most likely involve actions to incite civil unrest and ultimately resist consolidation of gains by continuing the conflict. Therefore, if they are to plan effectively, commanders require sufficient situational awareness to understand what needs to happen, what cannot happen, and what could happen in their AO. Establishing a sustainable position of relative advantage that facilitates achievement of military objectives is the goal, and all consolidate gains tasks should support that goal.

8-13. Army forces conduct continuous reconnaissance and, if necessary, gain or maintain contact with the enemy to defeat or preempt enemy actions and retain the initiative. Consolidating gains may include actions required to defeat any isolated or bypassed threat forces to increase area security and protect lines of communications. Commanders ensure that forces are properly task-organized and prepared to quickly defeat remaining conventional and irregular enemy forces. During the consolidation of gains, Army forces are responsible for accomplishing both the minimum essential stability tasks and the Army primary stability tasks. Commanders must quickly ensure the performance of minimum-essential stability tasks of providing security, food, water, shelter, and medical treatment. Once conditions allow, these tasks are a legal responsibility of Army forces.

8-14. Commanders may transition essential stability tasks to other forces or appropriate civilian organizations, if they are able to perform them. For example, there may be sufficient host-nation civilian or military governance in place to ensure that the population has adequate food and medical care. In that case, Army forces would continue consolidating gains by conducting the Army's primary stability tasks in their assigned areas of operation: establish civil security, establish civil control, restore essential services, support governance, support economic and infrastructure development, and conduct security cooperation. Army forces will retain the lead and perform tasks to establish civil security through the conduct of security force assistance (SFA). Eventually, the lead for the other four tasks will transfer to another military or civilian organization, although Army forces may retain a supporting role. (See ADRP 3-07 for more information on stability tasks.)

8-15. Activities to consolidate gains may occur over a significant amount of time and involve transitions in both focus and partners. Emphasis will shift from actions to ensure the defeat of remaining threat forces to measures that address the needs of the population and eventually to the transfer of responsibility from Army forces to a host-nation government, interagency partners, or other organizations.

8-16. A civilian population's perception of legitimacy may influence how it reacts to military forces. Forces establish credibility and legitimacy with local populations through the way they conduct operations. Adversaries are likely to seek legitimacy by assuming roles that gain favor with the population, such as resolving disputes, influencing key social leaders, providing essential services, or providing protection from criminal elements. Consolidation of gains activities may ultimately decide who possesses the ability to compel, control, influence, and gather support from a population. Therefore, throughout an operation, Army leaders have legal and moral responsibilities to establish area security and restore services while countering the efforts of those working against friendly goals. (See ADRP 3-07 for further information on legitimacy and other stability considerations.)

8-17. To win in the land domain requires Army forces to consolidate gains and transition an AO to a legitimate authority able to maintain security and public order. Regardless of the scale and scope of combat operations, detailed and continuous planning, task-organizing, and accounting for the effects of combat during all operations is essential to achieving the desired end state. Achieving a sustainable position of relative advantage across domains and in the information environment requires unified action and aggressive leadership.

THREATS TO THE CONSOLIDATION OF GAINS

The merit of an action lies in finishing it to the end.

Genghis Khan

8-18. An enemy or adversary attempts to prevent effective consolidation of gains in order to gain time for a favorable political settlement, set conditions for protracted resistance, and alter the nature of the conflict to suit its positions of relative advantage. An enemy or adversary will target both friendly forces and populations in a variety for ways.

8-19. Threat information warfare activities will focus on both altering the value of continued operations to the United States and altering the perceived value to other actors of continuing to act in a manner coincident to the interests of the United States. This effort seeks to undermine the authority and effectiveness of governing elements acting in concert with U.S. forces.

8-20. Threat forces will exploit any lack of cultural understanding displayed by U.S. forces. Threat forces will conduct information campaigns that portray a United States bent on political and economic global domination. They will paint U.S. military forces as brutal and unconstrained by the accepted rules of warfare. They will seek recruits from populations alienated by U.S. forces.

8-21. Threat forces will seek to shift the nature of the conflict in ways that favor their capabilities and discredit the stated goals of friendly forces. Threat forces will seek to prevent consolidation areas from becoming secure enough to permit substantial stability activities.

8-22. Threat forces will exploit social aspects of the environment for protection and freedom to maneuver. They can do this by using religious, medical, and other sensitive facilities as sanctuaries, concealing forces among populations, and influencing populations to oppose friendly forces.

8-23. Friendly forces should anticipate that operations to consolidate gains will be opposed. Threat forces are likely to exploit any opportunity as long as they retain the will to resist. Adversaries promote instability by encouraging competition for resources, promoting residual territorial claims, creating ethnic tension, supporting religious fanaticism, and positioning proxies in the government. Historically (for example, in the Philippines, Vietnam, Afghanistan, and Iraq) a protracted insurgency can reverse conventional military gains.

8-24. Enemy forces will likely continue to fight even after friendly forces attain their initial military objectives in the close area. Some enemy formations may be intentionally or unintentionally bypassed during close operations as friendly forces are focused on the decisive effort. An enemy is likely to maintain limited conventional capabilities, robust irregular capabilities, and an array of disruptive capabilities. An enemy may attempt to move remaining assets and capabilities into sanctuary locations. Equipment and ammunition stockpiles not destroyed during combat operations may be obtained by other actors, cached for later use, or moved to sanctuary areas. Regional adversaries may assist an enemy by providing proxy support through other regional actors.

CONSOLIDATION AREA FRAMEWORK

8-25. During large-scale combat operations, a consolidation area refers to an AO that extends from a higher echelon headquarters boundary to the boundary of forces in the close area. The consolidation area is where forces have established a level of control that allows the performance of tasks to consolidate gains. The consolidation area may or may not contain support areas that are focused on support to forward deployed units. Assigning a consolidation area to a subordinate headquarters allows the higher echelon headquarters to adequately focus resources on close operations and allows a subordinate headquarters to make progress towards consolidating gains and achieving desired objectives. It is the first step towards the overall consolidation of gains during large-scale combat operations. Figure 8-1 on page 8-6 shows corps and division consolidation areas within a geographic framework of large-scale combat operations.

8-26. If support areas are surrounded by a consolidation area, the higher echelon headquarters must clearly articulate the roles and responsibilities for controlling supply routes transitioning the area, the clearance of fires, and airspace control authorities. Commanders clearly define roles and responsibilities to allow sustainment units to focus on their primary functions.

8-27. Consolidation area activities require a balancing of area security and stability tasks. A unit assigned the responsibility for a consolidation area may not simultaneously perform all types of tasks. It may initially focus on security tasks and minimum essential stability tasks. Once the threat in the consolidation area has diminished, commanders may expand their efforts to the more deliberate conduct of primary stability tasks and, at some point in time, may facilitate transition of some activities to follow-on units or other unified action partners.

8-28. Divisions and corps assign subordinate units responsibility for consolidation areas to preserve the tempo of operations. Commanders assign consolidation areas to units in a follow and support role to avoid pulling combat power from the close and deep areas. Units assigned a consolidation area require the capability to conduct decisive action, which includes offensive, defensive, and stability tasks, to defeat enemy remnants or bypassed forces and preserve freedom of action for their higher headquarters. The consolidation area generally reduces the size of the support area and improves its security, allowing units in the support area to focus primarily on functional tasks and the local security of base clusters and routes.

> *Note.* A consolidation area requires additional combat power and is not intended to draw forces from the close or deep area. Theater armies, in coordination with corps, must plan for and request forces for the consolidation area.

8-29. Commanders should plan for consolidation areas early in the planning process and assign forces to the consolidation areas as soon as possible in order to allow combat forces to extend operations in the close and deep areas. A key component of this is requesting critical authorities to ensure effective application of such things as rules of engagement, direct liaison, cyberspace missions, and information operations. By smoothly transitioning consolidation areas to follow-on forces as close areas are secured, the commander enables combat forces in deep operations to maintain the initiative and maneuver without loss of momentum.

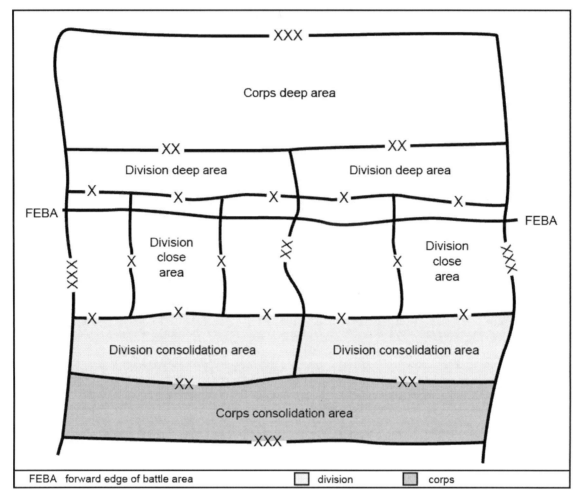

Figure 8-1. Consolidation area during large-scale combat operations

8-30. Boundaries for the division and corps consolidation areas may shift to reflect the capabilities of the units assigned them and the movement of friendly forces. Commanders purposefully task-organize units to consolidate gains based on security of the unit itself, defeat of enemy remnants within the consolidation area, security of the lines of communications, security of the support area, and minimal essential stability tasks. Once the boundaries of the division consolidation area become fixed for longer periods, units in the consolidation area will have opportunities to expand stability tasks within their AOs. A corps consolidation area may get larger with the intent of handing over responsibility for portions of its AO to host-nation or other forces as they become available.

8-31. Over time units assigned a consolidation area continuously improve the security and stability of their AOs. This is similar to the way units in the defense continuously improve their positions. This means that the consolidation area commander requests support and additional capabilities through the division or corps headquarters when the commander is required to make additional progress. Requests could include nongovernmental organization presence to accomplish specific tasks, or host-nation forces in order to facilitate a transition of control from the unit to a legitimate civilian authority.

8-32. Ultimately, most Army maneuver forces may be assigned consolidation areas when the overall focus of operations has transitioned from large-scale combat operations to consolidating gains. When this happens, the operational framework transitions from one suited for large-scale combat operations to one that accounts for the change of focus. Typically, consolidation areas will consist of contiguous unit boundaries that facilitate assigning responsibility for area security and control of physical terrain to subordinate echelons. Changing the operational framework may require significant repositioning of units, and potentially, changing task-organization.

8-33. Commanders may decide to place the support area within the consolidation area during large-scale combat operations to better secure it. When placing the support area inside a consolidation area, the higher echelon headquarters assigns clear roles and responsibilities for the affected subordinate headquarters. Commanders allocate sufficient combat power to address the requirements and the tasks associated with both areas. The advantages of placing the support area inside a consolidation area include unity of effort and alignment of unit capabilities with properly sized areas of operation. A BCT assigned a division consolidation area has organic combined arms and airspace control capabilities that allow the maneuver enhancement brigade (MEB) and sustainment units to focus on support to close and deep operations and not have to be concerned with bypassed enemy formations or clearance of fires. The maneuver enhancement brigade retains responsibility for security of the sustainment base clusters and routes upon which they transit in the close area, while the unit assigned the consolidation area maintains responsibility for the remainder of the area between the brigade and division rear boundaries.

8-34. During operations, there may be other corps and division assets positioned within the division consolidation area, but outside of a designated support area. Maintaining situational awareness of the various friendly and enemy formations likely to be located in the consolidation area is critical. As the terrain owner and manager, the commander responsible for the consolidation area is responsible for advising and deconflicting friendly terrain management with higher headquarters. Division may choose to establish a support area command post (SACP) (led by an assistant division commander) to help manage both the support and consolidation areas. (See chapter 2 for a discussion of command posts [CPs] and the division SACP.)

CONSOLIDATION OF GAINS ACTIVITIES

> *War is more than a true chameleon that slightly adapts its characteristics to the given case. As a total phenomenon its dominant tendencies always make war a paradoxical trinity— composed of primordial violence, hatred, and enmity, which are to be regarded as a blind natural forces; of the play of chance and probability within which the creative spirit is free to roam; and of its element of subordination, as an instrument of policy, which makes it subject to reason alone.*
>
> Carl von Clausewitz

8-35. Normally, most of a unit's efforts are focused on the performance of area security tasks at the onset of operations to consolidate gains. Area security includes activities to protect friendly forces, installations, and routes within a specific area. The protected forces and installations may be the civilian population, civil institutions, and civilian infrastructure. The weight of effort will shift toward the performance of stability tasks as operations progress. Ultimately, a military commander will transfer control of the area to a legitimate civilian authority. Therefore, when considering force structure, commanders should consider the progression of capabilities required over time as area security and minimal essential stability objectives are met within the consolidation area.

8-36. Units performing area security tasks require a force structure capable of performing the tasks required. This requires a task organization that addresses the warfighting functions as appropriate to the current operational variables within the assigned consolidation area. The unit conducting consolidation of gains refocuses its priorities toward the performance of the initial response tasks for the six primary stability tasks as the security situation stabilizes. (See ATP 3-07.5 for a discussion of these initial response tasks.)

8-37. This refocusing of priorities will require a force restructuring for more tactical units capable of horizontal and vertical engineering, CA, military information support, communications, logistics, protection assets, and other critical stability-related capabilities. Forces tasked with the consolidation of gains should be

Chapter 8

tailored (task-organized) for these tasks. For example, medical personnel may be necessary in the short term to provide immediate medical treatment. Engineer assets may be given on-order missions to improve civil infrastructure as the conditions improve. CA personnel will have a large role in assessing the situation and providing guidance to the commander on the prioritization of allocated resources.

AREA SECURITY OPERATIONS

8-38. By definition, area security operations focus on the protected force, installation, route, or area. Protected forces range from echelon headquarters through artillery and echelon reserves to the sustaining base. Protected installations can also be part of the sustaining base, or they can constitute part of the area's infrastructure. Areas to secure in the consolidation of gains may range from specific points (for example, bridges and defiles) and terrain features (including ridgelines and hills) to large civilian population centers and their adjacent areas. Population-centric area security missions are common across the range of military operations, but they are almost a requirement during irregular warfare. These population-centric area security operations typically combine aspects of the area defense and offensive tasks, like search and attack and cordon and search, to eliminate internal defense threats.

> **Operations to Consolidate Gains: Luzon**
>
> On 10 January 1945, elements of the U.S. Sixth Army landed on the island of Luzon in the Philippines and rapidly broke out of their beachhead towards the capital of Manila. After two months of heavy fighting, the ten American divisions committed to the campaign overcame all resistance in the densely-packed city, and the surviving Japanese fled to the mountainous and remote regions of the island.
>
> In order to consolidate these hard-won gains, Soldiers began at once to re-establish law and order, first under a military government but later transitioning to local authorities. The Japanese had exported most of the island's rice crop to their home islands, and the citizens required a massive infusion of food to prevent a famine. Allied prisoners of war liberated from camps across the island needed immediate medical care. Much of the island's infrastructure, especially around Manila, had to be rebuilt, including vital bridges and ports essential to moving supplies. At the same time, units of the Eighth Army were still engaged in combat in many of the outlying islands, attempting to reduce the last pockets of Japanese resistance.
>
> Despite these challenges, by 1946 the United States effectively passed control of the islands to the Philippine government. Commerce had resumed, and most U.S. combat forces had successfully redeployed. Though significant challenges remained, including the threat of an insurgency by the communist Huks, the islands had a functioning civil government strong enough to counter the threat.
>
> Historians attribute the success of U.S. forces to prior planning and effective execution of operations that deterred opposition forces and prevented significant violence and the breakdown of civil order.

8-39. Area security forces must retain readiness over long periods with intermittent enemy contact. The enemy normally tries to avoid engaging friendly forces except on favorable terms. Forces performing area security tasks should not develop a false sense of security, even if the enemy appears to have ceased operations in a particular area. Commanders must assume that the enemy is observing friendly operations and is seeking the opportunity to attack.

Search and Attack

8-40. As part of area security, the consolidation of gains force may have to identify and destroy remaining pockets of enemy forces. *Search and attack* is a technique for conducting a movement to contact that shares many of the characteristics of an area security mission (FM 3-90-1). Commanders conduct a search and attack for one or more of the following purposes:

- Destroy the enemy: render enemy units in the AO combat-ineffective.
- Deny the area: prevent the enemy from operating unhindered in a given area (for example, in any area the enemy is using for a base camp or for logistics support).
- Protect the force: prevent the enemy from massing to disrupt or destroy friendly military or civilian operations, equipment, property, and key facilities.
- Collect information: gain information about the enemy and the terrain to confirm the enemy course of action predicted by the intelligence preparation of the battlefield (IPB) process.

8-41. A commander employs the search and attack form of a movement to contact when the enemy is operating as small, dispersed elements whose locations cannot be determined to targetable accuracy by methods other than a physical search, or when the task is to deny the enemy the ability to move within a given area. A search and attack is conducted primarily by dismounted infantry forces and often supported by armored, mechanized, motorized and aviation forces (see figure 8-2). A search and attack normally occurs during the conduct of irregular warfare. However, it may also be necessary when conducting noncontiguous operations within major combat operations.

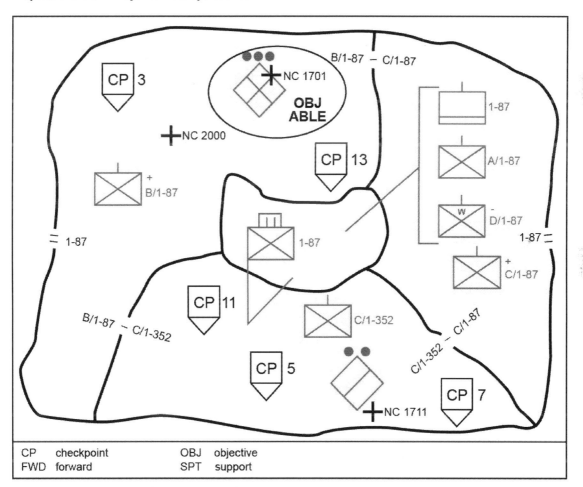

Figure 8-2. Search and attack

8-42. All units can conduct search and attack operations, although a division will rarely conduct search and attack operations simultaneously throughout its AO. BCTs, maneuver battalions, and companies normally conduct search and attack operations. However, during World War II, Germany and Japan, with their allies, conducted division and even corps-sized search and attacks designed to secure major lines of communications

Chapter 8

in Russia, the Balkans, and China. BCTs assist their subordinate maneuver battalions conducting search and attack by ensuring the availability of fires and other support.

8-43. A commander can task organize a unit into reconnaissance, fixing, and finishing forces, each with a specific purpose and task to accomplish. Alternatively, all units can be involved in the reconnaissance effort, with individual subordinate elements being tasked to perform the fixing and finishing functions based on the specifics of the situation.

8-44. Commanders establish control measures that allow for decentralized actions and small-unit initiative to the greatest extent possible. The minimum control measures for a search and attack are an AO, target reference points, objectives, checkpoints, and contact points. The use of target reference points facilitates responsive fire support once the reconnaissance force makes contact with the enemy. Commanders use objectives and checkpoints to guide the movement of subordinate elements. Contact points indicate a specific location and time for coordinating fires and movement between adjacent units. Commanders use other control measures, such as phase lines (PLs) and named areas of interest (NAIs), as necessary.

8-45. The search and attack plan places the finishing force, as the decisive operation, where it can best maneuver to destroy enemy forces or essential facilities once located by reconnaissance assets. Typically, the finishing force occupies a central location in the AO. However, the mission variables may allow the commander to position the finishing force outside the search and attack area. Commanders weight the decisive operation or main effort by using priority of fires and assigning priorities of support to available combat multipliers, such as engineer elements and helicopter lift support. Commanders establish control measures as necessary to consolidate units and concentrate the combat power of their forces before the attack. Once the reconnaissance force locates the enemy, the fixing and finishing forces can fix and destroy it. Commanders also develop contingency plans in the event that the reconnaissance force is compromised.

8-46. Each subordinate element operating in its own AO is tasked to destroy enemy forces within its capability. Units may enter an AO by infiltrating as an entire unit and then splitting out or by infiltrating as smaller units via ground, air, or water. Commanders should have in place previously established control measures and communications means between any closing elements to prevent fratricide and friendly fire incidents. The reconnaissance force conducts a zone reconnaissance to reconnoiter identified NAIs.

8-47. Once the reconnaissance force finds the enemy force, the fixing force develops the situation and executes one of two options based on the commander's guidance and the mission variables. The first option is to block identified routes that the detected enemy can use to escape or interdict with reinforcements. The fixing force maintains contact with the enemy and positions its forces to isolate and fix enemy forces before the finishing force attacks. The second option is to conduct an attack to fix enemy forces in their current positions until the finishing force arrives. The fixing force attacks, if attacking meets the commander's intent and if it can generate sufficient combat power against the detected enemy. Depending on the enemy's mobility and the likelihood of the reconnaissance force being compromised, commanders may need to position the fixing force before the reconnaissance force enters the AO. (See FM 3-90-1 for more information on the performance of the search and attack task.)

Cordon and Search

8-48. Once threats have been identified within a certain part of the consolidation of gains area, cordon and search is a technique that can be used for neutralizing the threat or confiscating material that could be used against friendly forces. *Cordon and search* is a technique of conducting a movement to contact that involves isolating a target area and searching suspected locations within that target area to capture or destroy possible enemy forces and contraband (FM 3-90-1). Cordon and search operations take place throughout the range of military operations. Commanders conducting a cordon and search organize their units into four elements: command, security, search or assault, and support. (See figure 8-3.) The security element must be large enough to establish both an inner and an outer cordon around the target area of the search. In that aspect, cordon and search operations are similar to encirclement operations. Cordon and search is normally conducted at the maneuver battalion echelon and below. (FM 3-90.2 discusses encirclement operations. ATP 3-06.20 discusses multi-Service tactics, techniques, and procedures for cordon and search operations.)

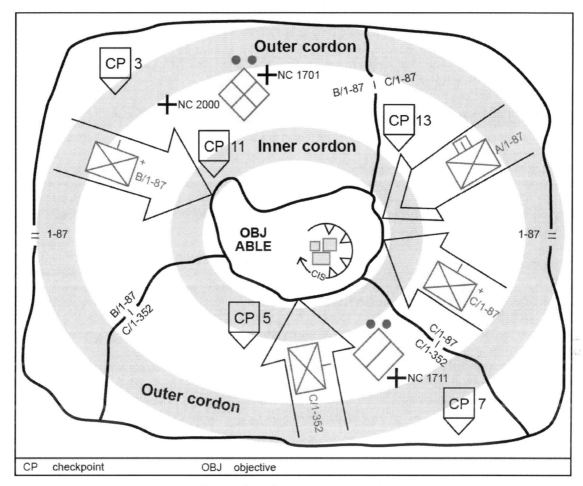

Figure 8-3. Cordon and search

STABILITY TASKS

8-49. Operations focused on stability seek to stabilize the environment enough so that the host nation can begin to resolve the root causes of conflict and state failure. During consolidation of gains, these operations focus on stability tasks to establish conditions that support the transition to legitimate authorities. Initially, this is accomplished by performing the minimum essential stability tasks of providing security, food, water, shelter, and medical treatment. Once conditions allow, these tasks are a legal responsibility of Army forces. However, commanders may not need to have Army forces conduct all of these essential tasks. Other military units or appropriate civilian organizations may be available to adequately perform them. As the operational environment (OE) and time allow, the effort will transition to the more deliberate of execution of the six stability tasks (See ADRP 3-07 for additional information on stability tasks.)

Establish Civil Security

8-50. Civil security is the provision of security for state entities and the population, including protection from internal and external threats. Establishing a safe, secure, and stable environment is crucial to obtaining local support for military operations. The primary task of establishing civil security may include SFA tasks, depending on the tasks assigned. As soon as a host-nation's security forces can perform this task, Army forces transition civil security responsibilities to them. Within the security sector, transformation tasks focus on developing legitimate, sustainable, and stable security institutions. Civil security sets the conditions for enduring stability and peace.

8-51. Transitional public security is the military forces' establishment, promotion, restoration, and maintenance of public order. It is an urgent task of military commanders after major combat ceases. Transitional public security enables the provision of water, food, and medical care. Provost marshals and MP units have experience with transitional public security tasks, as most transitional public security tasks are policing tasks, and they can advise commanders accordingly. (See JP 3-07 for more information on transitional public security.)

Establish Civil Control

8-52. Civil control fosters the rule of law. The rule of law means that all persons, institutions, and entities—public and private, including the state itself—are accountable to laws that are publicly promulgated, equally enforced, independently adjudicated, and consistent with international human rights principles. To strengthen civil control and the rule of law, Army units seek to improve the capability, capacity, and legitimacy of host-nation judicial and corrections systems by providing training and support to law enforcement and judicial personnel. Army units focus on implementing temporary or interim capabilities to lay the foundation for host-nation or inter-organizational development of judicial systems.

Restore Essential Services

8-53. Restoring services essential to local expectations of normalcy allows people to return to their daily activities and prevents further destabilization. Ideally, the host-nation's government and civilian relief agencies should restore and develop essential services. In most cases, local, international, and U.S. agencies have arrived in country long before U.S. forces. However, when partner organizations are not well established or lack capacity, Army units accomplish these tasks until other organizations can.

Support Governance

8-54. Governance is the set of activities conducted by a government or community organization to maintain societal order, define and enforce rights and obligations, and fairly allocate goods and services. Effective, legitimate governance ensures these activities are transparent, accountable, and involve public participation. Elections, while often an end state condition in planning, do not ensure these outcomes. In societies divided along ethnic, tribal, or religious lines, elections may further polarize factions. Generally, representative institutions based on universal suffrage offer the best means of fostering governance acceptable to most citizens. If a host-nation's government or community organizations cannot provide governance, some degree of military support may be necessary. In extreme cases where civil government or community organizations are dysfunctional or absent, international law requires military forces to provide basic civil administration.

Support Economic and Infrastructure Development

8-55. Long-term peace and stability require sustainable host-nation economic and infrastructure development. The end state is the creation of a sustainable economy. In post-conflict and fragile states, host-nation actors, interagency partners, and inter-organizational partners often have the most useful knowledge and skills regarding the restoration and facilitation of economic and infrastructure development. However, if security considerations or other factors restrict their ability to intervene, Army units should assist host-nation entities to foster sustainable economic and infrastructure development.

8-56. The Army financial management enterprise, which includes the financial management officer and banking officer, includes economic analysis planning in collaboration with unified action partners, U.S. national providers, and local authorities. This enterprise supports the host-nation's capability to regain the authority to govern and administer banking systems. (See FM 1-06 for additional information on financial management operations.)

Conduct Security Cooperation

8-57. Security cooperation, as part of consolidation of gains, enhances military engagement and builds the security capacity of partner states. Security cooperation is comprised of multiple activities, programs, and missions, and it is functionally and conceptually related to security assistance, SFA, internal defense and development, foreign internal defense, and security sector reform. As an example, security sector reform involves disarming, demobilizing, and reintegration of former warring factions in the aftermath of an insurgency, assists the host-nation reform its security forces (for example, military and police), bolsters rule of law through constitutional reform, and conducts advisory missions. Army forces may be granted special authorities and called upon to execute tasks in support of these programs that build partner capacity in support of broader national security interests. Security cooperation activities can be executed discretely or in concert with each other across the range of military operations, consolidating many requirements, authorities, and force structures. (See FM 6-22 for more information on security cooperation activities.)

CONSOLIDATE GAINS RESPONSIBILITIES BY ECHELON

A nation cannot be kept permanently interned.

Sir Basil Liddell Hart

8-58. Theater army, corps, divisions, and brigade combat teams all have different responsibilities when consolidating gains. These responsibilities are discussed in paragraphs 8-59 through 8-75.

THEATER ARMY

8-59. Theater army supports consolidation of gains through the execution of Title 10, United States Code (USC) and executive agent activities. During planning, the theater army headquarters anticipates and requests the additional Army combat forces, functional capabilities, and resources required to consolidate gains. It plans the mechanisms required to move those capabilities in and out of the theater of operations. Upon request from the joint force commander or the activation of the time-phased force and deployment data, the theater army requests forces to support the ARFOR in enabling local authorities across all phases of an operation. The theater army should provide the joint task force (JTF) with troops specifically task-organized to focus on area security and stability tasks as part of force tailoring. This requires refined logistic estimates, security cooperation plans, engineer units capable of infrastructure development, CA requirements, communications shortcomings, and other critical capabilities.

8-60. Enabling legitimate authorities to provide essential services is a friendly force objective. This requires coordination with unified action partners, as well as a favorable attitude among the population towards friendly forces supporting the effort. The goal is to replace U.S. combat forces with organizations performing security cooperation activities during the consolidation of gains. Theater army is responsible for coordinating these activities.

8-61. SFA is a key component of any plan to consolidate gains. SFA requires trained, educated professional officers and noncommissioned officers as trusted advisors to partner nation security forces. Talent management in assigning this critical aspect of consolidate gains should be explicit in the theater army request or assignment of forces. Joint force commanders build conventional force and special operations forces (SOF) interdependence into SFA plans to achieve objectives.

8-62. The theater army also manages Army support to other services (ASOS) forces across the area of responsibility (AOR). It is often beneficial to engage other regional actors to assist in promoting the stability of the region once the level of conflict subsides to where it is acceptable to the leaders and population of these regional actors. The theater army is critical in managing this support.

8-63. One of the most critical tasks is the final transition of authority from military forces to legitimate civilian authorities. In most cases, successful consolidation of gains requires the theater army to retrograde equipment, close the JOA, coordinate for the redeployment of Army forces, and manage a long-term security cooperation plan.

CORPS

8-64. The corps headquarters manages division and brigade consolidation of gains, while anticipating, planning, and shaping the next areas targeted for these operations. Corps headquarters request the additional forces necessary to consolidate gains and allocate them to divisions. This requires coordination and planning simultaneous with operations occurring in the corps deep and close areas. As part of this, consolidate gains activities may require the Army corps to employ forces to perform these tasks:

- Establish area security throughout the entire corps AO. This includes offense and defense tasks to destroy or neutralize remaining threats and protect the civilian population and infrastructure.
- Conduct stability tasks necessary to create conditions that allow for the eventual transition to a legitimate authority.
- Conduct security force assistance to build the capability and capacity of foreign security forces.
- Dissuade enemies and adversaries from reinitiating hostile or disruptive operations and persuading them to abide by sanctions, laws, or international dictates.
- Coordinate and influence the assumption of responsibility by host-nation or other authorities.
- Synchronize psychological actions aligned with friendly unit activities (including deeds, words, and images) to favorably influence civilian attitudes toward friendly security forces and the eventual transition to legitimate authority.

8-65. As large-scale combat operations conclude in a corps AO, the corps headquarters reorganizes the AO into areas appropriate for the operational and mission variables to facilitate the most rapid consolidation of gains. Operations primarily focus on providing area security in high threat areas, followed by the performance of stability tasks in lower threat areas. As shown in figure 8-4, a corps can simultaneously attack and consolidate gains during this transitional period. Subordinate divisions may still have BCTs conducting large-scale combat operations while committing one or more BCTs to conduct consolidation of gains in the consolidation area. Ideally, the corps headquarters task-organizes follow and support divisions to execute stability related tasks.

Operations to Consolidate Gains

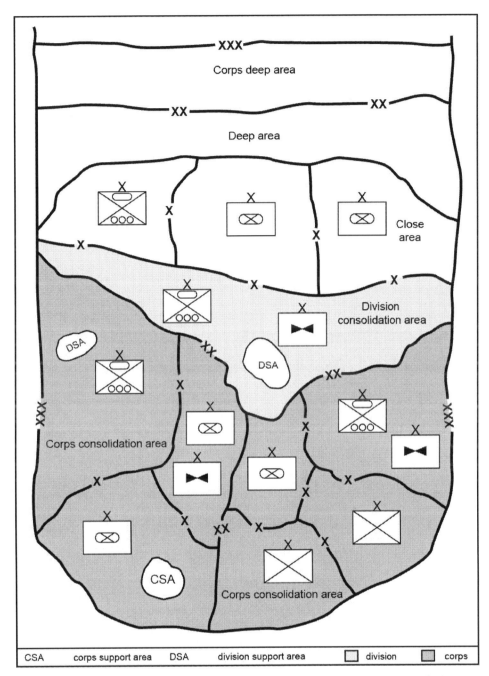

Figure 8-4. Consolidating gains after large-scale combat operations

DIVISIONS

8-66. Likewise, division commanders and staffs consider the impact of simultaneously conducting consolidation of gains activities and large-scale combat operations throughout all phases of their plans. Considerations include priority of assets, time phasing of operations, task organization changes, clearance of fires in the consolidation areas, and distance between the forward edge of the close area and the forward edge of consolidation areas. It is likely that concurrent, large-scale combat operations are ongoing across the AO and that additional combat forces are necessary to consolidate gains. Divisions must solve the problem of how to sustain those forces in combat operations concurrently with operations to consolidate gains. In addition, division support assets must be prepared to displace from established support areas, through areas

Chapter 8

where security may not yet be established, to new support areas to better support those forces in combat operations prior to area security being established.

8-67. Divisions manage BCT's as they establish area security and perform stability tasks throughout the division consolidation areas. This includes performing offensive and defensive tasks to destroy or neutralize remaining threats (usually company size or smaller) that were either by-passed in combat operations or have reconsolidated following major combat operations in the close area. Simultaneously, a task-organized, follow and assume BCT protects the civilian population and infrastructure from attacks by irregular forces and criminal elements within the division consolidation area.

8-68. Divisions perform the stability tasks necessary to create conditions within an AO that allow for transition to a legitimate civilian authority or a unified action partner to act as a transitional authority. Divisions also manage brigades as they conduct security force assistance to build the capability and capacity of foreign security forces. Finally, divisions dissuade adversaries from reinitiating hostile or disruptive operations and persuade them to abide by sanctions, laws, or international dictates.

8-69. Divisions are the first tactical echelon with a dedicated staff trained to synchronize actions of information-related capabilities (IRCs). These capabilities may be used to influence local governments and populations through compelling narratives that explain actions, discredit enemy propaganda, and highlight common goals, themes, and messages. Nothing makes narratives more compelling than witnessing positive physical security and stability activity on the ground.

8-70. Divisions enable subordinate BCTs to conduct detailed information collection. BCTs have limited organic information collection assets. BCTs often require additional assets to understand, shape, and influence the OE and consolidate positive gains that lead toward desired objectives during operations to consolidate gains. A division headquarters provides messaging themes aligned with friendly unit activities on the ground and focused on positively influencing civilian attitudes about the security forces and eventual transition to civilian authority.

8-71. Divisions work with host-nation or other authorities to assume responsibility in consolidating gains. They promote unified action partner efforts to ensure area security and stability. Division commanders extend their influence in the AO through the synchronization of efforts by various unified action partners.

BRIGADE COMBAT TEAMS AND TASK FORCES

8-72. Task-organized BCTs or task forces perform most area security tasks, and they facilitate the accomplishment of most stability tasks in concert with unified action partners. Army forces take specific actions upon culmination of large-scale combat operations against a conventional force. These actions include—
- Consolidation.
- Establish area security.
- Conduct stability tasks.

8-73. The transition from large-scale combat operations to area security tasks against a capable enemy entails risk. Units shifting focus from combat to consolidation are vulnerable to counterattack. Historical evidence suggests a force is most vulnerable after a hard-fought win. Commanders and staffs should plan for a quick transition and stress the importance of local security. (See FM 3-90-1 for additional information on consolidation.)

8-74. Forces first perform minimum-essential stability tasks, and then they maintain or reestablish a stable OE and provide essential governmental services, emergency infrastructure reconstruction, and humanitarian relief. BCTs further develop and reassess situations, perceptions, and opportunities through continuous information collection to maintain positive momentum to achieve additional gains.

8-75. BCTs conduct operations by seizing, retaining, and exploiting opportunities. BCTs leverage information collection, interactions with people and organizations, offensive and defensive tasks, IRCs, and cyberspace electromagnetic activities (CEMA) to achieve the security conditions essential to the effective performance of stability tasks.

TRANSITION

8-76. The transfer of an AO to a legitimate authority relieves the land force of area security and stability tasks and represents a transition from operations to consolidate gains to operations to shape or prevent. As the theater army headquarters coordinates the movement of most Army forces out of theater, it resumes the tasks necessary to sustain the gains consolidated during large-scale combat operations. Conditions on the ground and resources available determine what security and cooperation tasks have priority. It is possible that Army forces occupy long-term garrisons to prevent a recurrence of hostilities, as happened in Europe, Japan, and Korea in the 20th century. Alternatively, there may be a more robust security cooperation arrangement involving training of forces. Regardless of the tasks required in a specific AO, Army units will continue to perform the missions that reflect their strategic roles of shape, prevent, win, and consolidate gains in support of U.S. interests.

This page intentionally left blank.

Appendix A
Command and Support Relationships

This appendix discusses command and support relationships for joint and Army forces. Command and support relationships provide the basis for unity of command and unity of effort in operations.

FUNDAMENTAL CONSIDERATIONS

A-1. Establishing clear command and support relationships is a key aspect of any operation. Large-scale combat operations present unique and complex challenges that demand well defined command and support relationships among units. These relationships establish responsibilities and authorities between subordinate and supporting units. Some command and support relationships limit the commander's authority to prescribe additional relationships. Knowing the inherent responsibilities of each command and support relationship allows commanders to effectively organize their forces and helps supporting commanders understand their unit's role in the organizational structure.

JOINT COMMAND RELATIONSHIPS

A-2. As part of a joint force, Army commanders and staffs must understand joint command relationships. JP 1 specifies and details four types of joint command relationships:
- Combatant command (command authority) (COCOM).
- Operational control (OPCON).
- Tactical control (TACON).
- Support.

COMBATANT COMMAND (COMMAND AUTHORITY)

A-3. *Combatant command* is a unified or specified command with a broad continuing mission under a single commander established and so designated by the President, through the Secretary of Defense and with the advice and assistance of the Chairman of the Joint Chiefs of Staff (JP 1). Title 10, United States Code (USC), section 164 specifies this authority in law. Normally, the combatant commander exercises this authority through subordinate joint force commanders (JFCs), Service component commanders, and functional component commanders.

OPERATIONAL CONTROL

A-4. *Operational control* is the authority to perform those functions of command over subordinate forces involving organizing and employing commands and forces, assigning tasks, designating objectives, and giving authoritative direction necessary to accomplish the mission (JP 1). OPCON normally includes authority over all aspects of operations and joint training necessary to accomplish missions. It does not include directive authority for logistics or matters of administration, discipline, internal organization, or unit training. The combatant commander must specifically delegate these elements of COCOM. OPCON does include the authority to delineate functional responsibilities and operational areas of subordinate JFCs. In two instances, the Secretary of Defense may specify adjustments to accommodate authorities beyond OPCON in an establishing directive: when transferring forces between combatant commanders or when transferring members or organizations from the military departments to a combatant command. Adjustments will be coordinated with the participating combatant commanders.

Appendix A

TACTICAL CONTROL

A-5. *Tactical control* is the authority over forces that is limited to the detailed direction and control of movements or maneuvers within the operational area necessary to accomplish missions or tasks assigned (JP 1). TACON is inherent in OPCON. It may be delegated to and exercised by commanders at any echelon at or below the level of combatant command. TACON provides sufficient authority for controlling and directing the application of force or tactical use of combat support assets within the assigned mission or task. TACON does not provide organizational authority or authoritative direction for administrative and logistic support; the commander of the parent unit continues to exercise these authorities unless otherwise specified in the establishing directive.

SUPPORT

A-6. *Support* is the action of a force that aids, protects, complements, or sustains another force in accordance with a directive requiring such action (JP 1). Support is a command authority in joint doctrine. A supported and supporting relationship is established by a superior commander between subordinate commanders when one organization should aid, protect, complement, or sustain another force. Designating supporting relationships is important. It conveys priorities to commanders and staffs planning or executing joint operations. Designating a support relationship does not provide authority to organize and employ commands and forces, nor does it include authoritative direction for administrative and logistic support. Joint doctrine divides support into the categories listed in table A-1.

Table A-1. Joint support categories

Category	Definition
General support	That support which is given to the supported force as a whole and not to any particular subdivision thereof (JP 3-09.3).
Mutual support	That support which units render each other against an enemy, because of their assigned tasks, their position relative to each other and to the enemy, and their inherent capabilities (JP 3-31).
Direct support	A mission requiring a force to support another specific force and authorizing it to answer directly to the supported force's request for assistance (JP 3-09.3).
Close support	The action of the supporting force against targets or objectives that are sufficiently near the supported force as to require detailed integration or coordination of the supporting action. (JP 3-31)

A-7. Support is, by design, somewhat vague but very flexible. Establishing authorities ensure both supported and supporting commanders understand the authority of supported commanders. JFCs often establish supported and supporting relationships among components. For example, the maritime component commander is normally the supported commander for sea control operations; the air component commander is normally the supported commander for counter-air operations. An Army headquarters designated as the land component may be the supporting force during some campaign phases and the supported force in other phases.

A-8. The JFC may establish a support relationship between functional and Service component commanders. Conducting operations across a large operational area often involves both the land and air component commanders. The joint task force (JTF) commander places the land component in general support of the air component until the latter achieves air superiority. Conversely, within the land area of operations, the land component commander (LCC) becomes the supported commander and the air component commander provides close support. A joint support relationship is not used when an Army commander task organizes Army forces in a supporting role. When task-organized to support another Army force, Army forces use one of the four Army support relationships.

OTHER AUTHORITIES

A-9. Although discussed in joint doctrine, coordinating authority and direct liaison authorized are directly applicable to Army forces. These relationships can assist commanders in facilitating collaboration both within and outside their respective organizations, and they can promote information sharing concerning details of military operations.

Coordinating Authority

A-10. *Coordinating authority* is a commander or individual who has the authority to require consultation between the specific functions or activities involving forces of two or more Services, joint force components, or forces of the same Service or agencies, but does not have the authority to compel agreement (JP 1). In the event that essential agreement cannot be obtained, the matter shall be referred to the appointing authority. Coordinating authority is a consultation relationship, not an authority through which command may be exercised. Coordinating authority is more applicable to planning and similar activities than to operations. For example, a joint security commander exercises coordinating authority over area security operations within the joint security area. Commanders or leaders at any echelon at or below combatant command may be delegated coordinating authority. These individuals may be assigned responsibilities established through a memorandum of agreement between military and nonmilitary organizations. (See JP 1 for more information on coordinating authority.)

Direct Liaison Authorized

A-11. *Direct liaison authorized* is that authority granted by a commander (any level) to a subordinate to directly consult or coordinate an action with a command or agency within or outside of the granting command (JP 1). Direct liaison authorized is more applicable to planning than operations and always carries with it the requirement of keeping the commander granting direct liaison authorized informed. Direct liaison authorized is a coordination relationship, not an authority through which command may be exercised.

ARMY COMMAND AND SUPPORT RELATIONSHIPS

A-12. Army command and support relationships are similar but not identical to joint command authorities and relationships. Differences stem from the way Army forces task-organize internally and the need for a system of support relationships between Army forces. Another important difference is the requirement for Army commanders to handle the administrative support requirements that meet the needs of Soldiers. These differences allow for flexible allocation of Army capabilities within various Army echelons. Army command and support relationships are the basis for building Army task organizations. Certain responsibilities are inherent in the Army's command and support relationships.

ARMY COMMAND RELATIONSHIPS

A-13. Army command relationships define superior and subordinate relationships between unit commanders. By specifying a chain of command, command relationships unify effort and enable commanders to use subordinate forces with maximum flexibility. Army command relationships identify the degree of control of the gaining Army commander. The type of command relationship often relates to the expected longevity of the relationship between the headquarters involved, and it quickly identifies the degree of support that the gaining and losing Army commanders provide. Army command relationships include—
- Organic.
- Assigned.
- Attached.
- OPCON (see paragraph A-4).
- TACON (see paragraph A-5).

(See table A-2 on page A-4 for an illustration of Army command relationships.)

Appendix A

Table A-2. Army command relationships

If relationship is—	Then inherent responsibilities:							
	Have command relationship with—	May be task-organized by—	Unless modified, ADCON responsibility goes through—	Are assigned position or AO by—	Provide liaison to—	Establish and maintain communications with—	Have priorities established by—	Can impose on gained unit further command or support relationship of—
Organic	All organic forces organized with the HQ	Organic HQ	Army HQ specified in organizing document	Organic HQ	N/A	N/A	Organic HQ	Attached; OPCON; TACON; GS; GSR; R; DS
Assigned	Gaining unit	Gaining HQ	Gaining Army HQ	OPCON chain of command	As required by OPCON	As required by OPCON	ASCC or Service-assigned HQ	As required by OPCON HQ
Attached	Gaining unit	Gaining unit	Gaining Army HQ	Gaining unit	As required by gaining unit	Unit to which attached	Gaining unit	Attached; OPCON; TACON; GS; GSR; R; DS
OPCON	Gaining unit	Parent unit and gaining unit; gaining unit may pass OPCON to lower HQ[1]	Parent unit	Gaining unit	As required by gaining unit	As required by gaining unit and parent unit	Gaining unit	OPCON; TACON; GS; GSR; R; DS
TACON	Gaining unit	Parent unit	Parent unit	Gaining unit	As required by gaining unit	As required by gaining unit and parent unit	Gaining unit	TACON; GS GSR; R; DS

Note. [1] In NATO, the gaining unit may not task-organize a multinational force. (See TACON.)

ADCON	administrative control	HQ	headquarters
AO	area of operations	N/A	not applicable
ASCC	Army Service component command	NATO	North Atlantic Treaty Organization
DS	direct support	OPCON	operational control
GS	general support	R	reinforcing
GSR	general support–reinforcing	TACON	tactical control

Organic

A-14. *Organic forces* are those assigned to and forming an essential part of a military organization as listed in its table of organization for the Army, Air Force, and Marine Corps, and are assigned to the operating forces for the Navy (JP 1). Joint command relationships do not include organic because a JFC is not responsible for the organizational structure of units. That is a Service responsibility.

A-15. The Army establishes organic command relationships through organizational documents such as tables of organization and equipment and tables of distribution and allowances. If temporarily task-organized with another headquarters, organic units return to the control of their organic headquarters after completing the mission. To illustrate, within a brigade combat team (BCT), the entire brigade is organic. In contrast, within most functional and multifunctional brigades, there is a "base" of organic battalions and companies and a variable mix of assigned and attached battalions and companies.

Assigned

A-16. *Assign* is to place units or personnel in an organization where such placement is relatively permanent, and/or where such organization controls and administers the units or personnel for the primary function, or greater portion of the functions, of the unit or personnel (JP 3-0). Unless specifically stated, this relationship includes administrative control (ADCON).

Attached

A-17. *Attach* is the placement of units or personnel in an organization where such placement is relatively temporary (JP 3-0). A unit may be temporarily placed into an organization for the purpose of conducting a specific operation of short duration. Attached units return to their parent headquarters (assigned or organic) when the reason for the attachment ends. The Army headquarters that receives another Army unit through assignment or attachment assumes responsibility for the ADCON requirements, and particularly sustainment, that normally extend down to that echelon, unless modified by directives.

ARMY SUPPORT RELATIONSHIPS

A-18. Table A-3 on page A-6 lists Army support relationships. Army support relationships are not a command authority and are more specific than joint support relationships. Commanders establish support relationships when subordination of one unit to another is inappropriate. Army support relationships are—
- Direct support.
- General support.
- Reinforcing.
- General support-reinforcing.

A-19. Commanders assign a support relationship for several reasons. They include when—
- The support is more effective if a commander with the requisite technical and tactical expertise controls the supporting unit rather than the supported commander.
- The echelon of the supporting unit is the same as or higher than that of the supported unit. For example, the supporting unit may be a brigade, and the supported unit may be a battalion. It would be inappropriate for the brigade to be subordinated to the battalion; hence, the echelon uses an Army support relationship.
- The supporting unit supports several units simultaneously. The requirement to set support priorities to allocate resources to supported units exists. Assigning support relationships is one aspect of mission command.

Appendix A

Table A-3. Army support relationships

If relation-ship is—	Then inherent responsibilities—							
	Have command relation-ship with—	May be task-organized by—	Receive sustain-ment from—	Are assigned position or an area of operations by—	Provide liaison to—	Establish and maintain communi-cations with—	Have priorities established by—	Can impose on gained unit further support relationship of—
Direct support[1]	Parent unit	Parent unit	Parent unit	Supported unit	Supported unit	Parent unit; supported unit	Supported unit	See note[1]
Reinforc-ing	Parent unit	Parent unit	Parent unit	Reinforced unit	Reinforced unit	Parent unit; reinforced unit	Reinforced unit; then parent unit	Not applicable
General support-reinforc-ing	Parent unit	Parent unit	Parent unit	Parent unit	Reinforced unit and as required by parent unit	Reinforced unit and as required by parent unit	Parent unit; then reinforced unit	Not applicable
General support	Parent unit	Parent unit	Parent unit	Parent unit	As required by parent unit	As required by parent unit	Parent unit	Not applicable

Note. [1] Commanders of units in direct support may further assign support relationships between their subordinate units and elements of the supported unit after coordination with the supported commander.

A-20. Army support relationships allow supporting commanders to employ their units' capabilities to achieve results required by supported commanders. Support relationships are graduated from an exclusive supported and supporting relationship between two units—as in direct support—to a broad level of support extended to all units under the control of the higher headquarters—as in general support. Support relationships do not alter administrative control. Commanders specify and change support relationships through task organization.

A-21. **Direct support is a support relationship requiring a force to support another specific force and authorizing it to answer directly to the supported force's request for assistance.** A unit assigned a direct support relationship retains its command relationship with its parent unit, but it is positioned by and has priorities of support established by the supported unit. (Joint doctrine considers direct support a mission rather than a support relationship.) A field artillery unit in direct support of a maneuver unit is concerned primarily with the fire support needs of only that unit. The fires cell of the supported maneuver unit plans and coordinates fires to support the maneuver commander's intent. The commander of a unit in direct support recommends position areas and coordinates for movement clearances where the unit can best support the maneuver commander's concept of the operation.

A-22. General support is that support which is given to the supported force as a whole. It is not given to any particular subdivision of the force. Units assigned a general support relationship are positioned and have priorities established by their parent unit. A field artillery unit assigned in general support of a force has all of its fires under the immediate control of the supported commander or his designated force field artillery headquarters.

A-23. **Reinforcing is a support relationship requiring a force to support another supporting unit.** Only like units (for example, artillery to artillery) can be given a reinforcing mission. A unit assigned a reinforcing support relationship retains its command relationship with its parent unit, but it is positioned by the reinforced unit. A unit that is reinforcing has priorities of support established by the reinforced unit, then the parent unit. For example, when a direct support field artillery battalion requires more fires to meet maneuver force requirements, another field artillery battalion may be directed to reinforce the direct support battalion.

A-24. *General support—reinforcing* is a support relationship assigned to a unit to support the force as a whole and to reinforce another similar-type unit (ADRP 5-0). A unit assigned a general support-reinforcing support relationship is positioned and has its priorities established by its parent unit and secondly by the reinforced unit. For example, an artillery unit that has a general-support-reinforcing relationship supports the force as a whole and provides reinforcing fires for other artillery units.

ADMINISTRATIVE CONTROL

A-25. *Administrative control* is direction or exercise of authority over subordinate or other organizations in respect to administration and support (JP 1). ADCON is not a command or support relationship; it is a Service authority. It is exercised under the authority of and is delegated by the Secretary of the Army. ADCON is synonymous with the Army's Title 10 authorities and responsibilities.

A-26. ADCON responsibilities of Army forces involve the entire Army, and they are distributed between the Army institutional force and the operating forces. The institutional force consists of those Army organizations whose primary mission is to generate and sustain the operating force's capabilities for employment by JFCs. Operating forces consist of those forces whose primary missions are to participate in combat and the integral supporting elements thereof. Often, commanders in the operating force and commanders in the institutional force subdivide specific responsibilities. Army institutional force capabilities and organizations are linked to operating forces through co-location and reachback.

A-27. The Army Service component command (ASCC) is always the senior Army headquarters assigned to a combatant command. Its commander exercises command authorities as assigned by the combatant commander and ADCON as delegated by the Secretary of the Army. ADCON is the Army's authority to administer and support Army forces even while in a combatant command area of responsibility. COCOM is the basic authority for command and control of the same Army forces. The Army is obligated to meet the combatant commander's requirements for the operating forces. Essentially, ADCON directs the Army's support of operating force requirements.

A-28. Unless modified by the Secretary of the Army, administrative responsibilities normally flow from Department of the Army through the ASCC to those Army forces assigned or attached to that combatant command. ASCCs usually "share" ADCON for at least some administrative or support functions. "Shared ADCON" refers to the internal allocation of Title 10, U.S. Code, section 3013(b) responsibilities and functions. This is especially true for Reserve Component forces. Certain administrative functions, such as pay, stay with the Reserve Component headquarters, even after unit mobilization. Shared ADCON also applies to direct reporting units of the Army that typically perform single or unique functions. The direct reporting unit, rather than the ASCC, typically manages individual and unit training for these units. The Secretary of the Army directs shared ADCON.

This page intentionally left blank.

Appendix B
Risk Considerations

This appendix describes risks associated to large-scale ground combat. It then provides commanders and staffs considerations to mitigate those risks.

B-1. Risk, uncertainty, and chance are inherent in all military operations. Commanders seek to understand, balance, and take risks, rather than avoid risks. When commanders accept risk in large-scale combat operations, they create opportunities to seize, retain, and exploit the initiative and achieve decisive results. Opportunities come with risks. The willingness to incur risk is often the key to exposing enemy weaknesses that the enemy considers beyond friendly reach. Understanding risk requires assessments coupled with boldness and imagination. Successful commanders assess and mitigate risk continuously throughout an operation.

B-2. Commanders, supported by their staffs, make reasonable estimates and intentionally accept risks in combat. They focus on creating opportunities to win rather than simply preventing defeat—even when preventing defeat appears safer. Commanders carefully determine risks, analyze and minimize as many hazards as possible, and execute a plan that accounts for those hazards.

B-3. Commanders balance the tension between protecting the force and accepting risks in order to achieve military objectives. Thus, commanders must adequately plan and prepare for operations, and they are informed by a comprehensive understanding of an operational environment (OE). Commanders collaborate and dialog with subordinates when deciding how much risk to accept and how to minimize the effects of risk. Commanders also avoid delaying action while waiting for refined intelligence or perfect synchronization. Experienced commanders balance audacity and imagination with risk and uncertainty. They strike at a time and place and in a manner wholly unexpected by the enemy.

B-4. It is important to remember that accepting risk is a function of command, and it is a key planning consideration. The commander alone determines the level of risk that is acceptable with respect to aspects of operations. This level of risk should be expressed in the commander's guidance, incorporated into all plans and orders, and clearly understood by subordinate commanders.

B-5. Situational understanding is essential to managing risk. Seeing, understanding, and responding to windows of vulnerability or opportunity within each domain and the information environment can reduce risk to the force and enhance success in chaotic and high-tempo operations. Critical to understanding how much risk to incur is the regular review of assumptions that underlie particular courses of action. Challenging assumptions and demanding accurate staff estimates helps to ensure that when the commander decides to accept risk the decision is based on a reasonable understanding of how much risk is being taken. Table B-1 on page B-2 lists examples of risk considerations that commanders should review when making decisions on risk.

Appendix B

Table B-1. Risk considerations

Commander's risk considerations
How much combat power the main effort receives and why? (Commanders confirm the task organization annex. Subordinate units strictly follow the parameters of command or support relationships outlined in appendix A of this manual.)
Is sufficient guidance provided to empower, provide vision and purpose, and set conditions for the team to succeed? (Commanders drive the operations process through understanding, visualizing, describing, directing, leading, and assessing.)
Where does the commander best command the organization? (The commander's presence can mitigate confusion in combat and prevent the headquarters from losing focus.)
What authorities exist to empower subordinate commanders to build relationships of trust with unified action partners and local officials within their assigned area of operations? (Leaders need to build those relationships. Commanders should consider any obstacles to building those relationships, regardless of how slight, as having the potential to disrupt combat operations.)
How can commanders ensure that their staffs do not overclassify products? (Overclassification slows shared understanding with unified action partners.)
How does the unit move into positions of advantage in restricted terrain without being heavily disrupted by enemy assets? (Units observed by enemy forces can be engaged. Units engaged by an enemy on the enemy's terms can be destroyed.)
Are staffs at higher echelons enabling parallel planning across all echelon headquarters? (Staffs need to take full advantage of liaison officers. They should also use digital systems to share planning information with the staffs of subordinate, supporting, and supported staffs, if digital connectively exists.)
Are processes in place to allow staffs to interact with the unit commander and extract the commander's intent and overall visualization when problems are identified? (The staff battle rhythm needs to provide the staff with time to interact with the commander during the day.)
Are staffs developing and rehearsing systems to ensure the flow of information from all sources to the respective staff functions and synchronization cells? (Chiefs of staff and executive officers are responsible for ensuring all echelons have a feedback mechanism to the headquarters, including special operations.)
Do the efforts of special staff officers support the scheme of maneuver and the information narrative? (This should occur without placing excessive demands on subordinate commanders.)
Does the staff use simulation and gaming techniques to refine or synchronize operations as part of different rehearsal exercises across echelons and across domains? (This requires elements of the staff to operate as a red team to subject important parts of the plan to examination.)
Is the staff seeking synchronization across multiple domains? (Staffs determine windows when, either because of battle rhythm or effects, an enemy's collection or lethal effects capability is limited. Staffs recommend those windows of time for movements or attacks by subordinate units.)
What operations security considerations are most important for reducing risk to friendly forces? (This requires an accurate knowledge of enemy information collection means and capabilities on the part of the assistant chief of staff, intelligence (G-2) and battalion or brigade intelligence staff officer (S-2) and protection cell staffs.)
When and where could military deception operations best reduce friendly vulnerabilities while creating enemy dilemmas? (This requires an accurate knowledge of enemy information collection means and capabilities. It also requires an understanding of how the targeted enemy decision maker receives and processes information in addition to that leader's frame of reference.)

Source Notes

This division lists sources by page number. Where material appears in a paragraph, it lists both the page number followed by the paragraph number. All websites accessed 21 July 2017.

1-1 "War is thus an act...": Carl von Clausewitz, *On War*, edited and translated by Michael Howard and Peter Paret (Princeton, NJ: Princeton University Press, 2000), 75.

1-2 **Close Combat: Hürtgen Forest**: Army University Press staff, unpublished text, 2017.

1-3 **Lethality: Eastern Ukraine**: Army University Press staff, unpublished text, 2017.

1-4 "No matter how clearly...": Sir Michael Howard quoted in Developments Concepts and Doctrine Centre, *Future Character of Conflict: Strategic Trends Programme* (United Kingdom: Ministry of Defence, 2010), 2.

1-6 "Since men live upon...": Julian S. Corbett, *Some Principles of Maritime Strategy* (Annapolis, MD: Naval Institute Press, 1988), 16.

1-6 "Print is the sharpest...": Josef Stalin, Speech, April 19, 1923, *Bartlett's Familiar Quotations* (New York: Little, Brown and Company, 2002), 686.

1-9 "I know from personal...": General Mark Clark in "For 1957, I Resolve...," *Asbury Park Press* (Asbury Park, NJ: January 13, 1957), 24. Available at https://www.newspapers.com/newspage/144197456/.

1-12 "[S]eparate ground, sea, and...": Dwight D. Eisenhower, "Special Message to the Congress on Reorganization of the Defense Establishment," April 3, 1958. Online by Gerhard Peters and John T. Woolley, *The American Presidency Project*. http://www.presidency.ucsb.edu/ws/?pid=11340.

1-13 "I should like to...": Saladin in Beha Ed-din, *The Life of Saladin* (London: Committee of the Palestine Exploration Fund, 1897), 222.

1-14 "We in America have...": Ronald Regan, "Remarks at a Ceremony Commemorating the 40th Anniversary of the Normandy Invasion, D-day June 6, 1984," The Public Papers of President Ronald W. Reagan, Ronald Reagan Presidential Library. Available at https://www.reaganlibrary.archives.gov/archives/speeches/1984/60684a.htm.

1-15 "To be prepared for war...": George Washington, *Dictionary of Military and Naval Quotations*, compiled by Robert Debs Heinl, Jr. (Annapolis, MD: United States Naval Institute, 1988), 247.

1-15 "War's very object is...": General Douglas MacArthur, Speech to Congress in House of Representatives, April 19, 1951. Available at https://www.trumanlibrary.org/whistlestop/study_collections/koreanwar/index.php?action=pdf&documentid=ma-2-18.

1-15 "Loss of hope,...": B.H. Liddell Hart, in *Military Air Power: The CADRE Digest of Air Power Opinions and Thoughts*, compiled by Charles M. Westenhoff (Maxwell Air Force Base, AL: Air University Press, 1990), 36.

1-16 "The community of danger...": Xenophon, *Dictionary of Quotations (Classical)*, Thomas Benfield Harbottle (New York: Swan Sonnenschein & Co., Limited, 1906), 435.

1-17 "It is even better...": Carl von Clausewitz: *On War, Dictionary of Military and Naval Quotations*, compiled by Robert Debs Heinl, Jr. (Annapolis, MD: United States Naval Institute, 1988), 1.

1-18 "In war trivial ...": Julius Caesar: *De Bello Gallico*, i, 51 B. C., *Dictionary of Military and Naval Quotations*, compiled by Robert Debs Heinl, Jr. (Annapolis, MD: United States Naval Institute, 1988), 340.

Source Notes

1-19 "Never tell people…": General George Patton, "Quotes." Available at http://www.generalpatton.com/quotes/.

1-19 "[T]here is a wide…": Xenophon, *Memorabilia* 3.1.7. The Perseus Project, Tufts University. The Annenberg CPB/Project provided support for entering this text. Available at http://perseus.uchicago.edu/perseus-cgi/citequery3.pl?dbname=GreekFeb2011&query=Xen.%20Mem.%203.1.11&getid=1.

1-20. "Never forget that no military leader has ever become great without audacity.": Carl von Clausewitz.

1-22 "The more I see…": Field Marshal Wavell cited in *Speaking Generally* (London, 1946), 78–9, in Martin Van Creveld's *Supplying War: Logistics from Wallenstein to Patton* (New York: Cambridge University Press, 1991), 231–232.

1-23 "To remain separated for…": Field Marshal Helmuth von Moltke the Elder in T. Miller Maguire, "Campaign in Bohemia," *The United Service Magazine* (London: Will Clowes & Sons, Limited, 1866), 174–175.

1-30 "And to control many…": Sun Tzu, *The Art of War*, translated by Samuel B. Griffith (New York: Oxford University Press, 1963), 90.

Chapter 2

2-1 "[T]he real object of…": Elihu Root, *Annual Reports of the War Department*, Vol IV (Washington DC: Government Printing Office, 1903), 94.

2-1 "The land domain is...": JP 3-31, *Command and Control for Land Operations*. 24 February 2014. Available at http://www.dtic.mil/doctrine/new_pubs/jointpub.htm.

2-5 **Sustainment: OPERATIONS DESERT SHIELD and DESERT STORM**: vignette in FM 100-5 [obsolete], Operations (Washington DC: Government Printing Office, 1993), 12-2.

2-14 "Success in war can.": Major General Leonard Word in *Field Service Regulation* (Washington DC: Government Printing Office, 1918), 3.

2-21 "No army can be…": General William T. Sherman, *Memoirs of William T. Sherman by Himself* (New York: D. Appleton and Company, 1875), 386.

2-22 "I am doing all …": Field Marshal Earl Kitchener in "K," a speech by the Earl of Darby, *The Lord Kitchener Memorial Book*, edited by Sir Hedley Le Bas (New York: Hodder and Stoughton, 1917). Published on behalf of the Lord Kitchener national memorial fund. Available at http://www.archive.org/details/lordkitchenermem00leba.

2-23 "Unless you understand the…": Mao Tse Tung, *Selected Military Writings of Mao Tse-Tung* (Peking: Foreign Languages Press, 1967), 78.

2-24 "Lead me, follow me,…": General George Smith Patton Jr. http://www.generalpatton.com/quotes/.

2-24 "I have published under...": Field Marshal Slim, Lecture to the Staff College, 2 November 1967, Field-Marshal Viscount William Slim, *Defeat into Victory: Battling Japan in Burma and India, 1942—1945* (New York: Cooper Square Press, 2000), 210-211.

2-26 "There has never been…": President Harry Truman, "Going Forward with a Campaign of Truth," *State Department Bulletin* (Washington DC: Government Printing Office, May 1, 1950), 669.

2-28 "In wartime, truth is…": Winston S. Churchill, *The Second World War, Volume 5, Closing the Ring* (Boston: Houghton Mifflin Company, 1979), 383.

2-29 "The mind of the enemy…": Brigadier General Samuel B. Griffith II, USMC, "Introduction" to FMFRP 12-18, *Mao Tse-tung on Guerrilla Warfare* (Washington DC: Government Printing Office, 1989), 23.

2-29 **Messaging Against Iraqi Forces**: Army University Press staff, unpublished text, 2017.

2-39 "Diverse are the situations..." Field Marshal Helmuth von Moltke the Elder, "Moltkes Taktisch-strategische Aufsätze aus den Jahren 1867 bis 1871: Zur hundertjährigen Gedenkfeier der Geburt des General-Feldmarschalls Grafen von Moltke," published in *Verordnungen für die höheren Truppenführer*, 29 June 1969 in Major General Werner Widder (German Army),

"Auftragstaktik and Innere Führung: Trademarks of German Leadership," *Military Review* (September-October 2002): 4.

2-41 "Nine-tenths of tactics are …": T. E. Lawrence, "The Evolution of a Revolt," *The Army Quarterly, Volume 1* (London: William Clowes & Sons, Ltd., October 1920 and January 1921), 60.

2-42 "In war nothing is…": Dwight D. Eisenhower, "Dwight D. Eisenhower, 1953-61." Available at https://www.cia.gov/library/center-for-the-study-of-intelligence/csi-publications/books-and-monographs/our-first-line-of-defense-presidential-reflections-on-us-intelligence/eisenhower.html.

2-45 "Hard pounding this, gentlemen…": Duke of Wellington, "Anecdotes of the Duke of Wellington at the Battle of Waterloo," *The Scots Magazine and Edinburgh Literary Miscellany: Being a General Repository of Literature, History, and Politics, for 1816*, Volume 78 (Edinburgh: Archibald Constable and Company, 1816), 9-10.

2-47 "Even the bravest cannot…": Homer in *The Iliad*, xiii. c. 1000 B. C., *Dictionary of Military and Naval Quotations*, compiled by Robert Debs Heinl, Jr. (Annapolis, MD: United States Naval Institute, 1988), 102.

2-48 "Mobility is the true test…": B. H. Liddell Hart, *Quotes for Air Force Logistician*, Volume I (Maxwell Air Force Base, AL: Air Force Logistics Management Agency, 2006), 41.

2-49 "Petty geniuses attempt to…": Frederick the Great, "The Instruction of Frederick the Great for His Generals," vii, 1747, *Roots of Strategy: The 5 Greatest Military Classics of All Time*, edited by Thomas R. Phillips (Harrisburg, PA: Stackpole Books, 1985), 345.

2-52 "Thus a victorious army…": Sun Tzu, *The Art of War*, translated by Samuel B. Griffith (New York: Oxford University Press, 1963), 87.

2-53 **Readiness: Task Force Smith**: Department of Military History staff, unpublished text, 2017.

2-54 "To lead untrained people…": Confucius, quoted by E. C. Meyer, "Address to the Pre-Command Course 25 July 1980," Fort Leavenworth, KS, 110.

2-55 "The instruments of battle…": Charles Ardant du Picq, *Battle Studies: Ancient and Modern Battles*, translated by John N. Greely and Robert C. Cotton (New York: MacMillian Company, 1921), 68.

2-56 **Fight as You Train: Kasserine Pass**: Army University Press staff, unpublished text, 2017.

Chapter 3

3-1 "The history of failure…": General Douglas MacArthur quoted in William Manchester, *American Caesar: Douglas MacArthur 1880–1964* (Boston: Little, Brown and Company, 1978), 182-183.

3-3 **Security Force Assistance and Long-Term Consolidation of Gains: Colombia**: White Paper: "The Arc of Change: Operational Time to Influence Trajectories," United States Army Special Operations Command G-9 staff, Fort Bragg, NC.

3-3 "All planning, particularly strategic…": Carl von Clausewitz, *On War*, edited and translated by Michael Howard and Peter Paret (Princeton, NJ: Princeton University Press, 1989), 220.

3-8 **Humanitarian Assistance: Republic of Cameroon**: vignette in FM 100-5 [obsolete], *Operations* (Washington DC: Government Printing Office, 1993), 13-6.

3-8 "When near he makes…": Sun Tzu, *The Art of War*, translated by Samuel B. Griffith (New York: Oxford University Press, 1963), 41.

Chapter 4

4-1 "The ultimate determinant in…": Rear Admiral J. C. Wylie, *Military Strategy: A General Theory of Power Control* (Annapolis, MD: Naval Institute Press, 2014), 72.

4-1 "The price of greatness…": Winston Churchill: Speech, "The Gift of a Common Tongue, September 6, 1943, Harvard." The International Churchill Society. Available at https://www.winstonchurchill.org/resources/speeches/1941-1945-war-leader/the-price-of-greatness-is-responsibility.

Source Notes

4-2 **Prevent: OPERATION DESERT SHIELD**: Army University Press staff, unpublished text, 2017.

4-3 "[Y]ou don't have to...": Ronald Reagan, "Remarks at a Campaign Rally for Senator Robert W. Kasten, Jr., in Waukesha, Wisconsin, October 23, 1986" *The Public Papers of President Ronald W. Reagan*. Ronald Reagan Presidential Library. https://reaganlibrary.archives.gov/archives/speeches/1987/061287d.htm.

4-3 **Flexibility: 8th U.S. Army in Korea**: Army University Press staff, unpublished text, 2017.

4-14 "[I]t is an unfortunate...": John F. Kennedy: "Speech of Senator John F. Kennedy, Civic Auditorium, Seattle, WA," September 6, 1960. Online by Gerhard Peters and John T. Woolley, *The American Presidency Project*. http://www.presidency.ucsb.edu/ws/?pid=25654.

4-18 "It is discipline that makes...": Xenophon in "Speech to the Greek officers after the defeat of Cyrus at Cunaxa," 401 B.C., *Dictionary of Military and Naval Quotations*, compiled by Robert Debs Heinl, Jr. (Annapolis, MD: United States Naval Institute, 1988), 91.

4-21 **Deterrence: EXERCISE REFORGER**: Army University Press staff, unpublished text, 2017.

4-21 "He is best secure...": Publilius Syrus: *Dictionary of Military and Naval Quotations*, compiled by Robert Debs Heinl, Jr. (Annapolis, MD: United States Naval Institute, 1988), 291.

4-23 "Secret operations are essential...": Sun Tzu, *The Art of War*, translated by Samuel B. Griffith (New York: Oxford University Press, 1963), 149.

Chapter 5

5-1 "Therefore, the purpose of...": General Donn Starry in "Operational Concept for the Airland Battle," TRADOC Pam 525-5 [obsolete], *The Airland Battle and Corps 86* (Washington DC: Government Printing Office, 1981), 2.

5-1 "Jointness implies cross-Service...": JP 1, *Doctrine for the Armed Forces of the United States* (Washington DC: Government Printing Office, 25 March 2013), ix.

5-5 "A general in all of his...": Frederick the Great in *Instructions for His Generals*, iii. 1747, *Dictionary of Military and Naval Quotations*, compiled by Robert Debs Heinl, Jr. (Annapolis, MD: United States Naval Institute, 1988), 102.

5-6 "Limited resources and capabilities...": FM 31-23 [obsolete], *Stability Operations* (Washington DC: Government Printing Office, 1972), 4-2.

5-7 "In the past, combat operations...": FM 90-14 [obsolete], *Rear Battle* (Washington DC: Government Printing Office, 1985), 1-1.

5-9 "The purpose of reconnaissance...": FM 17-95 [obsolete], *Cavalry Operations* (Washington DC: Government Printing Office, 1981), 5-1.

5-12 **Ambiguous Environment: Fighting for Information**: Army University Press staff, unpublished text, 2017.

5-12 "Security includes all measures...": FM 7-19 [obsolete], *Combat Support Company: Infantry Division Battle Group Operations* (Washington DC: Government Printing Office, 1960), 92.

5-16 "The passage of lines is...": FM 71-100-3 [obsolete], *Air Assault Division Operations* (Washington DC: Government Printing Office, 1996), 6-6.

5-16 "East and north of Bastogne,...": Charles B. MacDonald, "Victory in the Ardennes," *The Last Offensive* (Washington DC: Center for Military History, 1993), 33.

5-20 "Airborne forces execute parachute...": FM 57-1 [obsolete], *U. S. Army/U. S. Air Force Doctrine for Airborne Operations* (Washington DC: Government Printing Office, 1967), 6.

5-21 **Forcible Entry: OPERATION JUST CAUSE**: Army University Press staff, unpublished text, 2017.

5-22 "In the case of ...": B. H. Liddell Hart, *Strategy* (New York: Signet Printing, 1974), 324–325.

Chapter 6

6-1 "In war, the defensive exists...": Alfred Thayer Mahan, *Naval Strategy: Compared and Contrasted With the Principles and Practice of Military Operations on Land; lectures delivered at U.S. Naval War College, Newport, R.I., between the years 1887-1911* (Boston: Little, Brown and Company, 1915), 150.

6-2 "No plan survives contact...": Helmuth von Moltke the Elder, 1800-1891, *Dictionary of Military and Naval Quotations*, compiled by Robert Debs Heinl, Jr. (Annapolis, MD: United States Naval Institute, 1988), 239.

6-2 **Malaya and Singapore: Faulty Assumptions—Defeat Across Multiple Domains**: Army University Press staff, unpublished text, 2017.

6-8 "In preparing for battle...": Dwight David Eisenhower, quoted in Richard Nixon, *Six Crises* (New York: Pyramid Book, 1968), 253.

6-15 "A sudden powerful transition...": Carl von Clausewitz, *On War*, edited and translated by Michael Howard and Peter Paret (Princeton, NJ: Princeton University Press, 1989), 370.

6-18 "The commander adopts a plan...": FM 7-100 [obsolete], *Infantry Division* (Washington DC: Government Printing Office, 1960), 158.

6-20 "Troops carry out the organization...": FM 100-5 [obsolete], *Field Service Regulations: Operations* (Washington DC: Government Printing Office, 1944), 173.

6-25 "The organization of supply...": *Field Service Regulations: United States Army* (Washington DC: Government Printing Office, 1923), 121.

6-26 "[T]he best protection against..." John Randolph Spears, *David G. Farragut* (Philadelphia, PA: George W. Jacobs & Company, 1905), 283.

6-32 "The primary mission of the reserve...": FM 100-5 [obsolete], *Field Service Regulations: Operations* (Washington DC: Government Printing Office, 1949), 87.

6-39 "Mobile defense is that method...": FM 100-5 [obsolete], *Field Service Regulations: Operations* (Washington DC: Government Printing Office, 1954), 16.

6-46 "A retreating force covers its...": *Field Service Regulations: United States Army* (Washington DC: Government Printing Office, 1924), 46.

Chapter 7

7-1 "On no account should we...": Carl von Clausewitz, *Principles of War*, 1812, *Dictionary of Military and Naval Quotations*, compiled by Robert Debs Heinl, Jr. (Annapolis, MD: United States Naval Institute, 1988), 19.

7-2 **Intermittent Communications, Continuous Mission Command: Grant and Sherman Deep Operational Maneuver 1864-1865**: Army University Press staff, unpublished text, 2017.

7-5 "In case of doubt,...": General George Patton, http://generalpatton.com/quotes/.

7-7 "Plans for the deep battle...": FM 100-5 [obsolete], *Operations* (Washington DC: Government Printing Office, 1982), 7-2.

7-9 "Agitate him and ascertain...": Sun Tzu, *The Art of War*, translated by Samuel B. Griffith (New York: Oxford University Press, 1963), 100.

7-11 "Orders should not attempt...": *Field Service Regulations: United States Army* (Washington DC: Government Printing Office, 1905), 30.

7-14 "To be at the head...": General William T. Sherman, *Memoirs of General William T. Sherman by Himself*, Bloomington, IN: Indiana University Press, 1957), 407.

7-16 **Movement and Maneuver: The 1973 Arab-Israeli War**: Combined Arms Doctrine Directorate staff, unpublished text, 2017.

7-18 "The bulk of the artillery...": FM 6-20 [obsolete], *Field Artillery Field Manual: Tactics and Techniques* (Washington DC: Government Printing Office, 1940), 128.

7-22 "The deep envelopment based...": General of the Army Douglas MacArthur: To planning conference before the Inchon landings, 23 August 1950, in *Joint Military Operations Historical Collection*, (Washington DC: Government Printing Office, 1997), II-4.

7-23 "Everyone knows the moral effects...": Carl von Clausewitz, *On War*, edited and translated by Michael Howard and Peter Paret (Princeton, NJ: Princeton University Press, 1989), 137.

Source Notes

7-24 "Are there not other…": Sir Winston Churchill, Letter to Prime Minister Asquith, December 29, 1914. Available at www.nationalchurchillmuseum.org/churchill-in-world-war-i-and-aftermath.html.

7-32 "During the advance to contact…": FM 100-5 [obsolete], *Field Service Regulations: Operations* (Washington DC: Government Printing Office, 1949), 96.

7-40 "The enemy say…": Winfield Scott: To the 11th Infantry at Chippewa, Canada, 5 June 1814, in *Dictionary of Military and Naval Quotations*, compiled by Robert Debs Heinl, Jr. (Annapolis, MD: United States Naval Institute, 1988), 49.

7-49 "In pursuit you must…": Field-Marshal Viscount Wavell of Cyrenacia and Winchester, *Allenby: Soldier and Statesman* (Toronto: George G. Harrap & Co. Ltd., 1944), 183.

7-52 "Ambushes are prepared by…": FM 7-5 [obsolete], *Infantry Field Manual: Organization and Tactics of Infantry: The Rifle Battalion* (Washington DC: Government Printing Office, 1940), 88.

7-53 "A demonstration is an operation…": FM 60-5 [obsolete], *Amphibious Operations: Battalion in Assault Landings* (Washington DC: Government Printing Office, 1951), 4.

Chapter 8

8-1 "It is no doubt a good…": Polybius, in Joseph Callo, *John Paul Jones: America's First Sea Warrior* (Annapolis, MD: Naval Institute Press, 2006), 114.

8-1 "Loss of hope, rather…": B.H. Liddell Hart, in *Military Air Power: The CADRE Digest of Air Power Opinions and Thoughts*, compiled by Charles M. Westenhoff (Maxwell Air Force Base, AL: Air University Press, 1990), 36.

8-4 "The merit of an action…": Genghis Khan: Advice to his sons, 1206, *A Dictionary of Military Quotations*, compiled by Trevor Royle (London: Routledge, 1990), 24.

8-7 "War is more than…": Carl von Clausewitz, *On War*, edited and translated by Michael Howard and Peter Paret (Princeton, NJ: Princeton University Press, 1989), 89.

8-8 **Operations to Consolidate Gains: Luzon**: Army University Press staff, unpublished text, 2017.

8-13 "A nation cannot…": B.H. Liddell Hart, *Dictionary of Military and Naval Quotations*, compiled by Robert Debs Heinl, Jr. (Annapolis, MD: United States Naval Institute, 1988), 191.

Appendix A

This appendix has no source notes.

Appendix B

This appendix has no source notes.

Glossary

The glossary lists acronyms and terms with Army or joint definitions. Where Army and joint definitions differ, (Army) precedes the definition. Terms for which FM 3-0 is the proponent are marked with an asterisk (*). The proponent publication for other terms is listed in parentheses after the definition.

SECTION I – ACRONYMS AND ABBREVIATIONS

A2	antiaccess
AADC	area air defense commander
AAMDC	Army air and missile defense command
ABCT	armored brigade combat team
AD	area denial
ADA	air defense artillery
ADCON	administrative control
ADP	Army doctrine publication
ADRP	Army doctrine reference publication
AMD	air and missile defense
AO	area of operations
AOR	area of responsibility
AR	Army regulation
ARCYBER	United States Army Cyber Command
ARFOR	See Terms.
ARSOF	Army special operations forces
ASCC	Army Service component command
ASOS	Army support to other Services
ATP	Army techniques publication
BCD	battlefield coordination detachment
BCT	brigade combat team
BHL	battle handover line
C2	command and control
CA	civil affairs
CAB	combat aviation brigade
CACOM	civil affairs command
CAS	close air support
CBRN	chemical, biological, radiological, and nuclear
CEMA	cyberspace electromagnetic activities
COCOM	combatant command (command authority)
CONUS	continental United States

COP	common operational picture
CP	command post
CSSB	combat sustainment support battalion
CTC	combat training center
CWMD	countering weapons of mass destruction
D3SOE	denied, degraded, and disrupted space operational environment
DA	Department of the Army
DCO	defensive cyberspace operations
DIVARTY	division artillery
DOD	Department of Defense
DODD	Department of Defense directive
DODIN	Department of Defense information network
DODIN-A	Department of Defense information network-Army
EA	engagement area
E-MIB	expeditionary-military intelligence brigade
EMS	electromagnetic spectrum
EOD	explosive ordnance disposal
ESC	expeditionary sustainment command
EW	electronic warfare
FARC	Revolutionary Armed Forces of Colombia
FDO	flexible deterrent option
FEBA	forward edge of the battle area
FID	foreign internal defense
FLOT	forward line of own troops
FM	field manual
FORSCOM	United States Army Forces Command
FPF	final protective fire
FRO	flexible response option
FSCL	fire support coordination line
FSCM	fire support coordination measure
FSF	foreign security forces
G-2	assistant chief of staff, intelligence
G-3	assistant chief of staff, operations
G-4	assistant chief of staff, logistics
G-6	assistant chief of staff, signal
GCC	geographic combatant commander
HNS	host-nation support
HPT	high-payoff target
IADS	integrated air defense system
IBCT	infantry brigade combat team
IFC	integrated fires command

Glossary

IPB	intelligence preparation of the battlefield
IRC	information-related capability
J-4	logistics directorate of a joint staff
JAGIC	joint air-ground integration center
JAOC	joint air operations center
JFACC	joint force air component commander
JFC	joint force commander
JOA	joint operations area
JP	joint publication
JSA	joint security area
JSOA	joint special operations area
JTF	joint task force
LCC	land component commander
LOA	limit of advance
LTC	lieutenant colonel
LTG	lieutenant general
MBA	main battle area
MEB	maneuver enhancement brigade
MEDCOM (DS)	medical command (deployment support)
METL	mission-essential task list
MG	major general
MIB-T	military intelligence brigade-theater
MILDEC	military deception
MISO	military information support operations
MP	military police
NAI	named area of interest
NATO	North Atlantic Treaty Organization
OCO	offensive cyberspace operations
OCS	operational contract support
OE	operational environment
OPCON	operational control
OPLAN	operation plan
OPORD	operation order
OPSEC	operations security
PACE	primary, alternate, contingency, and emergency
PL	phase line
POL	petroleum, oils, and lubricants
RSOI	reception, staging, onward movement, and integration
S-2	battalion or brigade intelligence staff officer
S-3	battalion or brigade operations staff officer
S-4	battalion or brigade logistics staff officer

Glossary

S-6	battalion or brigade signal staff officer
SACP	support area command post
SBCT	Stryker brigade combat team
SFA	security force assistance
SIGINT	signals intelligence
SOF	special operations forces
SSE	space support element
SSR	security sector reform
TAC-D	tactical deception
TACON	tactical control
TAI	target area of interest
TC	training circular
TCF	tactical combat force
TCP	theater campaign plan
TEC	theater engineer command
TM	technical manual
TRP	target reference point
TSC	theater sustainment command
TTSB	theater tactical signal brigade
UAS	unmanned aircraft system
U.S.	United States
USAF	United States Air Force
USC	United States Code
USCYBERCOM	United States Cyber Command
USTRANSCOM	United States Transportation Command
WMD	weapons of mass destruction

SECTION II – TERMS

administrative control

   Direction or exercise of authority over subordinate or other organizations in respect to administration and support. (JP 1)

adversary

   A party acknowledged as potentially hostile to a friendly party and against which the use of force may be envisaged. (JP 3-0)

air assault

   The movement of friendly assault forces by rotary-wing or tiltrotor aircraft to engage and destroy enemy forces or to seize and hold key terrain. (JP 3-18)

airborne assault

   The use of airborne forces to parachute into an objective area to attack and eliminate armed resistance and secure designated objectives. (JP 3-18)

airspace control

   Capabilities and procedures used to increase operational effectiveness by promoting the safe, efficient, and flexible use of airspace. (JP 3-52)

air tasking order
   A method used to task and disseminate to components, subordinate units, and command and control agencies projected sorties, capabilities and/or forces to targets and specific missions. (JP 3-30)

allocation
   The distribution of limited forces and resources for employment among competing requirements. (JP 5-0)

ambush
   An attack by fire or other destructive means from concealed positions on a moving or temporarily halted enemy. (FM 3-90-1)

area defense
   A defensive task that concentrates on denying enemy forces access to designated terrain for a specific time rather than destroying the enemy outright. (ADRP 3-90)

area of influence
   A geographical area wherein a commander is directly capable of influencing operations by maneuver or fire support systems normally under the commander's command or control. (JP 3-0)

area of interest
   That area of concern to the commander, including the area of influence, areas adjacent thereto, and extending into enemy territory. (JP 3-0)

area of operations
   An operational area defined by a commander for land and maritime forces that should be large enough to accomplish their missions and protect their forces. (JP 3-0)

area reconnaissance
   A form of reconnaissance that focuses on obtaining detailed information about the terrain or enemy activity within a prescribed area. (ADRP 3-90)

area security
   A security task conducted to protect friendly forces, installations, routes, and actions within a specific area. (ADRP 3-90)

ARFOR
   The Army component and senior Army headquarters of all Army forces assigned or attached to a combatant commander, subordinate joint force command, joint functional command, or multinational command. (FM 3-94)

Army personnel recovery
   The military efforts taken to prepare for and execute the recovery and reintegration of isolated personnel. (FM 3-50)

Army Service component command
   The command responsible for recommendations to the joint force commander on the allocation and employment of Army forces within a combatant command. (JP 3-31)

assault position
   A covered and concealed position short of the objective from which final preparations are made to assault the objective. (ADRP 3-90)

assign
   To place units or personnel in an organization where such placement is relatively permanent, and/or where such organization controls and administers the units or personnel for the primary function, or greater portion of the functions, of the unit or personnel. (JP 3-0)

attach
   The placement of units or personnel in an organization where such placement is relatively temporary. (JP 3-0)

Glossary

attack
> An offensive task that destroys or defeats enemy forces, seizes and secures terrain, or both. (ADRP 3-90)

attack by fire position
> The general position from which a unit conducts the tactical task of attack by fire. (ADRP 3-90)

attack position
> (Army) The last position an attacking force occupies or passes through before crossing the line of departure. (ADRP 3-90)

battle handover line
> A designated phase line on the ground where responsibility transitions from the stationary force to the moving force and vice versa. (ADRP 3-90)

battle position
> A defensive location oriented on a likely enemy avenue of approach. (ADRP 3-90)

boundary
> A line that delineates surface areas for the purpose of facilitating coordination and deconfliction of operations between adjacent units, formations, or areas. (JP 3-0)

breakout
> An operation conducted by an encircled force to regain freedom of movement or contact with friendly units. It differs from other attack only in that simultaneous defense in other areas of the perimeter must be maintained. (ADRP 3-90)

breach
> A synchronized combined arms activity under the control of the maneuver commander conducted to allow maneuver through an obstacle. (ATP 3-90.4)

checkpoint
> A predetermined point on the ground used to control movement, tactical maneuver, and orientation. (ADRP 1-02)

clearance of fires
> The process by which the supported commander ensures that fires or their effects will have no unintended consequences on friendly units or the scheme of maneuver. (FM 3-09)

clearing
> A mobility task that involves the elimination or neutralization of an obstacle that is usually performed by follow-on engineers and is not done under fire. (ATP 3-90.4)

close area
> The portion of a commander's area of operations assigned to subordinate maneuver forces. (ADRP 3-0)

close support
> The action of the supporting force against targets or objectives that are sufficiently near the supported force as to require detailed integration or coordination of the supporting action. (JP 3-31)

combat power
> (Army) The total means of destructive, constructive, and information capabilities that a military unit or formation can apply at a given time. (ADRP 3-0)

combatant command
> A unified or specified command with a broad continuing mission under a single commander established and so designated by the President, through the Secretary of Defense and with the advice and assistance of the Chairman of the Joint Chiefs of Staff. (JP 1)

Glossary

commander's visualization

The mental process of developing situational understanding, determining a desired end state, and envisioning an operational approach by which the force will achieve that end state. (ADP 5-0)

command group

The commander and selected staff members who assist the commander in controlling operations away from a command post. (FM 6-0)

command post

A unit headquarters where the commander and staff perform their activities. (FM 6-0)

common-user logistics

Materiel or service support shared with or provided by two or more Services, Department of Defense agencies, or multinational partners to another Service, Department of Defense agency, non-Department of Defense agency, and/or multinational partner in an operation. (JP 4-09)

consolidate gains

Activities to make enduring any temporary operational success and set the conditions for a stable environment allowing for a transition of control to legitimate authorities. (ADRP 3-0)

consolidation

The organizing and strengthening a newly captured position so that it can be used against the enemy. (FM 3-90-1)

consolidation area

The portion of the commander's area of operations that is designated to facilitate the security and stability tasks necessary for freedom of action in the close area and to support the continuous consolidation of gains. (ADRP 3-0)

contact point

In land warfare, a point on the terrain, easily identifiable, where two or more units are required to make contact. (JP 3-50)

control measure

A means of regulating forces or warfighting functions. (ADRP 6-0)

coordinated fire line

A line beyond which conventional surface-to-surface direct fire and indirect fire support means may fire at any time within the boundaries of the establishing headquarters without additional coordination. (JP 3-09)

coordinating authority

A commander or individual who has the authority to require consultation between the specific functions or activities involving forces of two or more Services, joint force components, or forces of the same Service or agencies, but does not have the authority to compel agreement. (JP 1)

cordon and search

A technique of conducting a movement to contact that involves isolating a target area and searching suspected locations within that target area to capture or destroy possible enemy forces and contraband. (FM 3-90-1)

counterattack

Attack by part or all of a defending force against an enemy attacking force, for such specific purposes as regaining ground lost, or cutting off or destroying enemy advance units, and with the general objective of denying to the enemy the attainment of the enemy's purpose in attacking. In sustained defensive operations, it is undertaken to restore the battle position and is directed at limited objectives. (FM 3-90-1)

countermobility operations

(Army/Marine Corps) Those combined arms activities that use or enhance the effects of natural and man-made obstacles to deny enemy freedom of movement and maneuver. (ATP 3-90.8)

Glossary

cover
> (Army) A security task to protect the main body by fighting to gain time while also observing and reporting information and preventing enemy ground observation of and direct fire against the main body. (ADRP 3-90)

cyberspace
> A global domain within the information environment consisting of the interdependent networks of information technology infrastructures and resident data, including the Internet, telecommunications networks, computer systems, and embedded processors and controllers. (JP 3-12[R])

cyberspace electromagnetic activities
> The process of planning, integrating, and synchronizing cyberspace and electronic warfare operations in support of unified land operations. (ADRP 3-0)

danger close
> In close air support, artillery, mortar, and naval gunfire support fires, the term included in the method of engagement segment of a call for fire that indicates that friendly forces are within close proximity of the target. (JP 3-09.3)

decisive action
> The continuous, simultaneous combinations of offensive, defensive, and stability or defense support of civil authorities tasks. (ADRP 3-0)

decisive operation
> The operation that directly accomplishes the mission. (ADRP 3-0)

deep area
> The portion of the commander's area of operations that is not assigned to subordinate units. (ADRP 3-0)

defeat mechanism
> A method through which friendly forces accomplish their mission against enemy opposition. (ADRP 3-0)

defensive task
> A task conducted to defeat an enemy attack, gain time, economize forces, and develop conditions favorable for offensive or stability tasks. (ADRP 3-0)

delay line
> A phase line where the date and time before which the enemy is not allowed to cross the phase line is depicted as part of the graphic control measure. (FM 3-90-1)

delaying operation
> An operation in which a force under pressure trades space for time by slowing down the enemy's momentum and inflicting maximum damage on the enemy without, in principle, becoming decisively engaged. (JP 3-04)

demonstration
> In military deception, a show of force similar to a feint without actual contact with the adversary, in an area where a decision is not sought that is made to deceive an adversary. (JP 3-13.4)

Department of Defense information network
> The set of information capabilities, and associated processes for collecting, processing, storing, disseminating, and managing information on-demand to warfighters, policy makers, and support personnel, whether interconnected or stand-alone, including owned and leased communications and computing systems and services, software (including applications), data, security services, other associated services, and national security systems. (JP 6-0)

Glossary

destroy
> A tactical mission task that physically renders an enemy force combat-ineffective until it is reconstituted. Alternatively, to destroy a combat system is to damage it so badly that it cannot perform any function or be restored to a usable condition without being entirely rebuilt. (FM 3-90-1)

direct liaison authorized
> That authority granted by a commander (any level) to a subordinate to directly consult or coordinate an action with a command or agency within or outside of the granting command. (JP 1)

*direct support
> (Army) A support relationship requiring a force to support another specific force and authorizing it to answer directly to the supported force's request for assistance.

disengagement line
> A phase line located on identifiable terrain that, when crossed by the enemy, signals to defending elements that it is time to displace to their next position. (ADRP 3-90)

disintegrate
> To disrupt the enemy's command and control system, degrading its ability to conduct operations while leading to a rapid collapse of the enemy's capabilities or will to fight. (ADRP 3-0)

dislocate
> To employ forces to obtain significant positional advantage, rendering the enemy's dispositions less valuable, perhaps even irrelevant. (ADRP 3-0)

dismounted march
> Movement of troops and equipment mainly by foot, with limited support by vehicles. (FM 3-90-2)

double envelopment
> This results from simultaneous maneuvering around both flanks of a designated enemy force. (FM 3-90-1)

early-entry command post
> A lead element of a headquarters designed to control operations until the remaining portions of the headquarters are deployed and operational. (FM 6-0)

electromagnetic spectrum
> The range of frequencies of electromagnetic radiation from zero to infinity. It is divided into 26 alphabetically designated bands. (JP 3-13.1)

encirclement operations
> Operations where one force loses its freedom of maneuver because an opposing force is able to isolate it by controlling all ground lines of communications and reinforcement. (ADRP 3-90)

enemy
> A party identified as hostile against which the use of force is authorized. (ADRP 3-0)

engagement area
> An area where the commander intends to contain and destroy an enemy force with the massed effects of all available weapons and supporting systems. (FM 3-90-1)

envelopment
> A form of maneuver in which an attacking force seeks to avoid the principal enemy defenses by seizing objectives behind those defenses that allow the targeted enemy force to be destroyed in their current positions. (FM 3-90-1)

feint
> In military deception, an offensive action involving contact with the adversary conducted for the purpose of deceiving the adversary as to the location and/or time of the actual main offensive action. (JP 3-13.4)

Glossary

final protective fire

An immediately available prearranged barrier of fire designed to impede enemy movement across defensive lines or areas. (JP 3-09.3)

fire superiority

That degree of dominance in the fires of one force over another that permits that force to conduct maneuver at a given time and place without prohibitive interference by the enemy. (FM 3-90-1)

fire support coordination line

A fire support coordination measure established by the land or amphibious force commander to support common objectives within an area of operation; beyond which all fires must be coordinated with affected commanders prior to engagement, and short of the line, all fires must be coordinated with the establishing commander prior to engagement. (JP 3-09)

fires warfighting function

The related tasks and systems that provide collective and coordinated use of Army indirect fires, air and missile defense, and joint fires through targeting processes. (ADRP 3-0)

flank attack

A form of offensive maneuver directed at the flank of an enemy. (FM 3-90-1)

flexible deterrent option

A planning construct intended to facilitate early decision making by developing a wide range of interrelated responses that begin with deterrent-oriented actions carefully tailored to create a desired effect. (JP 5-0)

flexible response

The capability of military forces for effective reaction to any enemy threat or attack with actions appropriate and adaptable to the circumstances existing. (JP 5-0)

force projection

The ability to project the military instrument of national power from the United States or another theater, in response to requirements for military operations. (JP 3-0)

force tailoring

The process of determining the right mix of forces and the sequence of their deployment in support of a joint force commander. (ADRP 3-0)

forcible entry

Seizing and holding of a military lodgment in the face of armed opposition or forcing access into a denied area to allow movement and maneuver to accomplish the mission. (JP 3-18)

foreign internal defense

Participation by civilian and military agencies of a government in any of the action programs taken by another government or other designated organization to free and protect its society from subversion, lawlessness, insurgency, terrorism, and other threats to its security. (JP 3-22)

forms of maneuver

Distinct tactical combinations of fire and movement with a unique set of doctrinal characteristics that differ primarily in the relationship between the maneuvering force and the enemy. (ADRP 3-90)

forward edge of the battle area

The foremost limit of a series of areas in which ground combat units are deployed, excluding the areas in which the covering or screening forces are operating, designated to coordinate fire support, the positioning of forces, or the maneuver of units. (JP 3-09.3)

forward passage of lines

Occurs when a unit passes through another unit's positions while moving toward the enemy. (ADRP 3-90)

Glossary

free-fire area

A specific area into which any weapon system may fire without additional coordination with the establishing headquarters. (JP 3-09)

frontal attack

A form of maneuver in which an attacking force seeks to destroy a weaker enemy force or fix a larger enemy force in place over a broad front. (FM 3-90-1)

general support

That support which is given to the supported force as a whole and not to any particular subdivision thereof. (JP 3-09.3)

general support-reinforcing

(Army) A support relationship assigned to a unit to support the force as a whole and to reinforce another similar-type unit. (ADRP 5-0)

graphic control measure

A symbol used on maps and displays to regulate forces and warfighting functions. (ADRP 6-0)

guard

A security task to protect the main force by fighting to gain time while also observing and reporting information and preventing enemy ground observation of and direct fire against the main body. Units conducting a guard mission cannot operate independently because they rely upon fires and functional and multifunctional support assets of the main body. (ADRP 3-90)

host-nation support

Civil and/or military assistance rendered by a nation to foreign forces within its territory during peacetime, crises or emergencies, or war based on agreements mutually concluded between nations. (JP 4-0)

infiltration

A form of maneuver in which an attacking force conducts undetected movement through or into an area occupied by enemy forces to occupy a position of advantage behind those enemy positions while exposing only small elements to enemy defensive fires. (FM 3-90-1)

information operations

The integrated employment, during military operations, of information-related capabilities in concert with other lines of operation to influence, disrupt, corrupt, or usurp the decision-making of adversaries and potential adversaries while protecting our own. (JP 3-13)

intelligence warfighting function

The related tasks and systems that facilitate understanding the enemy, terrain, weather, civil considerations, and other significant aspects of the operational environment. (ADRP 3-0)

interdiction

An action to divert, disrupt, delay, or destroy the enemy's military surface capability before it can be used effectively against friendly forces, or to achieve enemy objectives. (JP 3-03)

in-transit visibility

The ability to track the identity, status, and location of DOD units, and non-unit cargo (excluding bulk petroleum, oils, and lubricants) and passengers; patients; and personal property from origin to consignee or destination across the range of military operations. (AR 700-80)

isolate

A tactical mission task that requires a unit to seal off—both physically and psychologically—an enemy from sources of support, deny the enemy freedom of movement, and prevent the isolated enemy force from having contact with other enemy forces. (FM 3-90-1)

joint operations

Military actions conducted by joint forces and those Service forces employed in specified command relationships with each other, which of themselves, do not establish joint forces. (JP 3-0)

Glossary

kill box
A three-dimensional permissive fire support coordination measure with an associated airspace coordinating measure used to facilitate the integration of fires. (JP 3-09)

landpower
The ability—by threat, force, or occupation—to gain, sustain, and exploit control over land, resources, and people. (ADRP 3-0)

leadership
The process of influencing people by providing purpose, direction, and motivation to accomplish the mission and improve the organization. (ADP 6-22)

Level I threat
A small enemy force that can be defeated by those units normally operating in the echelon support area or by the perimeter defenses established by friendly bases and base clusters. (ATP 3-91)

Level II threat
An enemy force or activities that can be defeated by a base or base cluster's defensive capabilities when augmented by a response force. (ATP 3-91)

Level III threat
An enemy force or activities beyond the defensive capability of both the base and base cluster and any local reserve or response force. (ATP 3-91)

line of contact
A general trace delineating the location where friendly and enemy forces are engaged. (FM 3-90-1)

line of departure
(Army) A phase line crossed at a prescribed time by troops initiating an offensive task. (ADRP 3-90)

linkup
A meeting of friendly ground forces, which occurs in a variety of circumstances. (ADRP 3-90)

local security
A security task that includes low level security activities conducted near a unit to prevent surprise by the enemy. (ADRP 3-90)

lodgment
A designated area in a hostile or potentially hostile operational area that, when seized and held, makes the continuous landing of troops and materiel possible and provides maneuver space for subsequent operations. (JP 3-18)

main battle area
The area where the commander intends to deploy the bulk of the unit's combat power and conduct decisive operations to defeat an attacking enemy. (ADRP 3-90)

main command post
A facility containing the majority of the staff designed to control current operations, conduct detailed analysis, and plan future operations. (FM 6-0)

main effort
A designated subordinate unit whose mission at a given point in time is most critical to overall mission success. (ADRP 3-0)

maneuver
Employment of forces in the operational area through movement in combination with fires to achieve a position of advantage in respect to the enemy. (JP 3-0)

meeting engagement
A combat action that occurs when a moving force, incompletely deployed for battle, engages an enemy at an unexpected time and place. (FM 3-90-1)

military deception
Actions executed to deliberately mislead adversary military, paramilitary, or violent extremist organization decision makers, thereby causing the adversary to take specific actions (or inactions) that will contribute to the accomplishment of the friendly mission. (JP 3-13.4)

military engagement
The routine contact and interaction between individuals or elements of the Armed Forces of the United States and those of another nation's armed forces, or foreign and domestic civilian authorities or agencies to build trust and confidence, share information, coordinate mutual activities, and maintain influence. (JP 3-0)

mission command system
The arrangement of personnel, networks, information systems, processes and procedures, and facilities and equipment that enable commanders to conduct operations. (ADP 6-0)

mission command warfighting function
The related tasks and systems that develop and integrate those activities enabling a commander to balance the art of command and the science of control in order to integrate the other warfighting functions. (ADRP 3-0)

mobile defense
A defensive task that concentrates on the destruction or defeat of the enemy through a decisive attack by a striking force. (ADRP 3-90)

movement and maneuver warfighting function
The related tasks and systems that move and employ forces to achieve a position of relative advantage over the enemy and other threats. (ADRP 3-0)

movement to contact
(Army) An offensive task designed to develop the situation and establish or regain contact. (ADRP 3-90)

multiechelon training
A training technique that allows for the simultaneous training of more than one echelon on different or complementary tasks. (ADRP 7-0)

mutual support
That support which units render each other against an enemy, because of their assigned tasks, their position relative to each other and to the enemy, and their inherent capabilities. (JP 3-31)

named area of interest
The geospatial area or systems node or link against which information that will satisfy a specific information requirement can be collected, usually to capture indications of adversary courses of action. (JP 2-01.3)

offensive task
A task conducted to defeat and destroy enemy forces and seize terrain, resources, and population centers. (ADRP 3-0)

operation
A sequence of tactical actions with a common purpose or unifying theme. (JP 1)

operational approach
A broad description of the mission, operational concepts, tasks, and actions required to accomplish the mission. (JP 5-0)

operational art
The cognitive approach by commanders and staffs—supported by their skill, knowledge, experience, creativity, and judgment—to develop strategies, campaigns, and operations to organize and employ military forces by integrating ends, ways, and means. (JP 3-0)

Glossary

operational control

The authority to perform those functions of command over subordinate forces involving organizing and employing commands and forces, assigning tasks, designating objectives, and giving authoritative direction necessary to accomplish the mission. (JP 1)

operational environment

A composite of the conditions, circumstances, and influences that affect the employment of capabilities and bear on the decisions of the commander. (JP 3-0)

operational framework

A cognitive tool used to assist commanders and staffs in clearly visualizing and describing the application of combat power in time, space, purpose, and resources in the concept of operations. (ADP 1-01)

operational reach

The distance and duration across which a force can successfully employ military capabilities. (JP 3-0)

operations process

The major mission command activities performed during operations: planning, preparing, executing, and continuously assessing the operation. (ADP 5-0)

organic

Assigned to and forming an essential part of a military organization as listed in its table of organization for the Army, Air Force, and Marine Corps, and are assigned to the operating forces for the Navy. (JP 1)

passage of lines

An operation in which a force moves forward or rearward through another force's combat positions with the intention of moving into or out of contact with the enemy. (JP 3-18)

passage lane

A lane through an enemy or friendly obstacle that provides safe passage for a passing force. (FM 3-90-2)

passage point

A specifically designated place where the passing units pass through the stationary unit. (FM 3-90-2)

penetration

A form of maneuver in which an attacking force seeks to rupture enemy defenses on a narrow front to disrupt the defensive system. (FM 3-90-1)

phase

(Army) A planning and execution tool used to divide an operation in duration or activity. (ADRP 3-0)

position of relative advantage

A location or the establishment of a favorable condition within the area of operations that provides the commander with temporary freedom of action to enhance combat power over an enemy or influence the enemy to accept risk and move to a position of disadvantage. (ADRP 3-0)

preparation

Those activities performed by units and Soldiers to improve their ability to execute an operation. (ADP 5-0)

preparation of the environment

An umbrella term for operations and activities conducted by selectively trained special operations forces to develop an environment for potential future special operations. (JP 3-05)

protection warfighting function

The related tasks and systems that preserve the force so the commander can apply maximum combat power to accomplish the mission. (ADRP 3-0)

Glossary

raid
An operation to temporarily seize an area in order to secure information, confuse an enemy, capture personnel or equipment, or to destroy a capability culminating with a planned withdrawal. (JP 3-0)

rearward passage of lines
Occurs when a unit passes through another unit's positions while moving away from the enemy. (ADRP 3-90)

reconnaissance
A mission undertaken to obtain, by visual observation or other detection methods, information about the activities and resources of an enemy or adversary, or to secure data concerning the meteorological, hydrographic, or geographic characteristics of a particular area. (JP 2-0)

reconnaissance in force
A deliberate combat operation designed to discover or test the enemy's strength, dispositions, and reactions or to obtain other information. (ADRP 3-90)

reconnaissance objective
A terrain feature, geographic area, enemy force, adversary, or other mission or operational variable, such as specific civil considerations, about which the commander wants to obtain additional information. (ADRP 3-90)

*reinforcing
A support relationship requiring a force to support another supporting unit.

relief in place
An operation in which, by direction of higher authority, all or part of a unit is replaced in an area by the incoming unit and the responsibilities of the replaced elements for the mission and the assigned zone of operations are transferred to the incoming unit. (JP 3-07.3)

reserve
(Army) That portion of a body of troops which is withheld from action at the beginning of an engagement, in order to be available for a decisive movement. (ADRP 3-90)

retirement
A form of retrograde in which a force out of contact moves away from the enemy. (ADRP 3-90)

retrograde
(Army) A defensive task that involves organized movement away from the enemy. (ADRP 3-90)

route reconnaissance
A directed effort to obtain detailed information of a specified route and all terrain from which the enemy could influence movement along that route. (ADRP 3-90)

search and attack
A technique for conducting a movement to contact that shares many of the characteristics of an area security mission. (FM 3-90-1)

screen
A security task that primarily provides early warning to the protected force. (ADRP 3-90)

security area
That area that begins at the forward area of the battlefield and extends as far to the front and flanks as security forces are deployed. Forces in the security area furnish information on the enemy and delay, deceive, and disrupt the enemy and conduct counterreconnaissance. (ADRP 3-90)

security force assistance
The Department of Defense activities that contribute to unified action by the United States Government to support the development of the capacity and capability of foreign security forces and their supporting institutions. (JP 3-22)

Glossary

security forces

Duly constituted military, paramilitary, police, and constabulary forces of a state. (JP 3-22)

security cooperation

All Department of Defense interactions with foreign security establishments to build security relationships that promote specific United States security interests, develop allied and partner nation military and security capabilities for self-defense and multinational operations, and provide United States forces with peacetime and contingency access to allied and partner nations. (JP 3-20)

security operations

Those operations undertaken by a commander to provide early and accurate warning of enemy operations, to provide the force being protected with time and maneuver space within which to react to the enemy, and to develop the situation to allow the commander to effectively use the protected force. (ADRP 3-90)

security sector reform

A comprehensive set of programs and activities undertaken by a host nation to improve the way it provides safety, security, and justice. (JP 3-07)

shaping operation

An operation that establishes conditions for the decisive operation through effects on the enemy, other actors, and the terrain. (ADRP 3-0)

single envelopment

A form of maneuver that results from maneuvering around one assailable flank of a designated enemy force. (FM 3-90-1)

space domain

The space environment, space assets, and terrestrial resources required to access and operate in, to, or through the space environment. (FM 3-14)

special reconnaissance

Reconnaissance and surveillance actions conducted as a special operation in hostile, denied, or diplomatically and/or politically sensitive environments to collect or verify information of strategic or operational significance, employing military capabilities not normally found in conventional forces. (JP 3-05)

spoiling attack

A tactical maneuver employed to seriously impair a hostile attack while the enemy is in the process of forming or assembling for an attack. (FM 3-90-1)

stability mechanism

The primary method through which friendly forces affect civilians in order to attain conditions that support establishing a lasting, stable peace. (ADRP 3-0)

stability tasks

Tasks conducted as part of operations outside the United States in coordination with other instruments of national power to maintain or reestablish a safe and secure environment and provide essential governmental services, emergency infrastructure reconstruction, and humanitarian relief. (ADRP 3-07).

striking force

A dedicated counterattack force in a mobile defense constituted with the bulk of available combat power. (ADRP 3-90)

support

The action of a force that aids, protects, complements, or sustains another force in accordance with a directive requiring such action. (JP 1)

support area

The portion of the commander's area of operations that is designated to facilitate the positioning, employment, and protection of base sustainment assets required to sustain, enable, and control operations. (ADRP 3-0)

supporting effort

A designated subordinate unit with a mission that supports the success of the main effort. (ADRP 3-0)

survivability

(Army) A quality or capability of military forces which permits them to avoid or withstand hostile actions or environmental conditions while retaining the ability to fulfill their primary mission. (ATP 3-37.34)

survivability operations

Those military activities that alter the physical environment to provide or improve cover, camouflage, and concealment. (ATP 3-37.34)

sustaining operation

An operation at any echelon that enables the decisive operation or shaping operation by generating and maintaining combat power. (ADRP 3-0)

sustainment warfighting function

The related tasks and systems that provide support and services to ensure freedom of action, extend operational reach, and prolong endurance. (ADRP 3-0)

tactical combat force

A rapidly deployable, air-ground mobile combat unit, with appropriate combat support and combat service support assets assigned to and capable of defeating Level III threats including combined arms. (JP 3-10)

tactical command post

A facility containing a tailored portion of a unit headquarters designed to control portions of an operation for a limited time. (FM 6-0)

tactical control

The authority over forces that is limited to the detailed direction and control of movements or maneuvers within the operational area necessary to accomplish missions or tasks assigned. (JP 1)

target

An area designated and numbered for future firing. (JP 3-60)

target area of interest

The geographical area where high-value targets can be acquired and engaged by friendly forces. (JP 2-01.3)

target reference point

A predetermined point of reference, normally a permanent structure or terrain feature that can be used when describing a target location. (JP 3-09.3)

threat

Any combination of actors, entities, or forces that have the capability and intent to harm United States forces, United States national interests, or the homeland. (ADRP 3-0)

trigger line

A phase line located on identifiable terrain that crosses the engagement area that is used to initiate and mass fires into an engagement area at a predetermined range for all or like weapon systems. (ADRP 1-02)

troop movement

The movement of troops from one place to another by any available means. (ADRP 3-90)

turning movement

(Army) A form of maneuver in which the attacking force seeks to avoid the enemy's principle defensive positions by seizing objectives behind the enemy's current positions thereby causing the enemy force to move out of their current positions or divert major forces to meet the threat. (FM 3-90-1)

unified land operations

Simultaneous offensive, defensive, and stability or defense support of civil authorities tasks to seize, retain, and exploit the initiative to shape the operational environment, prevent conflict, consolidate gains, and win our Nation's wars as part of unified action. (ADRP 3-0)

warfighting functions

A group of tasks and systems united by a common purpose that commanders use to accomplish missions and training objectives. (ADRP 3-0)

withdrawal operation

A planned retrograde operation in which a force in contact disengages from an enemy force and moves in a direction away from the enemy. (JP 3-17)

zone reconnaissance

A form of reconnaissance that involves a directed effort to obtain detailed information on all routes, obstacles, terrain, and enemy forces within a zone defined by boundaries. (ADRP 3-90)

References

All URLs accessed on 22 August 2017.

REQUIRED PUBLICATIONS

Readers require these publications for fundamental concepts, terms, and definitions.

DOD Dictionary of Military and Associated Terms. August 2017.

ADRP 1-02. *Terms and Military Symbols.* 16 November 2016.

ADRP 3-0. *Operations.* 6 October 2017.

RELATED PUBLICATIONS

These publications are referenced in this publication.

JOINT AND DEPARTMENT DEFENSE PUBLICATIONS

Most Department of Defense publications are available at the Department of Defense Issuances website www.dtic.mil\whs\directives.

Joint publications are available at http://www.dtic.mil/doctrine/new_pubs/jointpub.htm.

DODD 4270.5. *Military Construction.* 12 February 2005.

JP 1. *Doctrine for the Armed Forces of the United States.* 25 March 2013.

JP 2-0. *Joint Intelligence.* 22 October 2013.

JP 2-01.3. *Joint Intelligence Preparation of the Operational Environment.* 21 May 2014.

JP 3-0. *Joint Operations.* 17 January 2017.

JP 3-01. *Countering Air and Missile Threats.* 21 April 2017.

JP 3-02. *Amphibious Operations.* 18 July 2014.

JP 3-03. *Joint Interdiction.* 9 September 2016.

JP 3-04. *Joint Shipboard Helicopter and Tiltrotor Aircraft Operations.* 6 December 2012.

JP 3-05. *Special Operations.* 16 July 2014.

JP 3-07. *Stability.* 3 August 2016.

JP 3-07.3. *Peace Operations.* 1 August 2012.

JP 3-09. *Joint Fire Support.* 12 December 2014.

JP 3-09.3. *Close Air Support.* 25 November 2014.

JP 3-10. *Joint Security Operations in Theater.* 13 November 2014.

JP 3-12[R]. *Cyberspace Operations.* 5 February 2013.

JP 3-13. *Information Operations.* 27 November 2012.

JP 3-13.1. *Electronic Warfare.* 8 February 2012.

JP 3-13.4. *Military Deception.* 14 February 2017.

JP 3-16. *Multinational Operations.* 16 July 2013.

JP 3-17. *Air Mobility Operations.* 30 September 2013.

JP 3-18. *Joint Forcible Entry Operations.* 11 May 2017.

JP 3-20. *Security Cooperation.* 23 May 2017.

JP 3-22. *Foreign Internal Defense.* 12 July 2010.

JP 3-30. *Command and Control of Joint Air Operations*. 10 February 2014.
JP 3-31. *Command and Control for Joint Land Operations*. 24 February 2014.
JP 3-33. *Joint Task Force Headquarters*. 30 July 2012.
JP 3-35. *Deployment and Redeployment Operations*. 31 January 2013.
JP 3-40. *Countering Weapons of Mass Destruction*. 31 October 2014.
JP 3-50. *Personnel Recovery*. 2 October 2015.
JP 3-52. *Joint Airspace Control*. 13 November 2014.
JP 3-60. *Joint Targeting*. 31 January 2013.
JP 3-61. *Public Affairs*. 17 November 2015.
JP 4-0. *Joint Logistics*. 16 October 2013.
JP 4-01. *The Defense Transportation System*. 18 July 2017.
JP 4-09. *Distribution Operations*. 19 December 2013.
JP 5-0. *Joint Planning*. 16 June 2017.
JP 6-0. *Joint Communications System*. 10 June 2015.

ARMY PUBLICATIONS

Army doctrinal publications are available at http://www.apd.army.mil/.
ADP 1-01. *Doctrine Primer*. 2 September 2014.
ADP 3-0. *Operations*. 6 October 2017.
ADP 3-07. *Stability*. 31 August 2012.
ADP 5-0. *The Operations Process*. 17 May 2012.
ADP 6-0. *Mission Command*. 17 May 2012.
ADP 6-22. *Army Leadership*. 1 August 2012.
ADRP 1. *The Army Profession*. 14 June 2015.
ADRP 1-03. *The Army Universal Task List*. 2 October 2015.
ADRP 2-0. *Intelligence*. 31 August 2012.
ADRP 3-05. *Special Operations*. 31 August 2012.
ADRP 3-07. *Stability*. 31 August 2012.
ADRP 3-09. *Fires*. 31 August 2012.
ADRP 3-37. *Protection*. 31 August 2012.
ADRP 3-90. *Offense and Defense*. 31 August 2012.
ADRP 4-0. *Sustainment*. 31 July 2012.
ADRP 5-0. *The Operations Process*. 17 May 2012.
ADRP 6-0. *Mission Command*. 17 May 2012.
ADRP 7-0. *Training Units and Developing Leaders*. 23 August 2012.
AR 11-31. *Army Security Cooperation Policy*. 21 March 2013.
AR 525-93. *Military Operations: Army Deployment and Redeployment*. 12 November 2014.
AR 570-9. *Manpower and Equipment Control: Host Nation Support*. 29 March 2006.
AR 700-80. *Logistics Army In-Transit Visibility*. 30 September 2015.
ATP 2-01. *Plan Requirements and Assess Collection*. 19 August 2014.
ATP 2-01.3/MCRP 2-3A. *Intelligence Preparation of the Battlefield/Battlespace*. 10 November 2014.
ATP 2-19.3. *Corps and Division Intelligence Techniques*. 26 March 2015.
ATP 3-04.1 *Aviation Tactical Employment*. 13 April 2016.
ATP 3-05.2. *Foreign Internal Defense*. 19 August 2015.

References

ATP 3-06.20/MCRP 3-30.5/NTTP 3-05.8/AFTTP 3-2.62. *Multi-Service Tactics, Techniques, and Procedures for Cordon and Search Operations*. 18 August 2016.

ATP 3-07.5. *Stability Techniques*. 31 August 2012.

ATP 3-09.13. *The Battlefield Coordination Detachment*. 24 July 2015.

ATP 3-09.24. *Techniques for the Fires Brigade*. 21 November 2012.

ATP 3-09.34/MCRP 3-25H/NTTP 3-09.2.1/AFTTP 3-2.59. *Kill Box Multi-Service Tactics, Techniques, and Procedures for Kill Box Planning and Employment*. 16 April 2014.

ATP 3-11.36/MCRP 3-37B/NTTP 3-11.34/AFTTP 3-2.70. *Multi-Service Tactics, Techniques, and Procedures for Chemical, Biological, Radiological, and Nuclear Aspects of Command and Control*. 1 November 2013.

ATP 3-34.23. *Engineer Operations—Echelons Above Brigade Combat Team*. 10 June 2015.

ATP 3-34.40/MCWP 3-17.7. *General Engineering*. 25 February 2015.

ATP 3-35. *Army Deployment and Redeployment*. 23 March 2015.

ATP 3-35.1. *Army Pre-Positioned Operations*. 27 October 2015.

ATP 3-37.10/MCRP 3-40D.13. *Base Camps*. 27 January 2017.

ATP 3-37.34/MCWP 3-17.6. *Survivability Operations*. 28 June 2013.

ATP 3-39.30. *Security and Mobility Support*. 30 October 2014.

ATP 3-52.2/MCRP 3-25F/NTTP 3-56.2/AFTTP 3-2.17. *TAGS Multi-Service Tactics, Techniques, and Procedures for the Theater Air-Ground System*. 30 June 2014.

ATP 3-55.3/MCRP 2-2A/NTTP 2-01.3/AFTTP 3-2.88. *ISR Optimization Multi-Service Tactics, Techniques, and Procedures for Intelligence, Surveillance, and Reconnaissance Optimization*. 14 April 2015.

ATP 3-55.4. *Techniques for Information Collection During Operations Among Populations*. 5 April 2016.

ATP 3-60. *Targeting*. 7 May 2015.

ATP 3-90.4/MCWP 3-17.8. *Combined Arms Mobility*. 8 March 2016.

ATP 3-90.8/MCWP 3-17.5. *Combined Arms Countermobility Operations*. 17 September 2014.

ATP 3-90.40. *Combined Arms Countering Weapons of Mass Destruction*. 29 June 2017.

ATP 3-91. *Division Operations*. 17 October 2014.

ATP 3-91.1. *The Joint Air Ground Integration Center*. 18 June 2014.

ATP 3-92. *Corps Operations*. 7 April 2016.

ATP 3-93. *Theater Army Operations*. 26 November 2014.

ATP 3-94.2. *Deep Operations*. 1 September 2016.

ATP 4-02.46. *Army Health System Support to Detainee Operations*. 12 April 2013.

ATP 4-10/MCRP 4-11 H/NTTP 4-09.1/AFMAN 10-409-O. *Multi-Service Tactics, Techniques, and Procedures for Operational Contract Support*. 18 February 2016.

ATP 4-10.1. *Logistics Civil Augmentation Program Support to Unified Land Operations*. 1 August 2016.

ATP 4-14. *Expeditionary Railway Center Operations*. 29 May 2014.

ATP 4-15. *Army Watercraft Operations*. 3 April 2015.

ATP 4-16. *Movement Control*. 5 April 2013.

ATP 4-32. *Explosive Ordnance Disposal (EOD) Operations*. 30 September 2013.

ATP 4-32.1. *Explosive Ordnance Disposal (EOD) Group and Battalion Headquarters Operations*. 24 January 2017.

ATP 4-32.3. *Explosive Ordnance Disposal (EOD) Company, Platoon, and Team Operations*. 1 February 2017.

ATP 4-43. *Petroleum Supply Operations*. 6 August 2015.

References

ATP 4-70. *Assistant Secretary of the Army for Acquisition, Logistics, and Technology Forward Support to Unified Land Operations.* 12 May 2014.

ATP 4-91. *Army Field Support Brigade.* 15 December 2011.

ATP 4-92. *Contracting Support to Unified Land Operations.* 15 October 2014.

ATP 4-93. *Sustainment Brigade.* 11 April 2016.

ATP 4-94. *Theater Sustainment Command.* 28 June 2013.

ATP 5-0.1. *Army Design Methodology.* 1 July 2015.

ATP 6-0.5. *Command Post Organization and Operations.* 1 March 2017.

FM 1-0. *Human Resources Support.* 1 April 2014.

FM 1-06. *Financial Management Operations.* 15 April 2014.

FM 3-01. *U.S. Army Air and Missile Defense Operations.* 2 November 2015.

FM 3-04. *Army Aviation.* 29 July 2015.

FM 3-04.120. *Air Traffic Services Operations.* 16 February 2007.

FM 3-05. *Army Special Operations.* 9 January 2014.

FM 3-06. *Urban Operations.* 26 October 2006.

FM 3-07. *Stability.* 2 June 2014.

FM 3-09. *Field Artillery Operations and Fire Support.* 4 April 2014.

FM 3-11/MCWP 3-37.1/NWP 3-11/AFTTP 3-2.42. *Multiservice Doctrine for Chemical, Biological, Radiological, and Nuclear Operations.* 1 July 2011.

FM 3-12. *Cyberspace and Electronic Warfare Operations.* 11 April 2017.

FM 3-13. *Information Operations.* 6 December 2016.

FM 3-14. *Army Space Operations.* 19 August 2014.

FM 3-22. *Army Support to Security Cooperation.* 22 January 2013.

FM 3-39. *Military Police Operations.* 26 August 2013.

FM 3-50. *Army Personnel Recovery.* 2 September 2014.

FM 3-52. *Airspace Control.* 20 October 2016.

FM 3-53. *Military Information Support Operations.* 4 January 2013.

FM 3-55. *Information Collection.* 3 May 2013.

FM 3-57. *Civil Affairs Operations.* 31 October 2011.

FM 3-61. *Public Affairs Operations.* 1 April 2014.

FM 3-63. *Detainee Operations.* 28 April 2014.

FM 3-81. *Maneuver Enhancement Brigade.* 21 April 2014.

FM 3-90-1. *Offense and Defense Volume I.* 22 March 2013.

FM 3-90-2. *Reconnaissance, Security, and Tactical Enabling Tasks Volume II.* 22 March 2013.

FM 3-94. *Theater Army, Corps, and Division Operations.* 21 April 2014.

FM 3-96. *Brigade Combat Team.* 8 October 2015.

FM 3-98. *Reconnaissance and Security Operations.* 1 July 2015.

FM 3-99. *Airborne and Air Assault Operations.* 6 March 2015.

FM 4-02. *Army Health System.* 26 August 2013.

FM 4-95. *Logistics Operations.* 1 April 2014.

FM 6-0. *Commander and Staff Organization and Operations.* 5 May 2014.

FM 6-02. *Signal Support to Operations.* 22 January 2014.

FM 6-22. *Leader Development.* 30 June 2015.

FM 7-0. *Train to Win in a Complex World.* 5 October 2016.

FM 27-10. *The Law of Land Warfare.* 18 July 1956.

TC 3-04.7. *Army Aviation Maintenance*. 2 February 2010.

TC 3-05.3. *Security Force Assistance Deployment Handbook*. 22 May 2015.

TC 7-100.2. *Opposing Force Tactics*. 9 December 2011.

TM 3-34.48-1. *Theater of Operations: Roads, Airfields, and Heliports—Road Design*. 29 February 2016.

TM 3-34.48-2. *Theater of Operations: Roads, Airfields, and Heliports—Airfield and Heliport Design*. 29 February 2016.

OBSOLETE PUBLICATIONS

This section contains references to obsolete historical doctrine. The Archival and Special Collections in the Combined Arms Research Library (CARL) on Fort Leavenworth in Kansas contains copies. These publications are obsolete doctrine publications referenced for citations only.

Field Service Regulation. Washington DC: Government Printing Office, 1918.

Field Service Regulations: United States Army. Washington DC: Government Printing Office, 1924.

Field Service Regulations: United States Army. Washington DC: Government Printing Office, 1923.

Field Service Regulations: United States Army. Washington DC: Government Printing Office, 1905.

FM 6-20 [obsolete]. *Field Artillery Field Manual: Tactics and Techniques*. Washington DC: Government Printing Office, 1940.

FM 7-5 [obsolete]. *Infantry Field Manual: Organization and Tactics of Infantry: The Rifle Battalion*. Washington DC: Government Printing Office, 1940.

FM 7-19 [obsolete]. *Combat Support Company: Infantry Division Battle Group Operations*. Washington DC: Government Printing Office, 1960.

FM 7-100 [obsolete]. *Infantry Division*. Washington DC: Government Printing Office, 1960.

FM 17-95 [obsolete]. *Cavalry Operations*. Washington DC: Government Printing Office, 1981.

FM 31-23 [obsolete]. *Stability Operations*. Washington DC: Government Printing Office, 1972.

FM 57-1 [obsolete]. *U. S. Army/U. S. Air Force Doctrine for Airborne Operations*. Washington DC: Government Printing Office, 1967.

FM 60-5 [obsolete]. *Amphibious Operations: Battalion in Assault Landings*. Washington DC: Government Printing Office, 1951.

FM 71-100-3 [obsolete]. *Air Assault Division Operations*. Washington DC: Government Printing Office, 1996.

FM 90-14 [obsolete]. *Rear Battle*. Washington DC: Government Printing Office, 1985.

FM 100-5 [obsolete]. *Field Service Regulations: Operations*. Washington DC: Government Printing Office, 1954.

FM 100-5 [obsolete]. *Field Service Regulations: Operations*. Washington DC: Government Printing Office, 1949.

FM 100-5 [obsolete]. *Field Service Regulations: Operations*. Washington DC: Government Printing Office, 1944.

FM 100-5 [obsolete]. *Operations*. Washington DC: Government Printing Office, 1993.

FM 100-5 [obsolete]. *Operations*. Washington DC: Government Printing Office, 1982.

FMFRP 12-18. *Mao Tse-tung on Guerrilla Warfare*. Washington DC: Government Printing Office, 1989.

TRADOC Pam 525-5 [obsolete]. *The Airland Battle and Corps 86*. Washington DC: Government Printing Office, 1981.

OTHER PUBLICATIONS

A Dictionary of Military Quotations. Compiled by Trevor Royle. London: Routledge, 1990.

Bartlett's Familiar Quotations. New York: Little, Brown and Company, 2002.

References

Callo, Joseph. *John Paul Jones: America's First Sea Warrior*. Annapolis, MD: Naval Institute Press, 2006.

Churchill, Winston S. Letter to Prime Minister Asquith, December 29, 1914. Available at www.nationalchurchillmuseum.org/churchill-in-world-war-i-and-aftermath.html.

Churchill, Winston S. *The Second World War, Volume 5, Closing the Ring*. Boston: Houghton Mifflin Company, 1979.

Churchill, Winston S. "The Gift of a Common Tongue, September 6, 1943, Harvard." The International Churchill Society. Available at https://www.winstonchurchill.org/resources/speeches/1941-1945-war-leader/the-price-of-greatness-is-responsibility.

Clausewitz, Carl von. *On War*. Edited and translated by Peter Paret and Michael E. Howard. Princeton, NJ: Princeton University Press, 2000.

Corbett, Julian S. *Some Principles of Maritime Strategy*. Annapolis, MD: Naval Institute Press, 1988.

Creveld, Martin Van. *Supplying War: Logistics from Wallenstein to Patton*. New York: Cambridge University Press, 1991.

Developments Concepts and Doctrine Centre. *Future Character of Conflict: Strategic Trends Programme*, United Kingdom: Ministry of Defence, 2010.

Dictionary of Military and Naval Quotations. Compiled by Robert Debs Heinl, Jr. Annapolis, MD: United States Naval Institute, 1988.

du Picq, Charles Ardant. *Battle Studies: Ancient and Modern Battles*. Translated by John N. Greely and Robert C. Cotton. New York: MacMillian Company, 1921.

Duke of Wellington. "Anecdotes of the Duke of Wellington at the Battle of Waterloo." *The Scots Magazine and Edinburgh Literary Miscellany: Being a General Repository of Literature, History, and Politics, for 1816*, Volume 78. Edinburgh: Archibald Constable and Company, 1816.

"Dwight D. Eisenhower, 1953-61." Available at https://www.cia.gov/library/center-for-the-study-of-intelligence/csi-publications/books-and-monographs/our-first-line-of-defense-presidential-reflections-on-us-intelligence/eisenhower.html.

Ed-din, Beha. *The Life of Saladin*. London: Committee of the Palestine Exploration Fund, 1897.

Eisenhower, Dwight D. "Special Message to the Congress on Reorganization of the Defense Establishment." April 3, 1958. Online by Gerhard Peters and John T. Woolley. *The American Presidency Project*. http://www.presidency.ucsb.edu/ws/?pid=11340.

"For 1957, I Resolve…." Asbury Park Press (January 13, 1957): 24. Available at https://www.newspapers.com/newspage/144197456/.

Frederick the Great. "The Instruction of Frederick the Great for His Generals." *Roots of Strategy: The 5 Greatest Military Classics of All Time*. Edited by Thomas R. Phillips. Harrisburg, PA: Stackpole Books, 1985.

Harbottle, Thomas Benfield. *Dictionary of Quotations (Classical)*. New York: Swan Sonnenschein & Co., Limited, 1906.

Hart, B. H. Liddell. *Strategy*. New York: Signet Printing, 1974.

Joint Military Operations Historical Collection. Washington DC: Government Printing Office, 1997.

Kennedy, John F. "Speech of Senator John F. Kennedy, Civic Auditorium, Seattle, WA, September 6, 1960." Online by Gerhard Peters and John T. Woolley. *The American Presidency Project*. Available at http://www.presidency.ucsb.edu/ws/?pid=25654.

Kitchener, Field Marshal Earl. "K." *The Lord Kitchener Memorial Book*. Edited by Sir Hedley Le Bas. New York: Hodder and Stoughton, 1917.

Lawrence, T. E. "The Evolution of a Revolt." *The Army Quarterly, Volume 1*. London: William Clowes & Sons, Ltd., October 1920 and January 1921.

MacArthur, General Douglas. "Speech to Congress in House of Representatives, April 19, 1951." Available at https://www.trumanlibrary.org/whistlestop/study_collections/koreanwar/index.php?action=pdf&documentid=ma-2-18.

MacDonald, Charles B. "Victory in the Ardennes." *The Last Offensive*. Washington DC: Center for Military History, 1993.

Maguire, T. Miller. "Campaign in Bohemia." *The United Service Magazine*. London: Will Clowes & Sons, Limited, 1866.

Mahan, Alfred Thayer. *Naval Strategy: Compared and Contrasted With the Principles and Practice of Military Operations on Land; lectures delivered at U.S. Naval War College, Newport, R.I., between the years 1887-1911*. Boston, Little, Brown and Company, 1915.

Manchester, William. *American Caesar: Douglas MacArthur 1880–1964*. Boston: Little, Brown and Company, 1978.

Mao Tse Tung. *Selected Military Writings of Mao Tse-Tung*. Peking: Foreign Languages Press, 1967.

McDonald, Charles B. The Last Offensive. U.S. Army Center of Military History Publication 7-9-1. Available at http://www.history.army.mil/catalog/pubs/7/7-9.html.

Meyer, E. C. "Address to the Pre-Command Course 25 July 1980." Fort Leavenworth, KS.

Military Air Power: The CADRE Digest of Air Power Opinions and Thoughts. Compiled by Charles M. Westenhoff. Maxwell Air Force Base, AL: Air University Press, 1990.

Nixon, Richard. *Six Crises*. New York: Pyramid Book, 1968.

Patton, General George S. "Quotes." Available at http://www.generalpatton.com/quotes/.

Quotes for Air Force Logistician, Volume I. Maxwell Air Force Base, AL: Air Force Logistics Management Agency, 2006.

Reagan, Ronald. "Remarks at a Campaign Rally for Senator Robert W. Kasten, Jr., in Waukesha, Wisconsin, October 23, 1986." *The Public Papers of President Ronald W. Reagan*. Ronald Reagan Presidential Library. Available at https://reaganlibrary.archives.gov/archives/speeches/1987/061287d.htm.

Regan, Ronald. "Remarks at a Ceremony Commemorating the 40th Anniversary of the Normandy Invasion, D-day June 6, 1984." The Public Papers of President Ronald W. Reagan, Ronald Reagan Presidential Library. Available at https://www.reaganlibrary.archives.gov/archives/speeches/1984/60684a.htm.

Root, Elihu. *Annual Reports of the War Department*, Vol IV. Washington DC: Government Printing Office, 1903.

Sherman, General William T. *Memoirs of William T. Sherman by Himself*. New York: D. Appleton and Company, 1875.

Slim, Field-Marshal Viscount William. *Defeat into Victory: Battling Japan in Burma and India, 1942—1945*. New York: Cooper Square Press, 2000.

Spears, John Randolph. *David G. Farragut*. Philadelphia: George W. Jacobs & Company, 1905.

Sun Tzu. *The Art of War*. Translated by Samuel B. Griffith. New York: Oxford University Press, 1963.

Truman, Harry. "Going Forward with a Campaign of Truth." *State Department Bulletin*. Washington DC: Government Printing Office, May 1, 1950.

United States Army Special Operations Command G-9. White Paper: "The Arc of Change: Operational Time to Influence Trajectories." Fort Bragg, NC.

Unpublished data. Army University Press. 2017.

Unpublished data. Combined Arms Doctrine Directorate staff. 2017.

Unpublished data. Department of Military History. 2017.

Widder, Major General Werner (German Army). "Auftragstaktik and Innere Führung: Trademarks of German Leadership." *Military Review* (September-October 2002): 4.

References

Wylie, Rear Admiral J. C. *Military Strategy: A General Theory of Power Control.* Annapolis, MD: Naval Institute Press, 2014.

Xenophon, *Memorabilia* 3.1.7. Available at http://perseus.uchicago.edu/perseus-cgi/citequery3.pl?dbname=GreekFeb2011&query=Xen.%20Mem.%203.1.11&getid=1.

UNITED STATES LAW

Most acts and public laws are available at http://thomas.loc.gov/home/thomas.php.

Title 10, United States Code. Armed Forces.

Title 22, United States Code. Foreign Relations and Intercourse.

PRESCRIBED FORMS

This section contains no entries.

REFERENCED FORMS

Unless otherwise indicated, DA forms are available on the Army Publishing Directorate Web site: http://www.apd.army.mil.

DA Form 2028. *Recommended Changes to Publications and Blank Forms.*

Index

Entries are by paragraph number.

A

administrative control, 2-9, A-25–A-28
    defined, A-25
administrative movement, 5-78
administrative responsibilities, corps, 2-60–2-61
    division, 2-67
    theater army, 2-9
adversary, defined, 1-37
air and missile defense, 2-231–2-232
    in the offense, 7-83–7-84
Air and Missile Defense Command, Army, 2-32–2-35
air assault, defined, 7-117
air defense, training considerations, 2-310
air movements, Army, 5-75
air tasking order, defined, 4-49
airborne assault, defined, 7-117
airspace control, 2-153–2-158
    defined, 2-153
    in consolidation and support areas, 5-51–5-54
    training considerations, 2-280–2-282
allocation, defined, 4-22
ambassador, United States diplomatic mission, 3-38–3-39
ambush, defined, 7-229
amphibious area of operations, 1-135
anticipated operational environments, 1-16–1-22
area, close, 1-146–1-149
    consolidation, 1-158–1-161
    deep, 1-150–1-154
    support, 1-155–1-157
area defense, 6-121–6-170
    conducting a counterattack during, 6-165–6-619
    conducting a spoiling attack in, 6-159–6-164
    consolidation of gains in, 6-170
    contiguous framework, 6-125–6-126
    control measures, 6-153–6-156
    defined, 6-121
    executing an, 6-157–6-158
    information collection in, 6-129–6-130
    main battle area forces in, 6-139–6-143
    noncontiguous framework, 6-127
    organization of forces for, 6-128–6-152
    reserve in, 6-144–6-152
    security forces in, 6-131–6-138
area of influence, defined, 1-138
area of interest, defined, 1-138
area of operations, defined, 1-136
    land force, 1-136–1-140
area of responsibility, 1-128–1-129
area reconnaissance, defined, 5-62
area security, defined, 5-67
    operations, 8-38–8-48
areas, close, deep, support, and consolidation, 1-141–1-161
Army, command relationships, A-13–A-17
    special operations forces, 2-52
    support relationships, A-18–A-24
    Air and Missile Defense Command, 2-32–2-35
    air movements, 5-75
    aviation maneuver (manned), 2-303
    aviation maneuver (unmanned), 2-304
    capabilities (combat power), 2-104–2-266
    command and support relationships, A-12–A-28
    cyberspace organizations, 3-53–3-56
    echelons, 2-4–2-103
    echelons, capabilities, and training, 2-1–2-319
    echelons, capabilities, and training, overview of, forces in large-scale combat operations, 5-14–5-29
    forces during operations to prevent, 4-70–4-93
    in joint operations, 1-61–1-67
    organizations, 3-42
Army personnel recovery, defined, 2-264
Army, pre-positioned stocks, 4-41–42
    rail and water movements, 5-76
Army Service component command, A-27–A-28
    defined, 2-7
Army, special operations forces, 3-57
    support to other Services and common user logistics, 2-10–2-12
    support to theater campaign planning, 3-44–3-47
assault position, defined, 7-107
assessments, operations, 3-8–3-10
assign, defined, A-16
attach, defined, A-17
attack, 7-171–7-195
    control measures for, 7-184–7-187
    defined, 7-171
    executing an, 7-188–7-195
    main body in, 7-178–7-180

Index

Entries are by paragraph number.

attack (*continued*)
   organization of forces for, 7-176–7-183
   reserve in, 7-181–7-183
   security forces for, 7-177
   subordinate forms, 7-228–7-232

attack position, defined, 7-107

attacking deeper into enemy territory, 6-223–6-224

aviation command, theater, 2-39–2-42

B

base security and defense, support of, 5-43–5-49

battle, multi-domain, 1-77–1-79

battle handover line, defined, 6-155

battlefield, multi-domain extended, 1-23–1-36

battlefield coordination detachment, 2-50–2-51

breach, defined, 5-88

breaching operations, 5-87–5-90
   in the offense, 7-244

breakout, defined, 6-221

brigade, combat aviation, 2-74–2-75
   expeditionary combat aviation, 2-76
   field artillery, 2-77–2-78
   maneuver enhancement, 2-79–2-81
   sustainment, 2-82–2-83

brigade combat teams, 2-68–2-71
   consolidate gains responsibilities, 8-72–8-75

brigades, 3-72–3-74
   functional, 2-84–2-103
   multifunctional, 2-73–2-83
   multifunctional and functional, 2-72–2-103
   sustainment, 2-21–2-23

C

capability, expeditionary, 1-110–1-113

chemical, biological, radiological, and nuclear brigade, 2-45
   defense operations, 2-256–2-257

civil affairs, corps, 4-107

operations, 2-159–2-162
   command, 2-28–2-30

civil affairs integration, training considerations, 2-283

civil control, establish, 8-52

civil security, establish, 8-50–8-51

civil-military operations, corps, 4-107

clearance of fires, defined, 7-81

clearing, defined, 5-92

clearing operations, 5-91–5-92

close area, 1-146–1-149
   defined, 1-146

close, deep, support, and consolidation areas, 1-141–1-161

cognitive, considerations, 1-126

combat aviation brigade, 2-74–2-75

combat engineering, forward aviation, 5-99–5-105

combat operations, characteristics, 1-4
   large-scale, 1-1–1-4

combat power, considerations while building, 4-116–4-122
   defined, 2-104
   elements of, 2-109–2-270

combat roads and trails, 5-95–5-98

combatant command, defined, A-3

combined training and exercises, 3-36

command and control nodes, enemy, 7-29–7-30

command and support relationships, A-1–A-28
   Army, A-12–A-28
   fundamental considerations, A-1

command group, 2-181–2-182
   defined, 2-181

command post, contingency, 2-170, 4-91–4-92
   defined, 2-165
   early-entry, 2-179–2-180
   effectiveness, 2-185–2-189
   main, 2-169
   operational, 2-171
   operations, 2-166–2-168

organization and employment considerations, 2-183–2-193
   support area, 2-174–2-178
   survivability, 2-190–2-193
   tactical, 2-172–2-173

command relationships, Army, A-13–A-17
   joint, A-2–A-11

commander and staff, roles, 2-125–2-126

commander's role, intelligence, 2-214–2-216

common-user logistics, and Army support to other Services, 2-10–2-12
   defined, 2-12

communications, planning considerations, 2-198

concept, Army operational, 1-68–1-85

conduct, information operations, 2-127–2-144
   large-scale ground combat, 1-65
   security cooperation, 8-57
   the operations process, 2-121–2-126

conducting a counterattack during an area defense, 6-165–6-619

conducting a spoiling attack in the area defense, 6-159–6-164

conflict continuum, range of military operations and, 1-2

considerations, cognitive, 1-126
   operational framework, 1-122
   physical, 1-123
   temporal, 1-124
   virtual, 1-125
   when preparing a mobile defense, 6-56–6-58
   when preparing an area defense, 6-50–6-55
   while building combat power, 4-116–4-122

consolidate gains, 1-66, 3-75–3-77
   defined, 8-1
   operations to, 8-1–8-76
   responsibilities by echelon, transition to, 5-115–5-118

Index

Entries are by paragraph number.

consolidate gains
 responsibilities, brigade
  combat teams, 8-72–8-75
  corps, 8-64–8-65
  divisions, 8-66–8-71
  task forces, 8-72–8-75
  theater army, 8-59–8-63
consolidation, defined, 8-10
consolidation area, 1-158–1-161
 defined, 1-158
 framework, 8-25–8-35
 operations, 5-41–5-54
 security planning, 7-47–7-53
consolidation of gains
 activities, 8-35–8-57
 in operations to prevent, 4-124–4-126
 in the area defense, 6-170
 in the defense, 6-226
 in the offense, 7-267
 threats to, 8-18–8-24
contact point, defined, 7-118
contiguous framework, area defense, 6-125–6-126
contingency command post, 2-170, 4-91–4-92
continuity of operations planning, 2-194–2-197
control measure, defined, 1-178
control measures, 1-178–1-181
 area defense, 6-153–6-156
 attack, 7-184–7-187
 defending encircled, 6-220
 delay, 6-200–6-201
 mobile defense, 6-187–6-188
 movement to contact, 7-142–7-145
 offensive encirclement, 7-249–7-250
 retirement, 6-212
 withdrawal, 6-207
conventional and special operations forces, interdependence, interoperability, and integration, 3-62–3-67
coordinating authority, defined, A-10
coordination, interagency, 3-37–3-41
cordon and search, defined, 8-48
corps, 2-54–2-61

administrative responsibilities, 2-60–2-61
civil affairs, 4-107
civil-military operations, 4-107
consolidate gains responsibilities, 8-64–8-65
cyberspace electromagnetic activities, 4-118
during operations to prevent, 4-94–4-122
employment, 4-105
engineer operations, 4-110
fires, 4-113
information operations, 4-108
intelligence operations, 4-111–4-112
military information support operations, 4-109
operational responsibilities, 2-56–2-59
operations security, 4-117
planning, 4-95–4-104
planning for the offense, 7-8–7-15
preparing for the offense, 7-54–7-94
protection, 4-114
reserve, 4-121
security operations, 4-106
space operations, 4-119
support area, 4-120
sustainment, 4-115
transition to the offense, 4-122
corps and below, organizations, 3-58–3-74
corps and division defensive tasks, planning, 6-8–6-48
corps and divisions preparing for the defense, 6-49–6-108
counterattack, defined, 6-165
counterfire training considerations, 2-309
countering weapons of mass destruction, 3-32
countermobility operations, 5-106–5-110
 defined, 5-106
country team, 3-40–3-41
cover, defined, 5-67
cyberspace and the electromagnetic spectrum, 1-31–1-36
 defined, 1-31

cyberspace and electronic warfare training considerations, 2-290–2-293
cyberspace electromagnetic activities, 2-145–2-151
 corps, 4-118
 defined, 2-145
cyberspace organizations, Army, 3-53–3-56

D

danger close, defined, 6-156
decisive action, 1-69–1-74
 defined, 1-69
decisive operation, defined, 1-162
decisive operations, planning, 7-33–7-37
decisive, shaping, sustaining operations, 1-162–1-164
deep area, 1-150–1-154
 defined, 1-150
deep operations as part of defensive preparations, 6-89–6-95
defeat and stability mechanisms, 1-93–1-100
defeat mechanism, defined, 1-94
defeat mechanisms, 1-94–1-98
defending encircled, 6-213–6-225
 attacking deeper into enemy territory, 6-223–6-224
 breakout from, 6-221
 control measures when, 6-220
 exfiltration, 6-222
 linkup, 6-225
 organization of forces when, 6-218–6-219
defense, consolidation of gains in, 6-226
 enemy, 7-6–7-7
 mobile, 6-171–6-191
defensive preparations, deep operations, 6-89–6-95
defensive task, defined, 6-1
defensive tasks, 6-109–6-211

Index

Entries are by paragraph number.

delay, 6-195–6-201
   control measures for, 6-200–6-201
   organization of forces for, 6-199
delay line, defined, 6-201
delaying operation, defined, 6-193
demonstration, defined, 7-230
Department of Defense information network, defined, 2-24
deployment, 4-46–4-69
destroy, defined, 1-95
detecting high-payoff targets, 6-86–6-88
detention operations, 2-259–2-263
deter, 1-56
direct liaison authorized, defined, A-11
direct support, defined, A-21
disintegrate, defined, 1-97
dislocate, defined, 1-96
dismounted march, defined, 5-73
division, administrative responsibilities, 2-67
   in operations to prevent, 4-123
   operational responsibilities, 2-63–2-66
   planning for the offense, 7-8–7-15
divisions, 2-62–2-67, 3-68–3-71
   consolidate gains responsibilities, 8-66–8-71
   preparing for the offense, 7-54–7-94
domain, 1-25–1-27
dominate, 1-58
double envelopment, defined, 7-97

E

early-entry command post, 2-179–2-180
   defined, 2-179
echelons, Army, 2-4–2-103
economic development support, 8-55–8-56
effectiveness, command post, 2-185–2-189

electromagnetic spectrum, and cyberspace, 1-31–1-36
   defined, 1-34
electronic and cyberspace warfare, training considerations, 2-290–2-293
elements, of operational art, 1-88–1-90
   of combat power, 2-109–2-270
employment, corps, 4-105
enable, civil authority, 1-60
   land forces, 4-88–4-99
encirclement operations, 7-245–7-266
   control measures, 7-249–7-250
   defined, 7-245
   defined, 5-85
   defined, 7-97
   executing, 7-251–7-266
   organization of forces, 7-246–7-248
enemy, attack, 6-5–6-7
   command and control nodes and logical entities, 7-29–7-30
   defense, 7-6–7-7
   defined, 1-37
   reserves, mobile, 7-27–7-28
   second tactical echelon, 7-25–7-26
engineer command, theater, 2-37–2-38
engineer operations, corps, 4-110
entry operations, 1-117–1-119
envelopment, 7-96–7-98
   defined, 7-96
environment, information, 1-28–1-30
essential services, restore, 8-53
establish, civil control, 8-52
   civil security, 8-50–8-51
execute, flexible deterrent options and flexible response options, 4-8–4-11
   executing, area defense, 6-157–6-158
   attack, 7-188–7-195
   encirclement, 7-251–7-266
   mobile defense, 6-189–6-191
   movement to contact, 7-158–7-170

executing a movement to contact, main body, 7-164–7-170
   security forces, 7-159–7-163
exfiltration, 6-222
expanded theater, 2-31
expeditionary capability, 1-110–1-113
   operation, 1-9
   combat aviation brigade, 2-76
   sustainment command, 2-17–2-20
exploit the initiative, 1-75–1-76
exploitation, fires in, 7-214–7-215
   maneuver in, 7-204–7-213
   mission command in, 7-199–7-203
   sustainment during, 7-216–7-217
exploitation and pursuit, 7-196–7-227
explosive ordnance disposal, 2-258
explosive ordnance disposal group, 2-47
external support contractors, 4-39

F

facilities, 4-29
feint, defined, 7-231
field artillery, 2-224–2-230
   brigade, 2-77–2-78
fighting for intelligence, 2-210–2-213
fire superiority, defined, 7-15
fire support, in the offense, 7-80–7-82
fire support coordination line, defined, 1-144
fires, 2-217–2-231
   corps, 4-113
   exploitation, 7-214–7-215
   preparing for in the defense, 6-80–6-95
   preparing for in the offense, 7-78–7-84
   training considerations, 2-307–2-310
fires warfighting function, defined, 2-217
fixing force, mobile defense, 6-182

Index

Entries are by paragraph number.

flank attack, 7-99–7-103
   defined, 7-99
flank security, 7-241–7-242
flexible deterrent options, defined, 4-8
flexible response options, defined, 4-8
force projection, 1-114–1-116
   defined, 4-25
forced marches, 5-77
forces, challenges, 1-5–1-15
forcible entry, 5-111–5-114
   defined, 5-111
foreign disclosure, 3-52
foreign internal defense, 3-22–3-24
   defined, 3-22
forms of maneuver, 7-95–7-119
   defined, 7-95
forward aviation combat engineering, 5-99–5-105
forward passage of lines, 7-234–7-236
   defined, 5-84
frontal attack, 7-104–7-108
   defined, 7-104
functional, brigades, 2-84–2-103
fundamental considerations, command and support relationships, A-1

G

gap-crossing operations, 5-93–5-93
general skills, 4-33–4-35
general support, defined, A-22
general support—reinforcing, defined, A-24
governance, support, 8-54
graphic control measure, defined, 1-180
guard, defined, 5-67

H

health service support, 2-245–2-246
high-payoff targets, detecting, 6-86–6-88
host-nation support, 4-43–4-45, 3-48–3-50
   defined, 3-50
humanitarian efforts, 3-33

I

infiltration, 7-109–7-114
   defined, 7-109
information, 2-113–2-115
information collection, defined, 2-209
   in area defense, 6-129–6-130
   in movement to contact, 7-148–7-150
information environment, 1-28–1-30
   defined, 1-28
information operations, conduct, 2-127–2-144
   corps, 4-108
   defined, 2-127
   shaping, 3-34–3-35
   training considerations, 2-284–2-285
information operations groups, theater, 2-48
information warfare, 1-41–1-43
infrastructure development, support, 8-55–8-56
integration, 4-67–4-69
intelligence, 2-205–2-216
   commander's role, 2-214–2-216
   fighting for, 2-210–2-213
   preparing for in the defense, 6-76–6-79
   preparing for in the offense, 7-75–7-77
   shaping activities, 3-29–3-31
   training considerations, 2-305–2-306
intelligence operations, corps, 4-111–4-112
intelligence warfighting function, defined, 2-205
interagency coordination, 3-37–3-41
interdependence, interoperability, and integration of conventional and special operations forces, 3-62–3-67
interdiction, defined, 1-144
in-transit visibility, defined, 4-54
isolate, defined, 1-98
isolation, 1-45–1-46

J

joint, command and control, theater army contributions to, 4-74–4-76
   command relationships, A-2–A-11
   large-scale combat operations, 5-1–5-13
   operational areas, 1-127
   operations, 1-51–1-160
   operations area, 1-132
   security area, 1-134
   special operations area, 1-133
joint operations, defined, 1-51
   the Army in, 1-61–1-67

K

kill box, defined, 6-92

L

land, control of, 1-7
land component headquarters in large-scale combat operations, 4-93
land force, area of operations, 1-136–1-140
landpower, defined, 5-14
large-scale combat operations, 5-1–5-118
   Army forces in, 5-14–5-29
   joint, 5-2–5-13
   land component headquarters in, 4-93
   stability in, 5-34–5-54
large-scale defensive operations, 6-1– 6-225
   overview of, 6-1–6-4
large-scale ground combat, conduct, 1-65
large-scale offensive operations, 7-1–7-267
   overview, 7-1–7-5
leadership, 2-110–2-112
   defined, 2-110
Level I threat, defined, 5-45
Level II threat, defined, 5-46
Level III threat, defined, 5-47
levels, of warfare, 1-19–1-20
linkup, 6-225
local security, defined, 5-67
lodgment, defined, 5-113
logical entities, enemy, 7-29–7-30
logistics, 2-237–2-243

Entries are by paragraph number.

M

main command post, defined, 2-169

main and supporting efforts, 1-165–1-167

main battle area, defined, 6-14

main battle area forces in an area defense, 6-139–6-143

main body, attack, 7-178–7-180
    executing a movement to contact, 7-164–7-170
    movement to contact, 7-137–7-140

main effort, defined, 1-165

main efforts, planning, 7-33–7-37

maintenance, 4-30–4-31

maneuver, defined, 1-80
    exploitation, 7-204–7-213
    forms of, 7-95–7-119

maneuver enhancement brigade, 2-79–2-81

medical command (deployment support), 2-25

medical support, theater army contributions to, 4-77–4-78

military deception, 2-133–2-137
    defined, 2-133

military engagement, defined, 3-16

military information support operations, 2-138–2-140
    corps, 4-109

military intelligence brigade-theater, 2-26–2-27

military police command, 2-36

mission command, 2-116–2-201, 1-83–1-85
    in exploitation, 7-199–7-203
    in pursuit, 7-221
    preparing for in the defense, 6-59–6-61
    preparing for in the offense, 7-55–7-62
    system, 2-165–2-182
    training considerations, 2-277–2-298

mission command system, defined, 2-120

mission command warfighting function, defined, 2-116
    tasks, 2-117–2-164

mission-essential tasks, 2-271

mobile defense, 6-171–6-191
    control measures, 6-187–6-188
    defined, 6-171
    executing, 6-189–6-191
    fixing force, 6-182
    organization of forces, 6-180–6-186
    reserve, 6-186
    striking force, 6-183–6-185

mobile enemy reserves, 7-27–7-28

mobility and countermobility, training considerations, 2-301–2-302

mobility operations, 5-86–5-110

mounted marches, 5-74

movement, 4-49–4-53
    administrative, 5-78
    troop, 5-71–5-78

movement and maneuver, 2-202–2-204
    operational, 1-108–1-109
    preparing for in the defense, 6-62–6-75
    preparing for in the offense, 7-63–7-74
    pursuit, 7-222–7-227
    training considerations, 2-299–2-304

movement and maneuver warfighting function, defined, 2-202

movement to contact, 7-123–7-170
    control measures for, 7-142–7-145
    defined, 7-123
    executing, 7-158–7-170
    information collection, 7-148–7-150
    main body, 7-137–7-140
    organization of forces, 7-128–7-141
    reserve, 7-141
    scheme of fires, 7-156–7-157
    scheme of maneuver, 7-151–7-155
    security elements, 7-130–7-136
    specific planning considerations, 7-146–7-157

multi-domain, battle, 1-77–1-79
    extended battlefield, 1-23–1-36

multifunctional, brigades, 2-73–2-83

N

noncontiguous framework, area defense, 6-127

O

offense, air and missile defense in, 7-83–7-84
    fire support in, 7-80–7-82

offensive encirclement control measures, 7-249–7-250

offensive task, defined, 7-1

offensive tasks, 7-120–7-227

onward movement, 4-65–4-66

operation, defined, 1-68
    expeditionary, 1-9

operational and strategic reach, 1-101–1-119
    command post, 2-171
    concept, 1-68–1-85
    movement and maneuver, 1-108–1-109
    responsibilities, 2-6–2-8

operational approach, defined, 1-90

operational art, 1-86–1-100
    defined, 1-86
    elements of, 1-88–1-90

operational contract support, 4-36, 3-51

operational control, defined, A-4

operational environments, anticipated, 1-16–1-22
    shape, 1-62–1-63
    trends in, 1-21–1-22

operational framework, 1-120–1-167
    considerations, 1-122
    defined, 1-120

operational reach, defined, 1-103

operational responsibilities, corps, 2-56–2-59
    division, 2-63–2-66

operations, area security, 8-38–8-48
    assessments, 3-8–3-10
    breaching, 5-87–5-90
    breaching in the offense, 7-244
    civil affairs, 2-159–2-162
    clearing, 5-91–5-92
    command post, 2-166–2-168

Index

Entries are by paragraph number.

operations (*continued*)
  countermobility, 5-106–5-110
  decisive, shaping, sustaining, 1-162–1-164
  encirclement, 7-245–7-266
  entry, 1-117–1-119
  gap-crossing, 5-93–5-94
  in the support and consolidation areas, 5-41–5-54
  information, 2-284–2-285
  joint, 1-51–1-160
  large-scale defensive, 6-1–6-225
  large-scale offensive, 7-1–7-267
  mobility, 5-86–5-110
  overview of Army, 1-1–1-187
  planning in depth, 7-21–7-30
  planning in depth in the defense, 6-21–6-28
  security, 5-66–5-70
  sequencing, 1-168–1-177
  theater of, 1-131
  to consolidate gains, 8-1–8-76
  to prevent, 4-1–4-126
  to shape, 3-1–3-77
  wet-gap crossing, 7-243
operations planning, continuity, 2-194–2-197
operations process, conduct, 2-121–2-126
  defined, 2-121
operations security, corps, 4-117
  training considerations, 2-288–2-289
organic, A-14–A-15
organic forces, defined, A-14
organization, theater army, 2-13–2-30
organization of forces, area defense, 6-128–6-152
  attack, 7-176–7-183
  defending encircled, 6-218–6-219
  delay, 6-199
  mobile defense, 6-180–6-186
  movement to contact, 7-128–7-141
  offensive encirclement, 7-246–7-248
  retirement, 6-211
  withdrawal, 6-206
organizations, corps and below, 3-58–3-74
  Army, 3-42
other tactical considerations, 7-233–7-266
overview, Army echelons, capabilities, and training, Army operations, 1-1–1-187
  large-scale combat operations, 5-2–5-54
  large-scale defensive operations, 6-1–6-4
  large-scale offensive operations, 7-1–7-5
  operations to consolidate gains, 8-1–8-17
  operations to prevent, 4-1–4-4
  operations to shape, 3-1–3-7

P

passage of lines, 5-82–5-84
  defined, 5-82
paths to victory, 1-182–1-187
penetration, 7-115–7-116
  defined, 7-115
personnel recovery, 2-264–2-266
personnel services, 2-244
phase, defined, 1-169
physical, considerations, 1-123
plan and conduct, space activities, 2-163–2-164
planning, 1-8
  corps, 4-95–4-104
  corps and division defensive tasks, 6-8–6-48
  corps and division in the offense, 7-8–7-15
  decisive and shaping operations or main and supporting efforts, 7-33–7-37
  operations in depth, 7-21–7-30
  operations in depth in the defense, 6-21–6-28
  reconnaissance and security operations, 7-31–7-32
  reconnaissance and security operations in the defense, 6-29–6-31
  reserve operations, 7-38–7-46
  reserve operations in the defense, 6-32–6-35
  retrograde operations in the defense, 6-36–6-39
  support and consolidation area security, 7-47–7-53
  support and consolidation area security in the defense, 6-40–6-48
planning considerations, communications, 2-198
position of relative advantage, defined, 1-80
positions of relative advantage, 1-80–1-82
preclusion, 1-44
preparation, defined, 6-49
preparation of the environment, defined, 3-57
preparing, area defense, considerations when, 6-50–6-55
  corps and divisions for the defense, 6-49–6-108
  corps and divisions for the offense, 7-54–7-94
  fires in the defense, 6-80–6-95
  fires in the offense, 7-78–7-84
  intelligence in the defense, 6-76–6-79
  intelligence in the offense, 7-75–7-77
  mission command in the defense, 6-59–6-61
  mission command in the offense, 7-55–7-62
  mobile defense, considerations when, 6-56–6-58
  movement and maneuver in the defense, 6-62–6-75
  movement and maneuver in the offense, 7-63–7-74
  protection in the defense, 6-102–6-108
  protection in the offense, 7-94
  sustainment in the defense, 6-96–6-101
  sustainment in the offense, 7-85–7-93
pre-positioned stocks, 4-41–42
prevent, activities, 4-7–4-25
  conflict, 1-64

Entries are by paragraph number.

primary, alternate, contingency, and emergency plans, 2-199–2-201
project the force, 4-25
protection, 2-247–2-266
  corps, 4-114
  during transit, 4-54–4-61
  preparing for in the defense, 6-102–6-108
  preparing for in the offense, 7-94
  theater army contributions to protection, 4-79–4-83
  training considerations, 2-314–2-320
protection warfighting function, defined, 2-247
psychological operation group, 2-43–2-44
public affairs, 2-141–2-144
pursuit, 7-218–7-227
  mission command in, 7-221
  movement and maneuver in, 7-222–7-227
  transition to, 7-220

Q-R

raid, defined, 7-232
rail, Army, 5-76
range of military operations, conflict continuum and, 1-2
rearward passage of lines, defined, 5-84
reception, 4-63
reception, staging, onward movement, and integration, 4-62–4-69
reconnaissance, 5-56–5-65
  defined, 5-56
reconnaissance and security operations in the defense, planning, 6-29–6-31
reconnaissance in force, defined, 5-64
reconnaissance objective, defined, 5-57
reconnaissance operations, planning, 7-31–7-32
regional, support groups, 2-49
regionally assigned and aligned forces, 3-60–3-61
reinforcing, defined, A-23
relationships, command and support, A-1–A-28

relative advantage, positions of, 1-80–1-82
relief in place, 5-79–5-81
  defined, 5-79
reserve, area defense, 6-144–6-152
  attack, 7-181–7-183
  corps, 4-121
  defined, 7-38
  mobile defense, 6-186
  movement to contact, 7-141
reserve operations, planning, 7-38–7-46
  planning in the defense, 6-32–6-35
responsibilities, administrative, 2-9
  consolidate gains by echelon, 8-58–8-75
  operational, 2-6–2-8
restore, essential services, 8-53
retain the initiative, 1-75–1-76
retirement, 6-208–6-212
  control measures for, 6-212
  defined, 6-193
  organization of forces for, 6-211
retrograde, 6-192–6-212
  defined, 6-192
retrograde operations in the defense, planning, 6-36–6-39
risk, 1-91–1-92
  considerations, B-1–B-5
risk management, 4-84–4-87
roads, combat, 5-95–5-98
roles, commander and staff, 2-125–2-126
route, defined, 7-135
route reconnaissance, defined, 5-61

S

sanctuary, 1-47–1-48
scheme of fires, movement to contact, 7-156–7-157
scheme of maneuver, 7-16–7-20
  defense, 6-13–6-20
  movement to contact, 7-151–7-155
screen, defined, 5-67
search and attack, 8-40–8-47
  defined, 8-40

second tactical echelon, enemy, 7-25–7-26
security area, defined, 6-30
security assistance, defined, 3-19
security cooperation, 3-17–3-18
  conduct, 8-57
  defined, 3-17
security elements, movement to contact, 7-130–7-136
security force assistance, 3-20–3-21
  brigade, 2-53
  defined, 3-20
security forces, attack, 7-177
  area defense, 6-131–6-138
  defined, 3-20
  executing a movement to contact, 7-159–7-163
security operations, 5-64–5-70
  corps, 4-106
  defined, 5-66
  planning, 7-31–7-32
security sector reform, 3-25–3-27
  defined, 3-22
seize initiative, 1-57
seize, retain, and exploit the initiative, 1-75–1-76
sequencing operations, 1-168–1-177
services and supplies, 4-28
set the theater, 4-12–4-16
shape, 1-55
  operational environments, 1-62–1-63
shaping, activities, 3-13–3-36
  additional, 3-28–3-36
shaping operation, defined, 1-163
shaping operations, planning, 7-33–7-37
signal support, theater-level, 2-24
single envelopment, defined, 7-97
space activities, plan and conduct, 2-163–2-164
space domain, 1-25–1-27
  defined, 1-25
space integration, training considerations, 2-294–2-298
space operations, corps, 4-119

Index

Entries are by paragraph number.

special operations and conventional forces, interdependence, interoperability, and integration of, 3-62–3-67

special operations forces, Army, 2-52, 3-57

special reconnaissance, defined, 5-65

specific planning considerations, movement to contact, 7-146–7-157

spoiling attack, defined, 6-159

stability, and defeat mechanisms, 1-93–1-100
    in large-scale combat operations, 5-34–5-54

stability mechanism, defined, 1-99

stability mechanisms, 1-99–100

stability tasks, 5-36–5-40, 8-49–8-57
    defined, 1-73

stabilize, 1-59

staging, defined, 4-64

strategic and operational reach, 1-101–1-119

strategic roles, 1-61

striking force, defined, 6-171
    mobile defense, 6-183–6-185

subordinate forms of the attack, 7-228–7-232

supplies and services, 4-28

support, base security and defense, 5-43–5-49
    defined, A-6
    governance, 8-54

support and consolidation area security in the defense, planning for, 6-40–6-48

support area, 1-155–1-157
    command post, 2-174–2-178
    corps, 4-120
    defined, 1-155
    operations, 5-41–5-54

support area security, planning, 7-47–7-53

support economic and infrastructure development, 8-55–8-56

support groups, regional, 2-49

support relationships, Army, A-18–A-24

supporting effort, defined, 1-167

supporting efforts, planning, 7-33–7-37

survivability, 2-254–2-255
    command post, 2-190–2-193
    defined, 2-254

survivability operations, defined, 2-254

sustaining operation, defined, 1-164

sustainment, 2-233–2-246
    brigade, 2-82–2-83
    brigades, 2-21–2-23
    corps, 4-115
    exploitation, 7-216–7-217
    preparation, 4-26–4-45
    preparing for in the defense, 6-96–6-101
    preparing for in the offense, 7-85–7-93
    training considerations, 2-311–2-313

sustainment command, expeditionary, 2-17–2-20
    theater, 2-15–2-23

sustainment warfighting function, defined, 2-233

system, mission command, 2-165–2-182

system support contracts, 4-40

systems warfare, 1-49–1-50

T

tactical command post, 2-172–2-173

tactical combat force, defined, 5-49
    in the support area, 5-50

tactical command post, defined, 2-172

tactical control, defined, A-5

tactical deception, training considerations, 2-286–2-287

tactical enabling tasks, 5-55–5-110

tailor Army forces, 4-17–4-24

target, defined, 6-26

targeting, defined, 2-219

task forces, consolidate gains responsibilities, 8-72–8-75

tasks, focus, 1-11

mission command warfighting function, 2-117–2-164

temporal, considerations, 1-124

terrain, use of, 7-237–7-240

theater, aviation command, 2-39–2-42
    engineer command, 2-37–2-38
    expanded, 2-31
    information operations groups, 2-48
    military intelligence brigade, 2-26–2-27

theater army, 3-43, 4-71–4-73, 2-4–2-5
    consolidate gains responsibilities, 8-59–8-63
    contributions to medical support, 4-77–4-78
    contributions to joint command and control, 4-74–4-76
    contributions to protection, 4-79–4-83
    organization, 2-13–2-30

theater, operations, 1-131
    of war, 1-130
    support contracts, 4-37–4-38
    sustainment command, 2-15–2-23

theater-level signal support, 2-24

threat, defined, 1-37
    in large-scale combat operations, 5-30–5-33
    planning for, 1-8

threats, 1-37–1-50, 1-1, 3-11–3-12, 4-5–4-6
    challenges by, 1-5
    to the consolidation of gains, 8-18–8-24

trails, combat, 5-95–5-98

training, 1-8
    considerations by warfighting function, 2-276–2-320
    techniques, 2-272–2-275

training and exercises, combined, 3-36

training considerations, air defense, 2-310
    airspace control, 2-280–2-282

Index

Entries are by paragraph number.

Army aviation maneuver (manned), 2-303
Army aviation maneuver (unmanned), 2-304
civil affairs integration, 2-283
counterfire, 2-309
cyberspace and electronic warfare, 2-290–2-293
electronic and cyberspace warfare, 2-290–2-293
fires, 2-307–2-310
intelligence, 2-305–2-306
mission command, 2-277–2-298
mobility and countermobility, 2-301–2-302
movement and maneuver, 2-299–2-304
operations security, 2-288–2-289
protection, 2-314–2-320
space integration, 2-294–2-298
sustainment, 2-311–2-313
tactical deception, 2-286–2-287
training for large-scale combat operations, 2-267–2-319
transit, protection during, 4-54–4-61
transition, 8-76
   to consolidate gains, 5-115–5-118
   to pursuit, 7-220
transition to the offense, corps, 4-122
transportation, 4-32
trends, operational environments, 1-21–1-22
troop movement, 5-71–5-78
   defined, 5-72
turning movement, 7-117–7-119
   defined, 7-117

U

unified action, 1-6
unified land operations, defined, 1-68
United States diplomatic mission, ambassador, 3-38–3-39
use of terrain, 7-237–7-240

V

vertical envelopments, defined, 7-97
victory, paths to, 1-182–1-187
virtual, considerations, 1-125

W

warfare, information, 1-41–1-43
   levels of, 1-19–1-20
   systems, 1-49–1-50
warfighting function, training considerations by, 2-276–2-320
warfighting functions, defined, 2-105
water movements, Army, 5-76
weapons of mass destruction, countering, 3-32
   enemy use of, 1-10
   protection from, 1-10
wet-gap crossing operations, 7-243
win, 1-67
withdrawal, 6-202–6-207
   control measures for, 6-207
   organization of forces for, 6-206
withdrawal operation, defined, 6-193

X-Y-Z

zone reconnaissance, defined, 5-62

Made in the USA
Lexington, KY
16 February 2018